Y0-AEY-654

CRC Series in The Biochemistry and Molecular Biology of the Cell Nucleus

Editor-in-Chief

Lubomir S. Hnilica, Ph.D.
Professor of Biochemistry and Pathology
Department of Biochemistry
Vanderbilt University School
of Medicine
Nashville, Tennessee

The Structure and Biological Function of Histones

Author
Lubomir S. Hnilica, Ph.D.

Chromosomal Nonhistone Proteins

Volume I: Biology
Volume II: Immunology
Volume III: Biochemistry
Volume IV: Structural Associations

Editor
Lubomir S. Hnilica, Ph.D.

Enzymes of Nucleic Acid Synthesis and Modification

Volume I: DNA Enzymes
Volume II: RNA Enzymes

Editor

Samson T. Jacob, Ph.D.
Professor
Department of Pharmacology
The Pennsylvania State University
The Milton S. Hershey Medical Center
Hershey, Pennsylvania

Enzymes of Nucleic Acid Synthesis and Modification

Volume II
RNA Enzymes

Editor

Samson T. Jacob, Ph.D.
Professor
Department of Pharmacology
The Pennsylvania State University
The Milton S. Hershey Medical Center
Hershey, Pennsylvania

CRC Series in the Biochemistry and Molecular Biology
of the Cell Nucleus

CRC Press, Inc.
Boca Raton, Florida

Library of Congress Cataloging in Publication Data
Main entry under title:

Enzymes of nucleic acid synthesis and
modification.

(CRC series in the biochemistry and molecular
biology of the cell nucleus)
Bibliography: p.
Includes index.
Contents: v. 1. DNA enzymes — v. 2. RNA
enzymes.
1. Nucleic acids—Synthesis. 2. Nucleic
acids—Metabolism. 3. Enzymes. I. Jacob,
Samson T. II. Series.
QP620.E58 574.87'328 82-4409
ISBN 0-8493-5517-6 (v. 1) AACR2
ISBN 0-8493-5518-4 (v. 2)

This book represents information obtained from authentic and highly regarded sources. Reprinted material is quoted with permission, and sources are indicated. A wide variety of references are listed. Every reasonable effort has been made to give reliable data and information, but the author and the publisher cannot assume responsibility for the validity of all materials or for the consequences of their use.

Direct all inquiries to CRC Press, Inc., 2000 Corporate Blvd., N.W., Boca Raton, Florida, 33431.

International Standard Book Number 0-8493-5517-6 (Volume I)
International Standard Book Number 0-8493-5518-4 (Volume II)

Library of Congress Card Number 82-4409
Printed in the United States

PREFACE

The past decade has witnessed remarkable advances in elucidation of the function, structure, and regulation of the enzymes responsible for nucleic acid synthesis and modification. To my knowledge, no single publication has covered the major enzymes involved in these important cellular events. It is, therefore, hoped that discussion of the current status of these enzymes, published as a single work, would be of considerable interest to a broad range of investigators.

The first volume gives an overview of the enzymes involved in DNA synthesis and modification; the second volume deals with the RNA-enzymes. Although the major emphasis of the book is on eukaryotic enzymes, a separate chapter dealing with prokaryotic DNA repair enzymes has been included to discuss the major advances in this field in recent years. There are two separate chapters on RNA polymerases to provide a comprehensive coverage of the enzymes from lower eukaryotes, plants, and higher eukaryotes.

As editor, I have been fortunate in securing as authors many of the leading investigators in the field of nucleic acid enzymes. It is particularly gratifying that they have in most instances provided new data from their laboratories. I express my deep appreciation to the authors for their prompt submission of the chapters, which facilitated rapid publication of this book.

Finally, I thank Drs. Aaron Shatkin, Bernard Moss, and Lawrence Loeb for useful suggestions and Ms. Edna Myeski for invaluable assistance in the preparation of this book.

Samson T. Jacob
1982

CONTRIBUTORS

R. L. P. Adams
Senior Lecturer
Department of Biochemistry
University of Glasgow
Glasgow, Scotland

Richard A. Bennett
Research Associate
Department of Physiology
The Milton S. Hershey Medical Center
The Pennsylvania State University
Hershey, Pennsylvania

R. H. Burdon
Professor of Biochemistry
University of Glasgow
Glasgow, Scotland

Mary Sue Coleman
Associate Professor
Department of Biochemistry
School of Medicine
University of Kentucky
Lexington, Kentucky

Martin R. Deibel, Jr.
Research Assistant Professor
Department of Biochemistry
School of Medicine
University of Kentucky
Lexington, Kentucky

Murray P. Deutscher
Professor
Department of Biochemistry
University of Connecticut Health Center
Farmington, Connecticut

S. J. Flint
Associate Professor
Department of Biochemical Sciences
Princeton University
Princeton, New Jersey

Michael Fry
Senior Lecturer
Unit of Biochemistry
Faculty of Medicine
Technion-Israel Institute of Technology
Haifa, Israel

Gary F. Gerard
Senior Scientist
Bethesda Research Laboratories
Gaithersburg, Maryland

Tom J. Guilfoyle
Associate Professor
Department of Botany
University of Minnesota
St. Paul, Minnesota

Samson T. Jacob
Professor
Department of Pharmacology
The Milton S. Hershey Medical Center
The Pennsylvania State University
Hershey, Pennsylvania

Jerry M. Keith
Associate Professor
Department of Biochemistry
College of Dentistry
New York University
New York, New York

Anthony E. Pegg
Professor
Department of Physiology
The Milton S. Hershey Medical Center
The Pennsylvania State University
Hershey, Pennsylvania

Kathleen M. Rose
Associate Professor
Department of Pharmacology
The Milton S. Hershey Medical Center
The Pennsylvania State University
Hershey, Pennsylvania

Dean A. Stetler
Assistant Professor
Department of Pharmacology
College of Medicine
The Pennsylvania State University
Hershey, Pennsylvania

Huber Warner
Professor of Biochemistry
Department of Biochemistry
University of Minnesota
St. Paul, Minnesota

ENZYMES OF NUCLEIC ACID SYNTHESIS AND MODIFICATION

Samson T. Jacob

Volume I

Volume II

TABLE OF CONTENTS

Volume II

Chapter 1

DNA-DEPENDENT RNA POLYMERASES OF PLANTS AND LOWER EUKARYOTES

Tom J. Guilfoyle

TABLE OF CONTENTS

I. INTRODUCTION

Eukaryotes possess three distinct classes of nuclear DNA-dependent RNA polymerases as well as specific organellar RNA polymerases (e.g., mitochondrial and chloroplast). The three classes of nuclear RNA polymerase as well as the organellar RNA polymerases have been purified from a variety of eukaryotes and characterized with respect to their chromatographic properties, molecular structures, immunological properties, capabilities of carrying out selective transcription in vitro, and regulation during growth and developmental transitions. The purpose of this review is to summarize our current understanding of plant and lower eukaryotic (e.g., fungi, algae, and protists*) DNA-dependent RNA polymerases. A review of animal RNA polymerases is the subject of another chapter in this book (see Chapter 2). For a more comprehensive treatment of historical literature and additional topics, the reader is referred to several recent reviews that specifically discuss RNA polymerases from higher plants[2-6] and lower eukaryotes[7-11] as well as more general reviews on eukaryotic RNA polymerases that emphasize the animal literature.[12-18]

In this chapter, the author discusses nuclear and organellar RNA polymerases from plants and lower eukaryotes in terms of sources of enzymes, purification methodology, catalytic properties, molecular structures, immunological relationships, transcriptional functions, selectivity of transcription in nuclear and reconstituted systems, and modulation during cell cycle and growth transitions. Since many of the above studies on plant and lower eukaryotic RNA polymerases are incomplete, the author will occasionally refer to literature on animal RNA polymerases where detailed studies have been conducted or where informative comparisons can be made among animal, plant, and lower eukaryotic enzymes. For condensation purposes in this chapter, references to animal literature are not always made even though the original results in a particular area of research may have been demonstrated with animal RNA polymerases.

II. RESOLUTION OF MULTIPLE NUCLEAR POLYMERASES

Nuclear RNA polymerases which have been solubilized from chromatin templates by sonication in the presence of high salt or those enzymes recovered in the soluble fraction of total tissue extracts can be separated into three distinct classes by chromatography on either anionic or cationic exchange resins. Resolution of three classes of nuclear RNA polymerase on the anionic exchange resin, diethylaminoethyl-sephadex® (DEAE-Sephadex®), has given rise to a nomenclature for eukaryotic RNA polymerases which is based on their order of elution from the resin when a gradient of increasing salt concentration is applied to the column; these enzymes are designated RNA polymerases I, II, and III.[19] RNA polymerases I, II, and III purified from a large number of animals, plants, and lower eukaryotes have similar elution patterns from DEAE-Sephadex® and DEAE-cellulose. With DEAE-Sephadex®, RNA polymerase I elutes at 0.10 to 0.15 *M*, RNA polymerase II at 0.17 to 0.25 *M*, and RNA polymerase

* The classification strategy of Whittaker and Margulis[1] is used throughout this chapter.

III at 0.20 to 0.35 *M* ammonium sulfate. When DEAE-cellulose is employed, the elution pattern of the nuclear RNA polymerases is similar to that observed with DEAE-Sephadex®, except that RNA polymerase III elutes at low salt concentrations in the range of 0.10 to 0.15 *M* ammonium sulfate.[15,20,21] One striking exception to the characteristic elution pattern of eukaryotic RNA polymerases from DEAE-Sephadex® is that reported for the slime mold, *Physarum polycephalum,* where RNA polymerase II or the α-amanitin-sensitive enzyme elutes prior to RNA polymerase I, the α-amanitin-resistant enzyme.[22-25]

With cationic exchange resins such as phosphocellulose, the order of elution is generally RNA polymerase II at about 0.10 *M*, RNA polymerase III at approximately 0.13 *M*, and RNA polymerase I at about 0.18 *M* ammonium sulfate.[15,20] Other cationic exchange resins such as carboxymethyl-Sephadex® (CM-Sephadex®) have also been utilized and while elution patterns similar to those observed with phosphocellulose are generally obtained for the three classes of nuclear RNA polymerase, the salt concentrations required for elution vary according to the cationic exchange resin employed.[15]

The relative position of elution for each class of RNA polymerase on ion exchange resins is readily determined by assaying aliquots of successive column fractions in the presence and absence of low (i.e., 1 μg/mℓ) and high (i.e., 100 to 1000 μg/mℓ) concentrations of the fungal toxin, α-amanitin. In general, column fractions containing RNA polymerase II activity will be inhibited by low concentrations of α-amanitin while fractions containing either RNA polymerase II or III activity will be inhibited at high concentrations of the fungal toxin. With yeast enzymes, however, fractions containing RNA polymerases I and II activities are inhibited by high concentrations of α-amanitin, while only those fractions containing RNA polymerase III activity are refractory to inhibition.[21,26]

III. REQUIREMENTS FOR ENZYMATIC ACTIVITY

To catalyze the in vitro synthesis of RNA, both prokaryotic and eukaryotic RNA polymerases require a DNA template, four ribonucleoside triphosphates (e.g., ATP, CTP, GTP, and UTP), and a divalent cation such as Mg^{++} or Mn^{++}. The requirement for a divalent cation probably reflects binding of nucleotides to the enzyme as metal chelates (i.e., MgATP).[27] The optimum divalent cation concentrations are generally in the range of 5 to 10 m*M* Mg^{++} or 1 to 2 m*M* Mn^{++}.[12] In addition to the above requirements, RNA polymerase activity is usually stimulated by the inclusion of monovalent cations supplied in the form of salts such as $(NH_4)_2SO_4$, NH_4Cl, KCl, or NaCl. Stimulation of in vitro transcription by monovalent cations is especially evident when engaged RNA polymerases are assayed in isolated nuclei or chromatin. With isolated plant nuclei or chromatin, as with isolated animal nuclei, α-amanitin-resistant RNA polymerase activity (e.g., RNA polymerases I + III) is most active at 0.05 to 0.10 *M* ammonium sulfate while α-amanitin-sensitive activity (e.g., RNA polymerase II) is optimized at 0.25 to 0.50 *M* ammonium sulfate.[28-32] The high salt optimum observed for RNA polymerase II on chromatin or nuclear templates appears to be a property of the chromatin transcription complex rather than an intrinsic property of the enzyme, since purified RNA polymerase II is most active at 0.025 to 0.10 *M* ammonium sulfate on deproteinized DNA templates.[20,32]

Purified RNA polymerases exhibit characteristic divalent cation preference and optima, ionic strength or monovalent cation optima, and template requirements. For example, purified yeast RNA polymerases I and II are optimally active at 0.03 to 0.05 *M* and 0.05 to 0.10 *M* ammonium sulfate, respectively, and RNA polymerase III displays a biphasic optimum at 0.04 to 0.10 *M* plus 0.18 to 0.30 *M* ammonium sulfate when the enzymes are assayed on native DNA templates.[21,26,33] All three classes of purified yeast RNA polymerases are more active with Mn^{++} than Mg^{++}.[33] However, these characteristics are not invariant properties of the three classes of nuclear RNA polymerase. Gissinger et al.[34] have demonstrated that the divalent cation requirements and ionic strength optima for each class of RNA polymerase depends on both the concentration of DNA and the integrity of the DNA utilized to assay the RNA polymerase activities.

RNA polymerases I and III utilize denatured and native DNA templates with about equal efficiency, but RNA polymerase II shows a strong preference for denatured DNA templates.[15] Dynan et al.[35] have reported that three potential artifacts may explain why wheat germ and other eukaryotic RNA polymerase II enzymes appear to prefer denatured over native DNA templates. These artifacts include: (1) greater self-absorption of tritium on filters containing precipitated native DNA compared to denatured DNA, (2) the use of limiting ribonucleoside triphosphate which favors transcription on denatured DNA in in vitro assays, and (3) contamination of commercial DNA preparations with deoxyribonucleases. When these artifacts are accounted for, wheat germ RNA polymerase II is more active with native than with denatured DNA templates. However, even when an RNA polymerase shows a preference for native DNA templates over denatured templates, initiation of RNA chains on the native templates is most likely occurring at single-stranded nicks or gaps in the duplex DNA (see Section X).

IV. PURIFICATION

A. Sources

Relatively large quantities of homogeneous enzyme are required for studies on RNA polymerase subunit structures, reconstitution of active enzyme from separated subunits, and in vitro transcription with reconstructed systems. A number of tissue sources have been identified that provide a large amount of enzyme for purification to homogeneity. Wheat germ (Phylum Tracheophyta) is an excellent source for RNA polymerases primarily because of the dehydrated nature of the tissue, the low level of proteolytic activity in the germ, and the commercial availability of the raw germ. The dehydrated condition of the germ is the major reason for the large amount of RNA polymerases per gram fresh weight. RNA polymerase II comprises 80 to 90% of the total RNA polymerase in wheat germ cell-free extracts, and this percentage of class II enzyme is observed for most other plant tissue extracts.[31,36-38] Because of the relatively large amounts of class II enzyme in plant tissues, this is the plant RNA polymerase most amenable to purification. A kilogram of raw wheat germ yields about 25 mg of homogeneous RNA polymerase II when the enzyme is purified by the method introduced by Jendrisak and Burgess.[39] Other plant embryonic axes (or germ) also provide good sources of RNA polymerase II. About 10 mg of RNA polymerase II can be recovered in pure form from 1 kg of dehydrated soybean embryonic axes with an overall recovery of 40%.[38,40] Germinated and growing plant organs appear to contain much less RNA polymerase II than ungerminated embryonic axes when calculations are based on tissue fresh weight; however, the lower quantities of enzyme present in growing organs are simply due to the hydrated state of the tissue, and the amount of enzyme based on equivalent amounts of protein or DNA in total tissue extracts is approximately equal in embryonic axes and growing organs.[38,40] Since large quantities of many different types of plant organs are easily obtained and since recoveries of RNA polymerase II can approach 40 to 50% even with hydrated tissues, a variety of plant sources have the potential to provide large quantities of pure RNA polymerase II.[38,40,41]

The amounts of RNA polymerases I and III in plant tissues are apparently much less than the amount of RNA polymerase II. RNA polymerases I and III comprise only about 10 to 20% of the total RNA polymerase activity in plant tissue extracts. About 3 mg of RNA polymerase I and 0.75 mg of RNA polymerase III can be purified from a kilogram of wheat germ with recoveries approaching 30 to 50% for each enzyme.[36,42] With hydrated plant organs, proliferating tissues are better sources than nonproliferating tissues for all three classes of RNA polymerase, but even with rapidly proliferating plant tissues only about 0.3 mg of pure RNA polymerase I and substantially less RNA polymerase III is recovered from a kilogram of tissue.[43,44]

Saccharomyces cerevisiae or yeast (Phylum Ascomycota or sac fungi) is an excellent source of lower eukaryotic RNA polymerases I, II, and III. Approximately 8 to 15 mg of RNA polymerase I, 20 mg of RNA polymerase II, and 3 to 10 mg of RNA polymerase III can be purified from a kilogram of yeast cells.[45-48] While yeast is a better source for RNA polymerases than other fungi and lower eukaryotes, certain other lower eukaryotic sources are worth noting since they are representative of enzymes from some additional phyla. A kilogram of *Aspergillus nidulans* mycelia (Form-Phylum Deuteromycota or fungi imperfecti) yields about 3 mg of pure RNA polymerase I[49] and 1 mg of pure RNA polymerase II.[50] *P. polycephalum* (Phylum Myxomycota or plasmodial slime molds) yields about 0.5 mg of pure RNA polymerase II per kilogram of microplasmodia.[51] Another protist, *Acanthamoeba castellanii* (Phylum Rhizopoda or naked amoebae), is a good source for all three classes of nuclear RNA polymerase. Approximately 0.75 mg of RNA polymerase I,[52] 2.5 mg of RNA polymerase II,[53] and 0.6 mg of RNA polymerase III[54] can be purified to homogeneity from a kilogram of cells.

B. Purification Methodology

The initial breakthrough in eukaryotic RNA polymerase purification resulted from the efficient solubilization of the enzymes by employing high salt-sonication procedures with isolated nuclei or whole cell extracts.[19] This procedure, which utilizes high salt (i.e., 0.3 *M* ammonium sulfate) to disrupt the nuclear chromatin and sonication to reduce the viscosity of the high salt extract, has been generally applicable to the purification of nuclear RNA polymerases from animals,[12] plants,[20,55-60] and lower eukaryotes.[24,45,61-66] Variations in this procedure such as sonication of nuclei or chromatin in low salt buffers (i.e., 0.05 *M* ammonium sulfate) have also been employed for purification of RNA polymerases from lower eukaryotes.[23,67] After solubilization of the RNA polymerases by the above procedures, purification of the enzymes can be achieved by conventional techniques such as ammonium sulfate fractionation, chromatography on anionic (i.e., DEAE-Sephadex® and DEAE-cellulose) and cationic (i.e., phosphocellulose and CM-Sephadex®) exchange resins, and sedimentation through glycerol or sucrose density gradients.[15] Application of these procedures has resulted in the purification of a number of plant and lower eukaryotic nuclear RNA polymerases to homogeneity;[20,23,45,56,58,62,65-69] however, overall yields of the purified RNA polymerases are generally less than 15%.

Several modifications have been introduced in the last few years that result in higher yields of purified RNA polymerases and these procedures have been especially applicable to the purification of plant and lower eukaryotic nuclear RNA polymerases. First, the isolation of nuclei for RNA polymerase purification is avoided. Even though the isolation of nuclei may provide a convenient purification method for nuclear-associated RNA polymerases (i.e., the isolation of nuclei may result in a several hundred to thousand-fold purification of the nuclear RNA polymerases), the available aqueous isolation procedures can result in nuclear damage and loss of RNA polymerases from the organelle by leaching during purification of the nuclei. It is also possible that the bulk of the RNA polymerase is not associated with nuclei or may leach from intact, undamaged nuclei upon cell disruption. In addition, the purification of isolated nuclei away from cellular debris and other organelles may result in large losses of nuclei. The isolation of nuclei is especially prohibitive with organisms that possess rigid cell walls composed of cellulose (e.g., plants) or chitin (e.g., fungi). Second, the RNA polymerases which are solubilized by the disruption of cells in low ionic strength buffer are purified by precipitation with and elution from Polymin P (polyethylenimine). This procedure was introduced by Jendrisak and Burgess[39] for the purification of RNA polymerase II from wheat germ and is applicable to any organism where RNA polymerases are released as soluble enzymes after cell disruption in low salt buffers. For cellular extracts from plants, the bulk of RNA polymerases II and III (and in some cases RNA polymerase I) is found to be soluble (e.g., the enzymes are recovered in 20,000 to 100,000 × g supernatants). The solubility of plant RNA polymerase II in tissue homogenates was first noted for maize seedlings[70,71] and subsequently found to be the case for a variety of plants.[40,72-75] The Polymin P procedure[39] has been applied to the purification of RNA polymerases I, II, and III from wheat germ[36,39,42,76] and *Acanthamoeba castellanii*,[52-54] RNA polymerases I and II from *Aspergillus nidulans*;[49,50] RNA polymerases II and III from yeast[48] and cauliflower;[41,74,77] and RNA polymerase II from soybean,[40,44,74] maize,[74] rye,[74] turnip,[32] pea,[78] *Agaricus bisporus* (a club fungus),[79] and *P. polycephalum*.[51] Third, affinity chromatography of RNA polymerases on heparin-Sepharose is employed as a purification step. RNA polymerases bind very tightly to heparin-Sepharose at high salt (i.e., 0.2 to 0.3 *M* ammonium sulfate) while most contaminating proteins do not. Heparin-Sepharose chromatography can be employed as an initial purification step for RNA polymerases that have been solubilized from nuclei or chromatin by the high salt-sonication procedure. This can be especially advantageous where dilution of the high salt extract results in aggregation of RNA polymerases with other chromatin components, since the solubilized RNA polymerases can be applied to the heparin-Sepharose column at high ionic strength where aggregation is inhibited.[43,59] Heparin-Sepharose chromatography has also been combined with the Polymin P technique to facilitate the purification of RNA polymerases I and III from *Acanthamoeba castellanii*,[52,54] RNA polymerases I, II, and III from wheat,[36] and RNA polymerase II from *P. polycephalum*.[51]

Application of conventional purification procedures (i.e., ion exchange chromatography and sedimentation on glycerol or sucrose density gradients) in combination with the Polymin P procedure and/or heparin-Sepharose chromatography can result in preparations of pure RNA polymerases at recoveries approaching 50%. Other methodology has been introduced in recent years that may also prove advantageous for purification of specific classes of nuclear RNA polymerase from particular organisms. Hydrophobic chromatography on ω- or δ-aminobutyl-Sepharose has been employed for the purification of wheat germ RNA polymerase III[80] and RNA polymerases II from rye germ.[57] Affinity chromatography on DNA-cellulose or DNA-Sepharose has been applied to the purification of a number of plant and lower eukaryotic RNA polymerases.[23,38,40,81,82] RNA polymerase III from yeast can be efficiently purified away from RNA polymerases I and II by chromatography on denatured DNA-cellulose.[48] The tight binding of yeast RNA polymerase III to denatured DNA-cellulose is thought to be a characteristic of the class III enzyme that is also reflected in the resistance of this enzyme to high salt inhibition of activity measured on native DNA.[26] Another procedure that has been employed in the purification of yeast RNA polymerases is ion-filtration or sieveorptive chromatography.[48] With this procedure, both gel filtration and ion exchange chromatography are combined in the same purification step and this has been employed for separating nucleic acids from RNA polymerases. Finally, the availability of new high capacity ion exchange resins has also improved the purification methodology for eukaryotic RNA polymerases. For example, DEAE-Sepharose CL-6B (Pharmacia) is a high capacity anionic exchange resin to which eukaryotic RNA polymerases bind with higher affinity than to other DEAE-substituted resins (i.e.,

DEAE-cellulose or DEAE-Sephadex®). Wheat germ RNA polymerases can be applied to a column of DEAE-Sepharose CL-6B at relatively high salt concentrations (i.e., 0.125 *M* ammonium sulfate). All of the applied RNA polymerase activity is retained on the column and RNA polymerase I is eluted at 0.25 *M* ammonium sulfate while RNA polymerases II and III are eluted at 0.5 *M* ammonium sulfate.[42] This resin has also been used for the purification of RNA polymerase II from *P. polycephalum*.[51] The greater affinity and higher capacity of DEAE-Sepharose CL-6D is thought to be due to the ability of the large RNA polymerase molecules to penetrate the beads of this high molecular weight sieving resin.[42]

V. SUBUNIT STRUCTURES

Nuclear RNA polymerases possess multimeric subunit structures with each class of enzyme consisting of some 7 to 15 polypeptides or putative subunits. The complexity of subunit structure has not always been fully realized due to analyses of enzymes under conditions which do not permit the complexity to be elucidated. To determine the complete subunit structures of the nuclear RNA polymerases, it is necessary to employ polyacrylamide gel systems with high resolving power since subunits range in size from about 10,000 daltons to greater than 200,000 daltons and many of the smaller subunits are clustered in the molecular weight range of 15,000 to 25,000. For example with wheat germ RNA polymerase II, gel systems must be used that allow the resolution of polypeptides ranging in size from 14,000 to 220,000 daltons, but also allow resolution of two polypeptides in each of the following molecular weight ranges: 40,000 to 42,000; 25,000 to 27,000; 20,000 to 21,000; 17,000 to 17,800; and 16,000 to 16,500.[83] In addition to the employment of high resolution polyacrylamide gel systems, it is also necessary to analyze sufficient amounts of enzyme to allow detection of the low molecular weight subunits which comprise little mass with respect to the overall mass of the enzyme complex. For example, when 1 μg of wheat germ RNA polymerase II is applied to a polyacrylamide gel containing dodecyl sulfate, the amount of 16,000 dalton subunit would be about 30 ng, an amount not readily detectable on polyacrylamide gels stained with Coomassie® Brilliant Blue.

In addition to having a sufficient amount of enzyme for analysis, it is important to have the enzyme sufficiently pure if an accurate subunit structure is to be obtained. Roeder[15] has pointed out several criteria that should be met before polypeptides associated with purified RNA polymerases can be designated as subunits. These criteria may be briefly summarized as:

1. The ratio of the amount of enzyme activity to the amount of each putative subunit should remain constant for individual gradient fractions if the pure enzyme is subjected to further analysis on ionic exchange resins.
2. The molar ratios of putative subunits should be integral values.
3. Individual fractions containing RNA polymerase activity collected from glycerol or sucrose density gradients should have identical subunit structures and subunit molar ratios.
4. Subunits should remain associated with the major protein band resolved on polyacrylamide gels subjected to electrophoresis under nondenaturing conditions.
5. The molecular weights of the putative subunits multiplied by their molar ratios should sum to the aggregate molecular weight of the native enzyme.
6. A similar subunit structure should be obtained for each class of enzyme purified from different tissues of the same or related organisms.

These criteria do not, however, provide definitive information on subunit structure since tightly bound polypeptides which are not required for enzymatic activity may, nevertheless, purify with the RNA polymerases. Elucidation of the structure of a "core" enzyme, which is the minimum structural composition required for enzymatic activity on simple templates (i.e., poly (dA-T)), will ultimately require either genetic analysis or reconstitution of the enzyme from individual subunits.

For the most part, only those subunit structures that have been analyzed under high resolution conditions and that have been subjected to tests that satisfy most of the criteria in the above paragraph will be discussed. Also included however, are brief references to results where only partial subunit structures were obtained.

A. RNA Polymerase I

Subunit structures for RNA polymerase I have been reported for a number of higher plants including soybean,[44,58] cauliflower,[20,43] wheat,[36] and parsley.[60] Subunit structures for the class I enzyme have also been reported for fungi including yeast,[7,45,46,62,84,85] *Mucor rouxii*,[81] and *Rhizopus stolonifer*,[86] for the fungi imperfecti *Aspergillus nidulans*,[49] and for the protists *Acanthamoeba castellanii*[52] and *P. polycephalum*.[23] Putative subunit structures of some representative RNA polymerase I enzymes are listed in Table 1. In general, the subunit structures of RNA polymerase I are similar across the plant, fungi, and protist kingdoms, and consist of two large subunits of approximately 190,000 and 120,000 daltons and several smaller subunits. Although both large subunits are generally present in approximately equimolar amounts, in some cases with higher plants the 190,000 dalton subunit may suffer limited proteolysis during purification which leads to the appearance of a 170,000 dalton polypeptide or several minor polypeptides between 190,000 and 120,000 daltons.[20,43,58] Proteolysis of the 190,000 dalton subunit is also observable in some yeast RNA polymerase I preparations.[45,87] Plant RNA polymerase I contains smaller subunits of approximately 40,000; 25,000; 20,000; as well as one or two polypeptides of approximately 17,000 daltons.[36,42-44] An additional polypeptide of about 50,000 daltons is present in wheat germ RNA polymerase I, but this polypeptide occurs at less than equimolar amounts.[36]

Yeast RNA polymerase I contains some 11 to 13 polypeptides in contrast to the 6 to 9 polypeptides associated with plant RNA polymerase I and the 10 polypeptides of *A. castellanii* class I enzyme (Table 1). With yeast RNA polymerase I, at least two polypeptides of 48,000 and 37,000 daltons can be dissociated from the enzyme by subjecting the purified enzyme to electrophoresis under nondenaturing conditions, phosphocellulose or blue dextran-Sepharose chromatography, or antibody affinity chromatography where antibody to the 190,000 subunit of yeast RNA polymerase I is covalently linked to Sepharose.[7,46,88-90]

B. RNA Polymerase II

For the class II RNA polymerases, subunit structures have been reported for a variety of plants, and these include soybean,[38,40,74] cauliflower,[41,43,74] wheat,[76,83] rye,[74] maize,[74,82] parsley,[75] and pea;[78] for a number of fungi including *Agaricus bisporus*,[79] yeast,[9,45,48,69] *R. stolonifer*,[86] and *M. rouxii*;[81] for a member of the fungi imperfecti *Aspergillus nidulans*;[50] and for the protists *Acanthamoeba castellanii*,[53] *P. polycephalum*,[25,51] and *Dictyostelium discoideum*.[67] Table 1 lists several representative class II RNA polymerase subunit structures.

The subunit structures of wheat, soybean, cauliflower, rye, maize, and yeast have been compared directly by analyzing the class II enzymes on the same slab gel or on identical gel systems.[74] For the plant enzymes, this has identified a general subunit structure consisting of two large subunits of 220,000 or 180,000 and 140,000 daltons and nine smaller subunits of about 40,000; 25,000; 21,000; 20,000; 17,500; 17,000; 16,300; 16,000; and 14,000 daltons. All subunits are present at nearly equimolar quantities except the 25,000 dalton polypeptide which is present at a molar ratio of approx-

Table 1
SUBUNIT STRUCTURES OF PLANT AND LOWER EUKARYOTIC NUCLEAR DNA-DEPENDENT RNA POLYMERASES

Cauliflower[a]			Wheat[b]			*Acanthamoeba*[c]			Yeast[d]		
I	II	III	I	II	III	I	II	III	I	II	III
190[e]	180		200	220		185	193		185	220	
		150			150			169			160
125	140	130	125	140	130	133	152	138	135	145	128
		70			94			82			82
		50			55	41.5		52	48		
38	40			40[f]			40[g]		44	46	
			$\underline{38}$		$\underline{38}$	$\underline{39}$		$\underline{39}$	$\underline{41}$		$\underline{41}$
						$\underline{37}$		$\underline{37}$			
						$\underline{35}$		34	36		
					30			30		33.5	34
								28.5			
		27		27 }	28						
$\underline{25}$[h,i]	$\underline{25}$	$\underline{25}$	24[h]	25 }	25	$\underline{22.5}$[h]	$\underline{22.5}$	$\underline{22.5}$	$\underline{28}$[h]	$\underline{28}$	$\underline{28}$
		24			24.5						
		23									
				21	20.5	17.5	18	17.5			
$\underline{22}$	$\underline{22}$	$\underline{22}$	$\underline{20}$	20	$\underline{20}$	$\underline{15.5}$	$\underline{15.5}$	$\underline{15.5}$	$\underline{24}$	$\underline{24}$	$\underline{24}$
	19	17.8			19.5				20	18	20
$\underline{17.5}$	$\underline{17.5}$	$\underline{17.5}$	$\underline{17.8}$	17.8	$\underline{17.8}$	$\underline{13.3}$	$\underline{13.3}$	$\underline{13.3}$	$\underline{14.5}$	$\underline{14.5}$	$\underline{14.5}$
	17		$\underline{17}$	17	$\underline{17}$		12.5		12.3	12.5	11
	16.2			16.3			12				
	16			16		10	10	10			
	14			14							

[a] References 40 and 43.
[b] References 36 and 42.
[c] References 52-54, 96, and 100.
[d] References 9 and 48.
[e] Molecular weights are given in kilodaltons. Accuracy in the estimation of molecular weight is probably ±10% of the values listed and may differ slightly to those discussed in the text.
[f] Actually two polypeptides of 42 and 40 kilodaltons sum to a molar ratio of about 1.
[g] A diffuse set of polypeptides sum to a molar ration of about 1.3.
[h] This subunit is basic in charge and is present at a molar ratio of about 2 in each organism.
[i] Common subunits in the RNA polymerases of each organism are underlined.

imately 2 and the 14,000 dalton polypeptide which is present at variable stoichiometry (generally less than a molar ratio of 1). A further exception to these equimolar ratios occurs in the 40,000 dalton subunits of wheat and rye where two polypeptides of 42,000 and 40,000 daltons sum to a molar ratio of 1, and the 25,000 dalton subunits of wheat, rye, and maize where polypeptides of 27,000 and 25,000 daltons sum to a molar ratio of about 2. The analogous subunits from each enzyme also migrate to the same relative positions on two dimensional polyacrylamide gels where subunits are separated primarily on the basis of charge in the first dimension and on the basis of molecular weight in the second dimension. Examples of direct comparisons of soybean, wheat, and yeast class II enzymes on one and two dimensional gels are shown in Figures 1 and 2.

The largest subunit of RNA polymerase II shows some heterogeneity in that one or both polypeptides of 220,000 and 180,000 daltons are associated with the purified enzyme. Figure 3 shows two different preparations of RNA polymerase II purified from soybean; one preparation contains the 220,000 dalton subunit while the other preparation contains the 180,000 dalton subunit. If only one of these polypeptides is present,

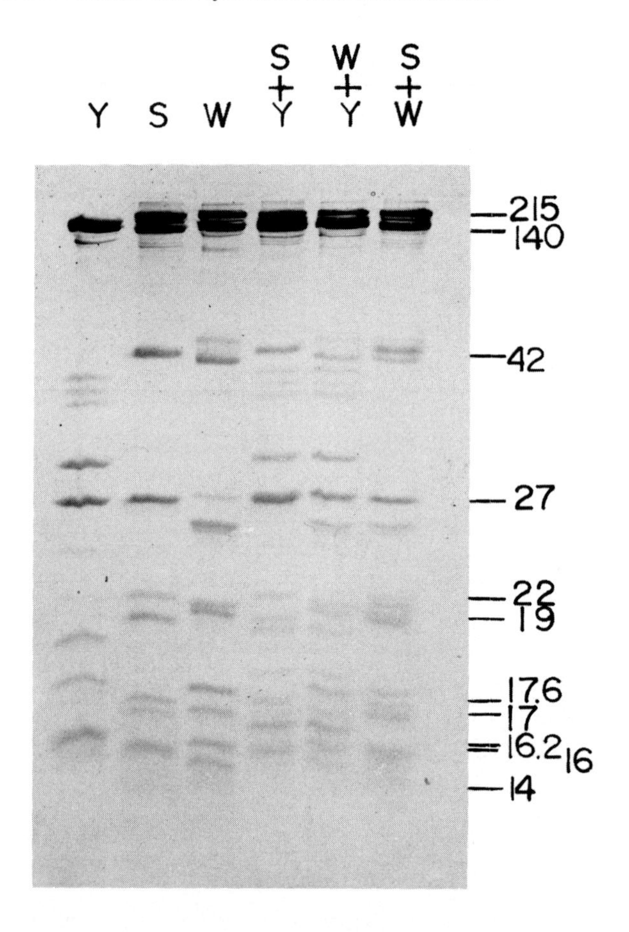

FIGURE 1. Polypeptide subunit patterns of yeast (Y), soybean (S), and wheat (W) class II RNA polymerases displayed on a 15% SDS polyacrylamide gel. Approximately 10 μg of each enzyme or mixtures of enzymes were applied to the gels. Numbers adjacent to the figure indicate molecular weights (kilodaltons) of soybean RNA polymerase II subunits. (Data from Guilfoyle, T. J. and Jendrisak, J. J., unpublished, 1978.)

then it is generally present at a molar ratio of 1; however, if both polypeptides are present, they generally sum to a molar ratio of 1. In some cases, even greater heterogeneity of the largest subunit has been observed, and in these cases several discrete polypeptides may be distinguished between 220,000 and 180,000 daltons.[36-38,51,91,92] This heterogeneity in the molecular weight of the largest subunit of RNA polymerase II has been found in animals,[15] plants,[6] fungi,[93] and protists.[51,53,67] There is evidence from peptide mapping that the 220,000 and 180,000 dalton subunits are related.[40,91] Figure 4 shows that the peptide pattern generated by limited proteolysis of soybean RNA polymerase II subunits is similar for the 220,000 and 180,000 dalton subunits, but that the patterns for these largest subunits differ from that of the 140,000 dalton subunit and other smaller subunits. Additional evidence for the relatedness of the 220,000 and 180,000 dalton subunits is provided by results that show that antibodies raised against the 180,000 dalton subunit cross-react with the 220,000 dalton subunit, but not with subunits of 140,000 daltons and smaller.[94]

The physiological significance of the heterogeneity observed in the largest subunit of RNA polymerase II remains questionable. Mixing experiments where tissues that contain only RNA polymerase II with the 220,000 dalton subunit are mixed with tissues

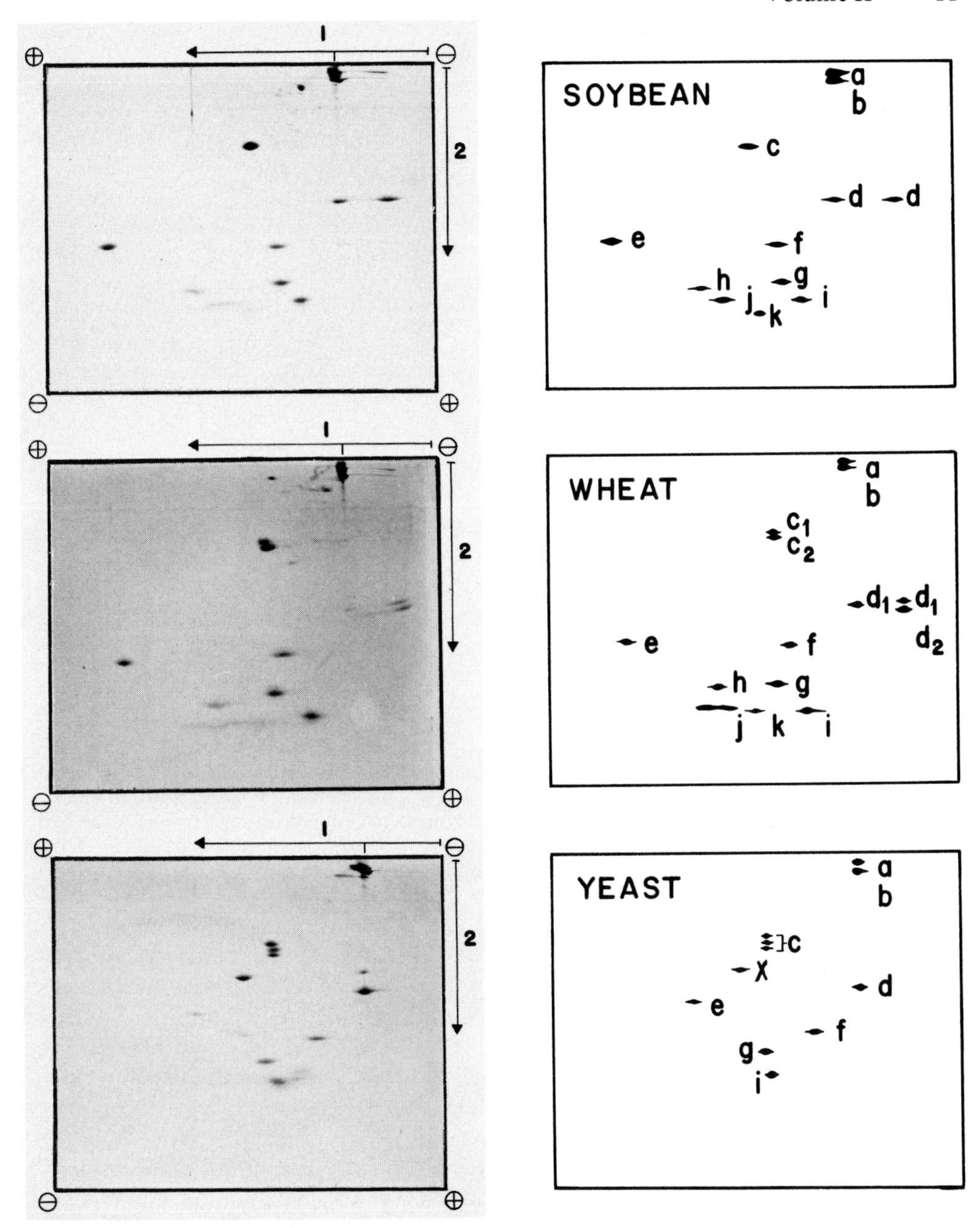

FIGURE 2. Polypeptide subunit patterns of soybean, wheat, and yeast class II RNA polymerases displayed on two dimensional polyacrylamide gels. The first dimension (horizontal) consisted of electrophoresis of individual enzymes on 7.5% polyacrylamide gels containing 8 *M* urea at pH 8.7 and the second dimension (vertical) consisted of electrophoresis of polypeptides out of the urea gels onto 3 mm thick, 15% SDS polyacrylamide slab gels. Approximately 25 μg of each enzyme were subjected to analysis. Direction of migration is indicated by the arrow. Letters correspond to the probable analogous subunits in each enzyme. For soybean the subunit molecular weights (kilodaltons) are (a) 215, (b) 138, (c) 42, (d) 27, (e) 22, (f) 19, (g) 17.6, (h) 17.0, (i) 16.2, (j) 16.1, and (k) 14; for wheat: (a) 220, (b) 140, (c) 42 and 40, (d) 27 and 25, (e) 21, (f) 20, (g) 17.8, (h) 17.0, (i) 16.3, (j) 16.0, and (k) 14; and for yeast. (a) 180, (b) 145, (c) 37 to 38, probably proteolytic degradation products of 46, (x) 33.5, (d) 28, (e) 24, (f) 20, (g) 14.5, and (i) 12.5. The splitting of the (d) subunits in soybean and wheat is an artifact of the gel system. (Data from Guilfoyle, T. J. and Jendrisak, J. J., unpublished, 1978).

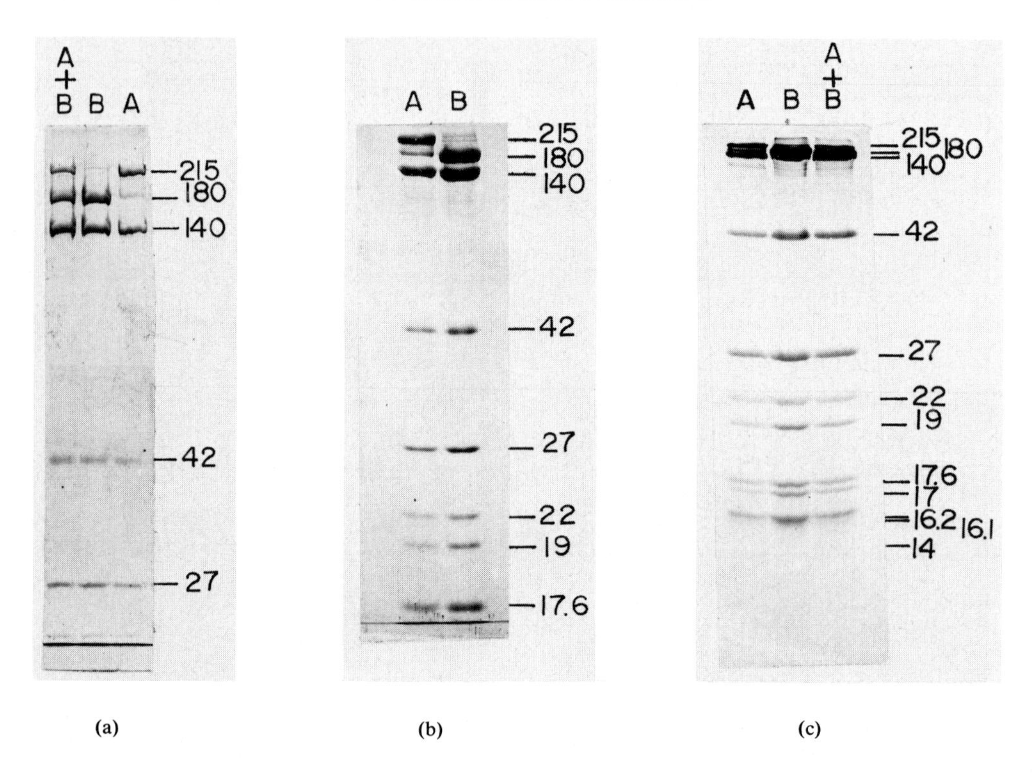

FIGURE 3. Polypeptide subunits of soybean RNA polymerases IIA (contains a largest subunit of 215,000 daltons) and IIB (contains a largest subunit of 180,000 daltons) separated on (a) 7.5%, (b) 10%, and (c) 15% SDS polyacrylamide gels. Enzymes were subjected to electrophoresis individually or as mixtures (2 to 10 μg amounts) on 0.75 mm thick slab gels. Molecular weights (kilodaltons) are indicated adjacent to the figures. (From Guilfoyle, T. J. and Jendrisak, J. J., *Biochemistry,* 17, 1860, 1978; © American Chemical Society. With permission.)

that contain only RNA polymerase II with the 180,000 dalton subunit (as judged by the form of the enzyme which is observed after purification of the enzymes to homogeneity) indicate that proteolysis of the 220,000 dalton subunit is not occurring during purification procedures since both forms of the enzyme are recovered at the expected ratio from such an experiment.[40] This result is, however, at variance with some other results which indicate that proteolysis of the 220,000 dalton subunit gives rise to the 180,000 dalton subunit during enzyme purification. When soybean RNA polymerase II is immunoprecipitated from partially purified enzyme preparations with antibody raised against the class II enzyme, both the 220,000 and 180,000 dalton subunits are observed on SDS polyacrylamide gels;[38] however, if the enzyme is purified to homogeneity from this same tissue, only the 180,000 dalton subunit is observed. With yeast, the 220,000 dalton subunit can be preserved during purification only if high amounts of the protease inhibitor, phenylmethylsulfonyl fluoride, are added to extraction and column buffers.[93] Yeast RNA polymerase II has recently been purified from a "protease strain" (presumably a strain of yeast low in one or more proteases) where 80% of the purified enzyme contains the 220,000 dalton subunit.[95] These latter experiments present a persuasive argument for the artifactual degradation of the 220,000 dalton subunit of RNA polymerase II to a 180,000 dalton polypeptide during enzyme purification.

It might be argued that the extreme complexity of subunit structures observed, for example in plants, results from the purification methods employed since all of the subunit structures reported here were obtained from enzymes purified by the method of Jendrisak and Burgess.[39] This method uses precipitation with and elution from Po-

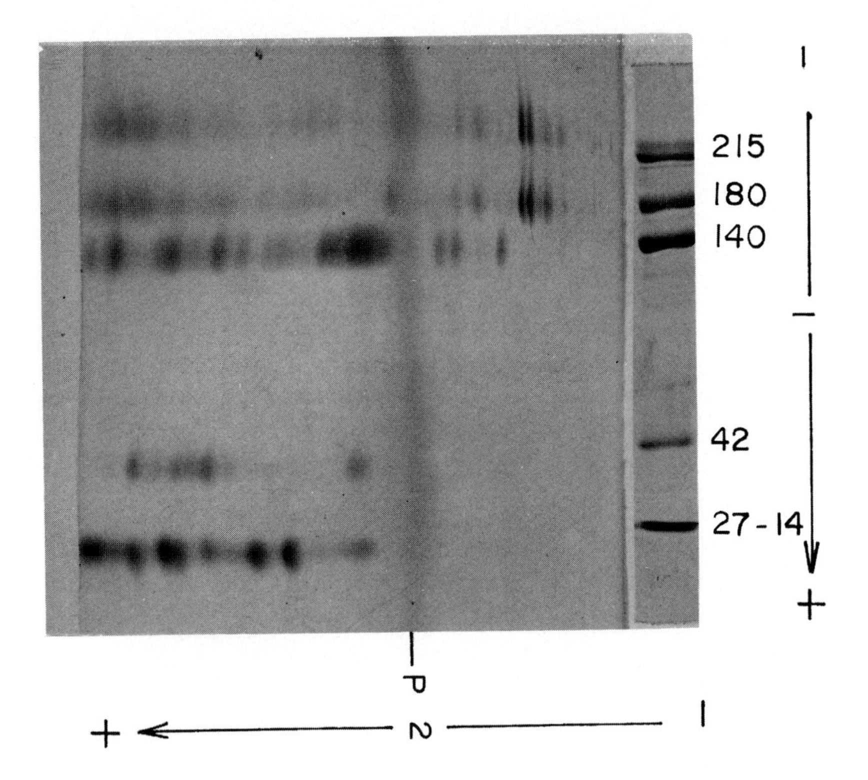

FIGURE 4. Peptide mapping of soybean RNA polymerases IIA and IIB high molecular weight subunits by limited proteolysis in 15% SDS polyacrylamide gels. The first dimension (1) represents a track from a 0.75 mm thick 5% SDS polyacrylamide slab gel to which was applied an equimolar mixture of soybean RNA polymerases IIA and IIB (25 μg of total protein). The first dimensional gel was positioned on top of the stacking gel of a 15% SDS polyacrylamide, 1.5 mm, slab gel. Buffer containing 100 μg/mℓ of *Staphylococcus aureus* V8 protease was layered over the first dimensional gel and electrophoresis was carried out at 12.5 mA until the marker dye reached the bottom of the second dimensional gel. Migration position of the protease is indicated by the letter P. Molecular weights (kilodaltons) of separated subunits in the first dimensional gel are indicated above the gel. (From Guilfoyle, T. J. and Jendrisak, J. J., *Biochemistry*, 17, 1860, 1978; © American Chemical Society. With permission.)

lymin P of soluble enzymes and is considered more gentle than high salt-sonication procedures required to solubilize nuclear-associated RNA polymerases (see Section IV.B). However, Figure 5 shows that cauliflower RNA polymerase II enzymes purified by the Polymin P method[39] or by the high salt-sonication method[19,20] have identical subunit structures. There appears to be no variability in RNA polymerase II subunit structure whether different purification methods are employed, whether the enzyme is purified from soluble or chromatin fractions of the cell, or whether different tissues in the same organism are used as a source of enzyme.

C. RNA Polymerase III

Due to the relatively low quantities of RNA polymerase III present in most tissues, few class III enzymes have been purified sufficiently to allow analysis of subunit structure. However, subunit structures have been reported for wheat,[42,80] cauliflower,[43] yeast,[26,48] *M. rouxii*,[81] and *A. castellanii*.[54] Several of these subunit structures are listed in Table 1. All class III enzymes examined contain two large subunits of about 150,000 and 130,000 daltons plus an additional intermediate sized subunit of 70,000 to 90,000 daltons, and as many as 11 smaller subunits are associated with the enzyme.

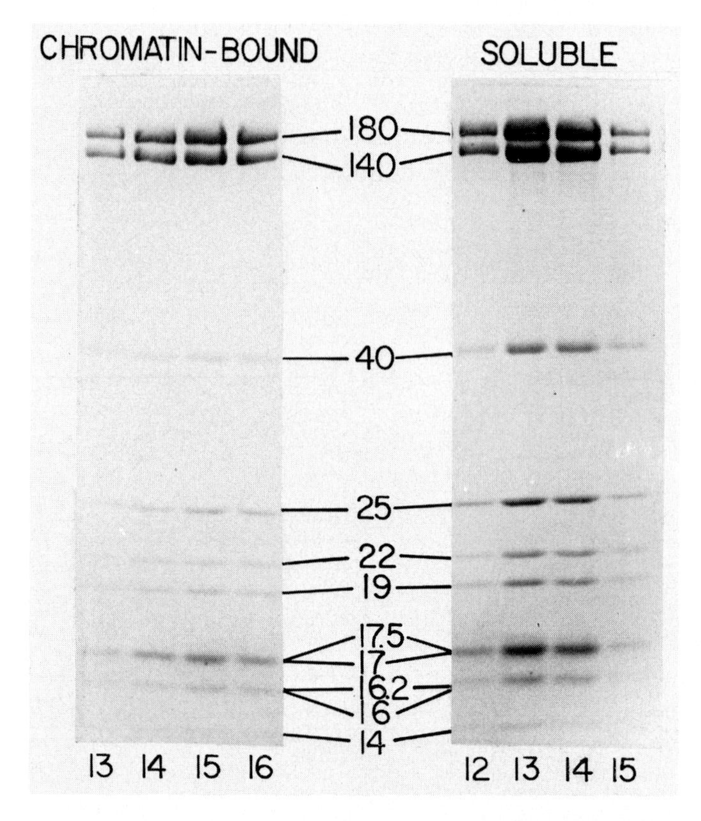

FIGURE 5. Polypeptide subunit structures of soybean RNA polymerase II enzymes purified from chromatin (by the high salt-sonication procedure[20]) and from cytosol (by the Polymin P procedure[74]). Individual lanes are successive fractions collected from a glycerol gradient. Fractions were subjected to electrophoresis on 7.5 to 15% gradient SDS polyacrylamide gels. Molecular weights (kilodaltons) are indicated adjacent to the figures. (Data from Guilfoyle, T. J., unpublished, 1981).

D. Common Subunits

This topic has recently been reviewed by Paule[96] and will only be briefly summarized here. Figure 6 shows a one dimensional gel of cauliflower RNA polymerases I, II and III where subunits of 25,000, 22,000, and 17,500 daltons are observed within each class of enzyme. In cauliflower, the class I and II enzymes have been analyzed by two dimensional polyacrylamide gel electrophoresis and the analogous subunits were shown to have identical mobilities.[43] In addition, the three common subunits of each class of enzyme cross-react when probed with antibody raised against cauliflower or soybean RNA polymerase II[43] (see Section VI and Figure 7). The common subunits of cauliflower, wheat, yeast, and *A. castellanii* RNA polymerases I, II, and III are indicated in Table 1. Each class of enzyme contains at least three common subunits in yeast, *Acanthamoeba,* wheat, and cauliflower, and two to three additional subunits are common to RNA polymerases I and III in yeast and *Acanthamoeba.* One of the common subunits (the 28,000 dalton subunit of yeast, the 22,500 dalton subunit of *Acanthamoeba,* and the 25,000 dalton subunit of cauliflower) within each organism is highly basic in charge (i.e., pI = 9) and is present at a molar ratio of about 2. This analogous subunit in soybean (molecular weight of 27,000) is found in RNA polymerases I and II; however, the structure of soybean RNA polymerase III has not been determined.[94] With wheat, the picture is less clear since the 25,000 and 27,000 dalton polypeptides

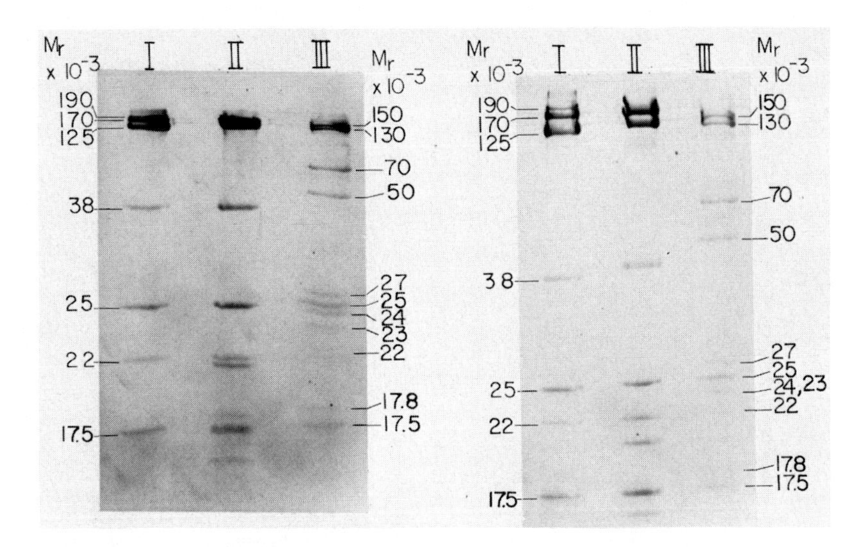

FIGURE 6. Subunit structures of cauliflower RNA polymerases I, II, and III displayed on (A) 15% and (B) 10 to 16% gradient SDS polyacrylamide gels. Approximately 2 μg of RNA polymerases I and II, and 0.5 μg of RNA polymerase III were applied to the gels. Numbers adjacent to the gels indicate molecular weights (kilodaltons) of subunits in RNA polymerases I and III. Molecular weights of RNA polymerase II subunits are found in Table 1. Subunits with identical mobilities in RNA polymerases I, II, and III are 25,000; 22,000; and 17,500 daltons. (From Guilfoyle, T. J., *Biochemistry*, 19, 5966, 1980; © American Chemical Society. With permission.)

of RNA polymerase II are both basic in charge and sum to a molar ratio of 2, but polypeptides of identical molecular weight are not associated with RNA polymerase I and only a 25,000 dalton polypeptide is associated with RNA polymerase III.

Other common subunits present in RNA polymerases I, II, and III from each organism are acidic in charge and present at molar ratios of approximately 1. When data are compiled from yeast, *Acanthamoeba*, cauliflower, and wheat germ, there is strong evidence for the existence of at least three subunits common to each class of RNA polymerase. This evidence is based on genetics,[97] one and two dimensional polyacrylamide gel analysis,[7,36,43,85,98-100] peptide mapping,[7,85] and immunology.[43,87] These common subunits probably perform similar functions in each class of RNA polymerase and may form a "core" enzyme along with the two largest subunits from each class of enzyme.[96]

VI. IMMUNOLOGICAL PROPERTIES

The immunological relationship of RNA polymerases I and II from yeast was first reported by Hildebrandt et al.[101] Antibodies raised against yeast RNA polymerase I inhibited in vitro RNA synthesis catalyzed by either RNA polymerase I or II from yeast and formed precipitan lines with the class I and II enzymes from yeast. Likewise, reciprocal experiments with antibody directed against yeast RNA polymerase II substantiated the immunological cross-reactivity of the yeast RNA polymerases. Buhler et al.[7,85] extended these immunological studies with yeast RNA polymerases by demonstrating that although antibodies raised against native RNA polymerase I cross-react with RNA polymerase II, antibodies directed against the largest subunit of yeast RNA polymerase I do not cross-react with other yeast RNA polymerase I subunits or with any yeast RNA polymerase II subunits.

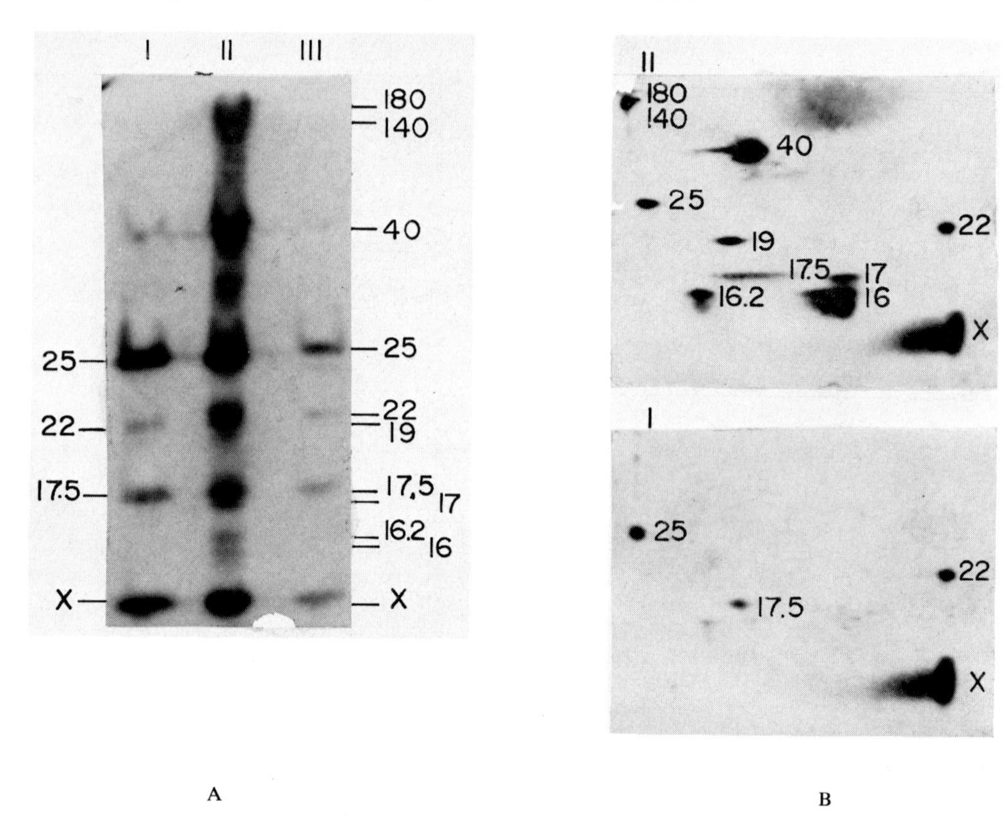

FIGURE 7. Binding of anticauliflower RNA polymerase II antibody to subunits of various RNA polymerases. RNA polymerase subunits were separated on SDS polyacrylamide gels, transferred to diazobenzyloxymethyl paper, incubated with anti-RNA polymerase II, and then incubated with ^{125}I-protein A. The autoradiograms displayed are (A) subunits of cauliflower RNA polymerases I, II, and III separated on 15% SDS polyacrylamide gels: (B) subunits of cauliflower RNA polymerases I and II separated on two dimensional 15% SDS polyacrylamide gels; and (C) subunits of cauliflower, soybean, yeast, and calf thymus RNA polymerase II enzymes and soybean RNA polymerase I separated on 15% SDS polyacrylamide gels. Molecular weights are indicated in kilodaltons. X is an unidentified antigen in the RNA polymerases. (From Guilfoyle, T. J., unpublished, 1981. Calf thymus RNA polymerase II was provided by Dr. R. Burgess.

A detailed immunological study of *Saccharomyces cerevisiae* RNA polymerases has recently been published.[87] Antibodies were prepared against native yeast RNA polymerases I and II as well as to each subunit of yeast RNA polymerase I. The immunological cross-reactivity of subunits associated with RNA polymerase I, II, and III from yeast as well as other species was evaluated by transferring subunits from SDS polyacrylamide gels to cellulose acetate membranes and probing these "protein blots" with ^{125}I-immunoglobulins or probing subunit-immunoglobulin complexes with ^{125}I-protein A from *Staphylococcus aureus.* The common subunits in class I, II, and III enzymes from yeast showed the expected cross-reactivity. In addition, the large subunits of each enzyme showed some faint cross-reactivity. Immunological cross-reactivity was also detected with the large subunits of RNA polymerase I from two distantly related yeast species, *Candida tropicalis* and *Endomycopsis fibuligera,* when these enzyme subunits were probed with antibody directed against *Saccharomyces cerevisiae* RNA polymerase I. However, no cross-reactivity was detected in a third species of yeast, *Schizosaccharomyces pombe.* Additional studies were carried out with antibodies raised against yeast RNA polymerase II. With antibody raised against the class II enzyme from yeast, binding of ^{125}I-immunoglobulins was observed for the largest subunit and one small subunit of wheat germ RNA polymerase II, but little or no binding was noted with

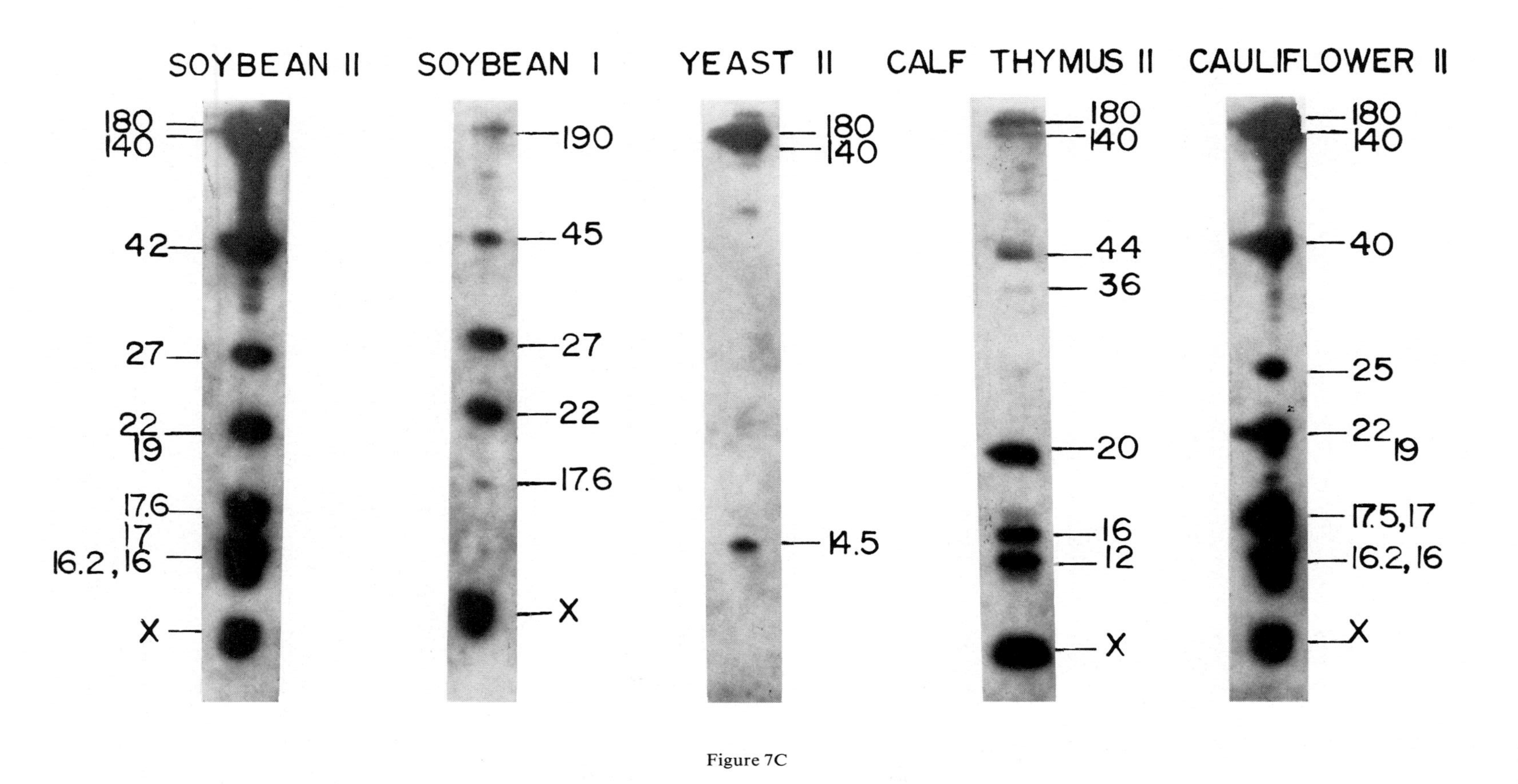
SOYBEAN II
180
140
42
27
22
19
17.6
17
16.2, 16
X
SOYBEAN I
190
45
27
22
17.6
X
YEAST II
180
140
14.5
CALF THYMUS II
180
140
44
36
20
16
12
X
CAULIFLOWER II
180
140
40
25
22
19
17.5,17
16.2,16
X

Figure 7C

subunits of calf thymus RNA polymerase II. No cross-reactivity was observed for subunits of *Escherichia coli* RNA polymerase when subunits of bacterial enzyme were probed with antibodies to either yeast RNA polymerases I or II.

Antibodies have also been raised against soybean RNA polymerases I and II,[38,44] cauliflower RNA polymerases II and III,[43] and the largest subunit (180,000 daltons) of cauliflower RNA polymerase II,[94] and these antibodies have been used to probe for cross-reacting subunits in a variety of class I, II, and III enzymes. Antibodies directed against native cauliflower RNA polymerase II bind to all the subunits of cauliflower, wheat, and soybean RNA polymerase II, and to the 25,000, 20,000, and 17,500 dalton subunits common to cauliflower RNA polymerases I and III.[43,94] Figure 7 shows some one and two dimensional gels of RNA polymerase subunits probed with antibodies directed against cauliflower RNA polymerase II. Antibodies raised against either cauliflower or soybean RNA polymerase II react with the largest subunits of calf thymus and yeast RNA polymerase II as well as with several smaller subunits in each enzyme.[94] Antibody raised against the 180,000 dalton subunit of cauliflower RNA polymerase II cross-reacts with the 180,000 and 220,000 dalton subunits of wheat and soybean RNA polymerase II, but with no other smaller subunits in any class II enzyme.[94]

The results obtained with antibodies directed against yeast and plant RNA polymerases and their subunits support the subunit status of polypeptides in each class of enzyme and the association of common subunits within each class of RNA polymerase. The observation that antibodies directed against individual subunits of a particular class of enzyme do not bind to other subunits in that class of enzyme indicates that the smaller subunits found in the nuclear RNA polymerases are not proteolytic degradation products of the larger subunits (the only exceptions being the 180,000 and 220,000 dalton subunits of RNA polymerase II and some proteolytic degradation products between the 190,000 and 120,000 dalton subunits of RNA polymerase I).[87,94] Finally, the immunological studies indicate that there is some evolutionary conservation in the large subunits of each class of enzyme within an individual species and within the same class of enzyme from divergent species. It is apparent that the largest subunits of plant RNA polymerase II enzymes share antigenic determinants even though peptide mapping of these subunits has indicated large differences in their respective fingerprints.[102]

VII. SUBUNIT FUNCTIONS

The elucidation of what subunits and other factors (i.e., enzyme-bound Zn^{++}) are required for RNA polymerase activity would be definitive if reconstitution of nuclear RNA polymerases could be achieved. To date, however, reconstitution of nuclear RNA polymerases from dissociated subunits has been unsuccessful. Therefore, other approaches such as the use of chemical probes and genetic mutations have been the only alternatives available to determine what function a subunit or other factor plays in the nuclear RNA polymerases.

That enzyme-bound Zn^{++} is required for nuclear RNA polymerase activities is suggested by experiments that demonstrate inhibition of RNA polymerase activity with the metal chelator 1,10-phenanthroline.[103,104] The detection of Zn^{++} by analytical procedures in wheat RNA polymerase II,[105] *Euglena gracilis* RNA polymerases I and II,[104,106] and yeast RNA polymerases I, II, and III[107-110] provides supporting evidence for the Zn^{++} requirement in nuclear RNA polymerases. Values for zinc content range from 1 g equivalent in yeast RNA polymerase II[107] to 7 g equivalents in wheat RNA polymerase II.[105] Enzyme-bound Zn^{++} is believed to function by promoting the activation of the 3′-hydroxyl group in the sugar moiety of RNA to facilitate nucleophilic attack of the α-phosphate group of the entering nucleotide.[27] Although Zn^{++} is clearly required for nuclear RNA polymerase activities, the subunits in these enzymes that bind Zn^{++} have not been identified.

Chemical reagent probes (i.e., pyridoxal 5′-phosphate) have been employed to investigate the roles of subunits in nuclear RNA polymerases. Pyridoxal 5′-phosphate which is thought to act as a nucleotide analogue inhibits yeast RNA polymerase I through the formation of a Schiff base with lysyl amino groups that may be located at the active site of the enzyme.[111] When yeast RNA polymerase I is reacted with pyridoxal 5′-phosphate and the Schiff base which forms is subsequently reduced with ^3H-$NaBH_4$, the tritium label is found associated with the 185,000; 137,000; 48,000; and 36,000 dalton subunits of the enzyme.[112] The addition of nucleotides protects the enzyme from inactivation and inhibits the binding of pyridoxal 5′-phosphate to the 185,000 and 137,000 dalton subunits. The addition of DNA also protects yeast RNA polymerase I from inactivation by pyridoxal 5′-phosphate and inhibits the binding to the 185,000; 137,000; 48,000; and 36,000 dalton subunits of the enzyme. These results suggest that both large subunits are involved with the interaction or binding of nucleotides and DNA and that the 48,000 and 36,000 dalton subunits are involved in DNA binding. More recent results indicate that the 48,000 and 36,000 dalton subunits can be removed from the enzyme by chromatography on blue dextran-Sepharose, a resin which interacts with polypeptides containing nucleotide binding sites.[90] The inclusion of ATP in the chromatography buffer prevents dissociation of the polypeptides from yeast RNA polymerase I. In these experiments, the yeast class I enzyme was completely inactivated by removal of the 48,000 and 36,000 dalton subunits; however, removal of these subunits by other procedures (see Section V.A) does not result in complete inactivation of the enzyme, but does result in inefficient transcription of native DNA templates, reduced binding to DNA templates, and reduced affinity for nucleotide substrates.[89,113] Additional information will be required before definitive assignment of transcriptional roles played by yeast RNA polymerase I subunits can be made.

Another approach to defining subunit function of nuclear RNA polymerases is based on genetics. Mutations in individual subunits of the RNA polymerases would aid in determining the functional role of each subunit. A mutant strain of yeast, *rpo* B_I, has been identified by its reduced RNA polymerase activity in crude cell-free extracts[114,115] and has been isolated.[114] The 32,000 and 16,500 dalton subunits are absent from the class II enzyme in this mutant, but it appears that the 220,000 dalton subunit has suffered the mutation.[116] The mutated enzyme possesses altered temperature and salt optima and is relatively deficient in catalyzing RNA chain initiation and elongation. Since removal of the 32,000 and 16,500 dalton subunits from the wild type enzyme can be achieved by electrophoresis or mild urea treatment without producing the enzymatic characteristics of the mutant enzyme, these small subunits are believed not to be the subunits that are mutated. Instead, a mutation in the 220,000 dalton subunit is supported by the different peptide maps obtained for the largest subunit of the wild type and *rpo* B_I strains of yeast. The absence of the 32,000 and 16,500 dalton subunits in the mutant enzyme is believed to result from the lack of interaction of these small subunits with the altered 220,000 dalton subunit in the *rpo* B_I strain. Based on the above results, it has been postulated that the 220,000 dalton subunit of yeast RNA polymerase II may be functionally equivalent to the β subunit of bacterial RNA polymerases.

VIII. α-AMANITIN INHIBITION

The most selective and best characterized inhibitor of eukaryotic RNA polymerases is the fungal toxin, α-amanitin. This octapeptide interacts directly with RNA polymerase II[13,117] and inhibits the propagation of RNA chains in in vitro transcription reac-

Table 2
α-AMANITIN SENSITIVITIES OF PURIFIED NUCLEAR RNA POLYMERASES

Classification and organism	Concentration of α-amanitin (μg/mℓ) required for 50% inhibition of RNA polymerase activity — RNA polymerase I	II	III	Ref.
Kingdom Animalia				
Phylum Chordata				
Mus musculus (mouse)	>500 (0)	0.01—0.05	20	15,119
Xenopus laevis (toad)	>500 (0)	0.01—0.05	20	15,119
Phylum Arthropoda				
Bombyx mori (moth)	>1000 (0)	0.01—0.05	>1000 (5)	15,119
Artemia salina (brine shrimp)	>1000 (10)	0.01	>1000 (30)	10,120
Kingdom Plantae				
Phylum Tracheophyta (vascular plants)				
Brassica oleracea (cauliflower)	>2000 (0)	0.05	1000—2000	20,43,74
Glycine max (soybean)	>100 (0)	0.05	ND	40,58,74
Pisum sativum (pea)	ND	0.05	ND	78
Petroselinum crispum (parsley)	>100 (0)	0.05	ND	60,91
Triticum aestivum (wheat)	>100 (0)	0.05	5	36,74
Secale cereale (rye)	>1000 (0)	0.05	5	57,74
Zea mays (maize)	ND	0.05	ND	74
Kingdom Fungi				
Phylum Basidiomycota (club fungi)				
Agaricus bisporus	ND	6.5	ND	79
Amanita brunnescens	ND	8.3	ND	124
Phylum Ascomycota				
Saccharomyces cerevisiae	300—600	1	2400 (5—35)	21,26
Podospora anserina	>1000 (0)	50	3200	126
Form-Phylum Deuteromycota (fungi imperfecti)				
Aspergillus nidulans	>400 (0)	>400 (0)	ND	49,50
Kingdom Protista				
Phylum Myxomycota (plasmodial slime molds)				
Physarum polycephalum	>200 (0)	0.05—0.1	ND	23,127
Dictyostelium discoideum	>10 (0)	0.05	ND	67
Phylum Rhizopoda (naked amoebae)				
Acanthamoeba castellanii	>500 (0)	0.009	20	65,66,68,128

Note: A > symbol indicates that 50% inhibition was not obtained, but that the concentration required for 50% inhibition is greater than the highest concentration tested; number in parenthesis is the % inhibition obtained. ND = not determined.

tions.[118] In general, plant and lower eukaryotic RNA polymerases show similar sensitivities to inhibition by α-amanitin as the mammalian RNA polymerases;[15] however, there are some notable exceptions. Table 2 presents a summary of α-amanitin sensitivities of RNA polymerases for a variety of higher and lower eukaryotic organisms.

All higher plant RNA polymerase I enzymes examined are refractory to α-amanitin inhibition[6] (Figure 8). Plant RNA polymerase II is inhibited by 50% at about 0.05 μg/mℓ of the fungal toxin.[74] The concentration of α-amanitin required for 50% inhibition of plant RNA polymerase III ranges from about 5 μg/mℓ for wheat[42] and rye[57] to 1000 to 2000 μg/mℓ for cauliflower.[20,43] This range in sensitivity for the RNA polymerase III enzymes purified from plants is not unlike that observed for animals where vertebrate class III enzymes are half maximally inhibited at about 10 to 20 μg/mℓ,[15] but the one insect[119] enzyme and one crustacean[120] enzyme examined are highly resistant to the fungal toxin (Table 2).

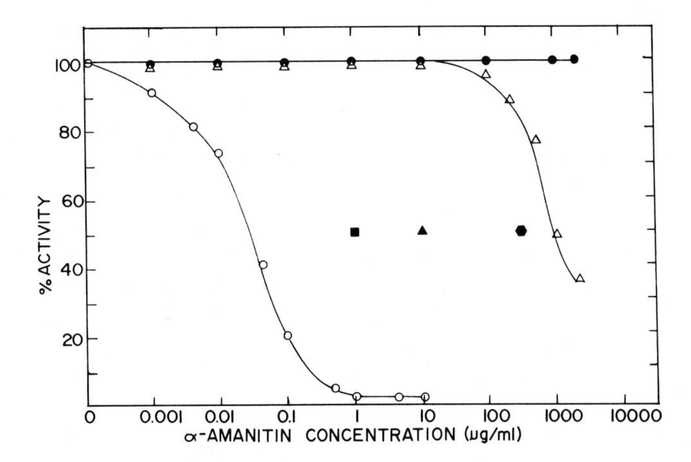

FIGURE 8. Effect of α-amanitin on the activities of RNA polymerases. (●) cauliflower RNA polymerase I, (○) cauliflower RNA polymerase II, (△) cauliflower RNA polymerase III, (▲) wheat RNA polymerase III,[42] (■) yeast RNA polymerase II,[21] and (⬢) yeast RNA polymerase I.[21] (From Guilfoyle, T. J., *Biochemistry,* 19, 5966, 1980; © American Chemical Society. With permission.)

The α-amanitin sensitivities of RNA polymerases purified from protists are similar to those reported for higher animals and plants. Enzymes which have been titrated with α-amanitin are listed in Table 2. RNA polymerase I enzymes are resistant to inhibition, RNA polymerase II enzymes are 50% inhibited in the range og 0.01 to 0.1 μg/mℓ, and the one protist (*A. castellanii*) class III enzyme examined is inhibited 50% at 20 μg/mℓ of α-amanitin.[66] Although α-amanitin inhibition curves are not available for other protists, RNA polymerase I is reported to be resistant and RNA polymerase II is reported to be inhibited at low concentrations of α-amanitin in the water mold (Phylum Oomycota), *Achlya bisexualis,*[121] and the chytrids (Phylum Chytridiomycota), *Blastocladiella emersonii*[61] and *Allomyces arbuscula.*[122]

In contrast to the general patterns of α-amanitin sensitivities observed for animal, plant, and protist RNA polymerases I, II, and III, the fungal RNA polymerases display a number of uncharacteristic sensitivities to the toxin (Table 2). The Basidiomycota (club fungi) RNA polymerase II enzymes are highly resistant to inhibition by α-amanitin. Since α-amanitin is produced in a number of club fungi,[123] it is not entirely surprising that α-amanitin-accumulating species possess RNA polymerase II enzymes highly resistant to the toxin. When α-amanitin sensitivity is assayed in isolated nuclei from the *Amanita* species that accumulate the toxin, 50% inhibition of the total sensitive RNA polymerase activity is approached at about 1200 to 1300 μg/mℓ of α-amanitin; however, even nonaccumulating species require about 10 to 30 μg/mℓ toxin for 50% inhibition.[79,121,123] RNA polymerase II enzymes purified from two nonaccumulating species, *Agaricus bisporus*[79] and *Amanita brunnescens,*[124] are inhibited 50% by 6 to 11 μg/mℓ of α-amanitin. Thus, class II enzymes which have been purified from the club fungi, including those species that do not accumulate the toxin, require at least 100- to 200-fold higher concentrations of α-amanitin for 50% inhibition compared to animals and plants.

At least one species in the conjugation fungi (Phylum Zygomycota), *R. stolonifer,* possesses a class II RNA polymerase that is 50% inhibited at 20 μg/mℓ of α-amanitin.[86,125] Whether RNA polymerase II from *M. rouxii,* a second species in this phylum, is more resistant than that characteristic of animals and plants is difficult to assess with available data, since only a single concentration (4.5 μg/mℓ) of α-amanitin was tested and greater than 85% inhibition was obtained.[81]

The α-amanitin sensitivities of RNA polymerases I, II, and III have been determined for two species in the Phylum Ascomycota or sac fungi (Table 2). RNA polymerase II purified from *Podospora anserina* is some 1000- to 5000-fold more resistant to α-amanitin than animal or plant class II enzymes.[126] While RNA polymerase I is resistant to 1000 to 2000 μg/mℓ of α-amanitin, 3200 μg/mℓ of the toxin produces 25% inhibition. RNA polymerase III from *Podospora* is inhibited 30% at 2000 μg/mℓ and 45% at 3200 μg/mℓ of the toxin. A second species in this phylum, *Saccharomyces cerevisiae,* contains RNA polymerase II that is 20- to 100-fold more resistant to α-amanitin than RNA polymerase II of plants and animals.[21,26,33] The class III enzyme from yeast is resistant to α-amanitin, similar to enzymes purified from *Bombyx mori, Artemia salina,* and cauliflower. Yeast RNA polymerase I has an α-amanitin sensitivity unique to any of the other class I enzymes; it is the only class I enzyme (with the possible exception of *Podospora*) which has been shown to be inhibited by the toxin. Although pure yeast RNA polymerase I is inhibited 50% by 300 to 600 μg/mℓ of α-amanitin, dissociation of two polypeptides of 48,000 and 37,000 from the enzyme results in an increased sensitivity to the toxin so that 50% inhibition is obtained with 100 to 200 μg/mℓ of α-amanitin.[7,88]

Aspergillus nidulans in the Form-Phylum Deuteromycota (fungi imperfecti) possesses the most α-amanitin resistant RNA polymerase II that has been purified and characterized.[50] This enzyme is refractory to inhibition by 400 μg/mℓ α-amanitin, the highest concentration tested.

From a comparison of the α-amanitin sensitivities of RNA polymerases I, II, and III across four kingdoms and some ten phyla (as discussed in the above paragraphs and summarized in Table 2), several generalizations can be made.

1. The animal, plant, and protist kingdoms contain species that possess RNA polymerases I, II, and III with characteristic sensitivities to α-amanitin where RNA polymerase I is refractory, RNA polymerase II is 50% inhibited at 0.01 to 0.05 μg/mℓ, and RNA polymerase III is 50% inhibited at 5 to 20 μg/mℓ.
2. Animal (at least Arthropoda) and plant class III enzymes may deviate from the typical sensitivity of 5 to 20 μg/mℓ of α-amanitin required for 50% inhibition to almost total resistance to inhibition.
3. Fungal RNA polymerase II is, in general, some 100- to 1000-fold more resistant to α-amanitin inhibition than animal, plant, or protist class II enzyme. This increased resistance of fungal RNA polymerase II is observed across three phyla and one form-phylum.
4. Yeast RNA polymerase I is the only class I enzyme sensitive to α-amanitin inhibition that has been described and the only enzyme observed to date that can be made more sensitive to the fungal toxin by altering its subunit structure (at least without mutagenesis).

IX. INTRACELLULAR LOCALIZATION AND TRANSCRIPTIONAL ROLES

Most of the information on the intracellular localization and transcriptional roles of nuclear RNA polymerases, I, II, and III has come from studies with animals.[15,18] RNA polymerase I is localized in the nucleolus and transcribes the ribosomal DNA cistrons producing precursor rRNA. RNA polymerase II is localized in the nucleoplasm and transcribes DNA sequences which code for precursors to mRNA. Certain small stable RNAs also appear to be synthesized by RNA polymerase II.[129] RNA polymerase III is also localized in the nucleoplasm and transcribes DNA sequences which code for 5S rRNA and precursors to tRNAs as well as other small stable RNAs. The meager amount of data available from plants and lower eukaryotes supports the results obtained with animals and this data is summarized below.

With plants, RNA polymerase I from soybean has been localized in the nucleolus[30] and has been shown to transcribe primarily ribosomal DNA sequences in isolated nuclei and chromatin as determined by competition-hybridization experiments with in vitro synthesized RNA.[130] RNA polymerase II has been shown to transcribe cauliflower mosaic virus DNA in isolated nuclei from infected turnip[32] and T-DNA in crown gall tumor nuclei.[131] Both of these DNAs presumably code for mRNA in vivo. The only other studies on the transcriptional roles of RNA polymerases from plants are experiments where the in vivo synthesis of specific classes of RNA were analyzed from wheat germinated in the presence and absence of α-amanitin.[132] These in vivo experiments indicate that class III RNA polymerase from plants catalyzes the synthesis of 5S rRNA and tRNA.

Other experiments that support the general conclusions of the similarity of transcriptional roles of RNA polymerases I, II, and III in animals, plants, and lower eukaryotes have been conducted with yeast. Yeast RNA polymerase III transcribes sequences coding for 5S rRNA and precursors to tRNA in isolated nuclei and chromatin.[133-136] The selective transcription of these DNA sequences has been demonstrated with endogenous nuclear RNA polymerase III[133,134] as well as with purified yeast RNA polymerase III which was added to yeast nuclei or chromatin.[135,136] Support for faithful transcription of these sequences by yeast RNA polymerase III in isolated nuclei or chromatin comes from size analysis, competition-hybridization, and RNA fingerprinting of in vitro transcription products. In addition to the selective transcription of DNA sequences that code for these small RNAs, the yeast class III enzyme appears to catalyze the synthesis of some large, unidentified RNAs in vitro.[133,134]

X. SELECTIVE BINDING AND INITIATION ON DEPROTEINIZED DNA TEMPLATES

Before proceeding to a discussion of results obtained for selective binding and initiation of RNA chains by eukaryotic RNA polymerases on deproteinized DNA templates, it is worthwhile for background purposes to briefly review the events required for faithful initiation of RNA chains by prokaryotic RNA polymerases. Bacterial RNA polymerase holoenzymes demonstrate a high degree of specificity in promoter site selection and initiation of RNA chains on deproteinized DNA templates.[137] The sequence of events for promoter site selection and RNA chain initiation by *Escherichia coli* RNA polymerase on templates such as T7 DNA and fd RF DNA have been described in some detail and a summation of these events follows. First, the bacterial RNA polymerase holoenzyme locates and binds selectively and stably to DNA promoter sequences forming an "I" or "closed" complex. This binary complex can be formed at low temperatures (i.e., 0°C) and is stable in the absence of nucleoside triphosphates with a half-time for dissociation of about 40 min. The "I" complex is, however, relatively sensitive to inhibition by polyanionic inhibitors such as heparin or drugs such as rifampicin. Second, the holoenzyme is thought to facilitate localized denaturation in the promoter region of the DNA duplex. This leads to the conversion of "I" complexes to "RS" (rapid start) or "open" complexes. In contrast to the "I" complex, the binary "RS" complex is stable for hours in the absence of nucleoside triphosphates, does not form at low temperatures, and escapes inhibition by polyanionic inhibitors and rifampicin when RNA synthesis is initiated. Third, in the presence of nucleoside triphosphates, the "RS" complexes catalyze the rapid formation of the first phosphodiester bond, thus starting the synthesis of a new RNA chain. Formation of the first phosphodiester bond results in the conversion of the binary RNA polymerase-DNA complex into a ternary complex consisting of RNA polymerase-DNA-nascent

RNA chain. Ternary complexes are extremely stable even at high ionic strength. The formation of ternary complexes results in dissociation of the sigma subunit of RNA polymerase and the core enzyme catalyzes the extension of RNA chains.

Like prokaryotic RNA polymerases, eukaryotic RNA polymerases transcribe both double stranded and single stranded DNA templates. Single stranded templates generally promote higher rates of transcription with the eukaryotic enzymes. Yeast RNA polymerases I and II transcribe unnicked double stranded DNA very inefficiently, but the introduction of unpaired gaps into the duplex DNA by treatment with pancreatic deoxyribonuclease and exonuclease III stimulates transcription by the yeast enzymes.[138] Likewise, the template activity of duplex calf thymus DNA is dramatically reduced after treating the template with endonuclease S_1, an enzyme that removes single stranded regions from duplex DNA. These and additional experiments indicate that yeast RNA polymerases I and II require a gap of some 25 to 50 nucleotides in length to initiate RNA chains on duplex DNA.[138] In contrast to the gap required for yeast RNA polymerase initiation, animal and plant RNA polymerases are capable of initiating transcripts at nicks, but they are equally inefficient as yeast enzymes in initiating on an unnicked DNA template.[13,14,139] Wheat germ RNA polymerase II binds specifically and stably to site specific nicks introduced into simian virus-40 (SV-40) DNA.[140] This binding protects about 38 base pairs of DNA against deoxyribonuclease I digestion and about 34 base pairs of DNA against micrococcal nuclease digestion. Binding at nicks is not only stable, but transcription initiated at nicks by wheat germ RNA polymerase II is resistant to inhibition by heparin.[139]

RNA transcripts that appear to initiate at nicks or gaps on a duplex DNA template may not always result from authentic initiations of new RNA chains with a 5′ terminal triphosphate.[141] Wheat germ RNA polymerase II is capable of catalyzing both the initiation of new RNA chains and the covalent linkage of RNA transcripts to the 3′ termini of the nicked DNA template. In the latter case, the DNA template is acting as a primer for the extension of RNA chains. The priming reaction is strongly inhibited when Mn^{++} is substituted for Mg^{++} as the divalent cation in the in vitro transcription assay.

Wheat germ RNA polymerases I, II, and III have been reported to form heparin resistant complexes in vitro on cauliflower mosaic virus DNA template.[142] These complexes are, however, formed primarily at random nicks, as demonstrated by others with different templates, or at site specific discontinuities in the cauliflower mosaic virus DNA (the site specific discontinuities are a peculiarity of the cauliflower mosaic virus genome).

Selective and sequence specific binding, unrelated to nicks in the DNA, has been reported with wheat germ RNA polymerase II and deproteinized adenovirus DNA,[143,144] SV-40 DNA,[145,146] and a recombinant plasmid containing the entire genome of cauliflower mosaic virus DNA.[147] Binary complexes formed with wheat germ RNA polymerase II and adenovirus 2 DNA were localized by electron microscopy to 12 to 14 specific strong sites and 2 to 3 weaker sites.[144] All but one of these binding sites corresponded to similar binding sites when human placenta RNA polymerase II was substituted for the plant enzyme and six of the binding sites overlapped DNA regions that function as promoters in vivo. It was suggested that wheat germ RNA polymerase II is capable of forming two types of binary complexes analogous to the "I" and "RS" complexes formed by bacterial RNA polymerases, since the wheat enzyme formed stable binary complexes at both 0 and 37°C. Similar to the bacterial enzymes, complexes formed at 37°C were resistant to inhibition by polyanions and a derivative of rifampicin, rifamycin AF/013. In addition, Witney et al.[144] showed that the localization of ternary complexes with wheat germ RNA polymerase II and adenovirus 2 DNA template mapped in regions similar to where the binary complexes mapped.

With superhelical SV-40 DNA, a fraction of wheat germ RNA polymerase molecules forms stable binary complexes with a half-time for dissociation of about 30 min.[145] This contrasts with results where calf thymus RNA polymerase II is substituted for the plant enzyme; these complexes are very unstable. It was suggested that the wheat enzyme might contain additional factors that facilitate stable binding of the enzyme to the duplex DNA. Both enzymes are able to form stable ternary complexes in the presence of a specific dinucleotide and a single nucleoside triphosphate on the SV-40 template. Stability of wheat germ RNA polymerase II-SV-40 binary complexes is greater for superhelical DNA than with either linear or relaxed circular DNA. Greater binding stability and transcription by wheat germ RNA polymerase II on superhelical DNA compared to linearized or relaxed circular DNA suggest that underwinding of the DNA template provides the necessary activation energy for localized denaturation of the duplex DNA allowing for greater binding and initiation of RNA transcripts.[148]

With linearized SV-40 DNA, three major binding sites were mapped by electron microscopy to similar regions on the duplex DNA with wheat germ RNA polymerase II, calf thymus RNA polymerase II, and *E. coli* RNA polymerase.[146] However, inclusion of the dinucleotide primer, ApA, and a nucleoside triphosphate, ATP, resulted in the specific binding of the prokaryotic enzyme to a single site on the duplex DNA template while both eukaryotic enzymes bound at many unrelated sites. Using a similar approach, Lescure et al.[95] have demonstrated that yeast RNA polymerase II can selectively bind and initiate transcription on the yeast alcohol dehydrogenase I portion of a recombinant plasmid containing the yeast structural gene. This selectivity is dependent on the presence of a dinucleotide, UpA, and UTP, and binding is upstream from the coding region of the gene.

It is clear from the above studies that purified eukaryotic RNA polymerases are capable of some degree of selective binding and initiation of RNA chains on deproteinized DNA templates; however, in no case has the selectivity been definitively correlated with an in vivo promoter or initiation site. From a number of studies with animal cell-free extracts, it is apparent that other factors are required along with RNA polymerase to obtain a high degree of faithful transcription in vitro (see Section XI).

XI. SELECTIVE TRANSCRIPTION OF SPECIFIC GENES

In vitro transcription studies with animal RNA polymerases and deproteinized DNA templates have demonstrated that ancillary factors are required in addition to the RNA polymerases for faithful transcription.[15,18] A possible contradiction to this generalization has been reported for the in vitro transcription of ribosomal DNA cistrons in the yeast genome with purified yeast RNA polymerase I. Originally, Hollenberg[149] reported that a partially purified RNA polymerase I from *S. cerevisiae* showed little selectivity for transcription of ribosomal cistrons with deproteinized genomic DNA as template; however, Cramer et al.[150] concluded that with yeast genomic DNA enriched for ribosomal DNA sequences, purified yeast RNA polymerase I, but not II, exhibited some asymmetric transcription of the ribosomal DNA cistrons. With another species of yeast, *S. carlsbergensis,* Van Keulen et al.[84,151] demonstrated that purified RNA polymerase I catalyzed selective and asymmetric transcription of ribosomal DNA provided that the DNA template had a high molecular weight. When high molecular weight yeast genomic DNA was used as template with low enzyme to DNA ratios, 85% of the RNA synthesized in vitro with purified yeast RNA polymerase I was rRNA. The size of the in vitro transcription product was reported to be approximately 42 S, similar to the size of precursor yeast rRNA, and the direction of transcription was reported to be from the 5′ side of 17 S rRNA to 26 S rRNA, in agreement with the direction of transcription determined in vivo. In addition to the requirement for template integrity, Holland et al.[47,152] showed that the choice of divalent cation (e.g., Mg^{++} vs. Mn^{++}) as well as the method of enzyme purification was important in obtaining selective transcription of yeast ribosomal DNA by the homologous class I enzyme. However, more recent studies[135] from the same laboratory have failed to consistantly detect selective transcription of ribosomal cistrons when a recombinant plasmid containing a yeast ribosomal gene is used as template. In contrast, Sawadogo et al.[153] have reported the selective and asymmetric transcription by purified yeast RNA polymerase I of yeast ribosomal DNA in a hybrid plasmid; however, the initiation site used in vitro appeared to be upstream from the presumptive in vivo initiation site. These results taken together indicate that although yeast RNA polymerase I shows some selectivity for transcription of the coding strand of deproteinized yeast ribosomal genes, other factors are probably required for accurate transcription of these genes in vitro.

Selective transcription by yeast RNA polymerase II has only recently been investigated with recombinant plasmids containing yeast structural genes. Lescure et al.[95] have demonstrated that under specialized assay conditions, purified yeast RNA polymerase II selectively initiates transcription on the yeast portion of a recombinant plasmid containing the yeast structural gene for alcohol dehydrogenase I. Selective transcription by purified yeast RNA polymerase II on cloned 2 μm yeast plasmid DNA has also been reported.[154] That additional factors will be required for accurate in vitro transcription with purified RNA polymerase II is expected from results obtained with animal systems.[155,156]

Cell-free extracts derived from plants and lower eukaryotes that promote accurate in vitro transcription have yet to be described. Faithful in vitro transcription in nonanimal systems has thus far been documented only for the transcription of 5 S rRNA genes where purified yeast RNA polymerase III is added to isolated yeast nuclei or chromatin.[135,136]

XII. RNA POLYMERASES AS MULTIFUNCTIONAL ENZYMES

Because of the complex subunit structures of nuclear RNA polymerases, there has been some interest in whether these enzymes may represent multienzyme complexes capable of catalyzing reactions in addition to transcription of DNA templates. Yeast RNA polymerase I contains two polypeptides that possess ribonuclease H activity.[157,158] Ribonuclease H catalyzes the degradation of RNA-DNA hybrids and it has been suggested that RNA-DNA hybrids in chromatin could possibly regulate the activity of that template.[158] Renaturation of RNA polymerase I subunits separated on SDS polyacrylamide gels reveals the presence of ribonuclease H activities associated with the 48,000 and 40,000 dalton polypeptides of the class I enzyme. The 48,000 dalton subunit of yeast RNA polymerase I can be readily dissociated from the enzyme complex (i.e., by phosphocellulose chromatography; see Section V.A) and this also dissociates ribonuclease H activity. Further evidence for the 48,000 dalton subunit of yeast RNA polymerase I being a ribonuclease H enzyme comes from peptide mapping; a 48,000 dalton ribonuclease H enzyme purified from chromatin and the 48,000 dalton subunit of the class I enzyme produce similar peptide maps when subjected to limited proteolysis on sodium dodecyl sulfate (SDS) polyacrylamide gels. Although the above results indicate that yeast RNA polymerase I possesses ribonuclease H activity as well as polymerase activity, the subunit status of the 48,000 dalton polypeptide of yeast RNA polymerase I is somewhat questionable since this polypeptide can be readily separated from the class I enzyme by relatively mild chromatographic and electrophoretic procedures.

Another study which suggests that nuclear RNA polymerases may be capable of catalyzing reactions other than transcription of DNA templates has recently been reported. Purified RNA polymerase II from tomato or wheat germ is capable of synthesizing full length copies of viroid RNA templates in vitro.[159] Viroids are single stranded covalently closed circular RNA molecules made up of about 360 nucleotides. The in vitro synthesis of full length viroid RNA with plant class II enzymes along with in vivo results that indicate viroid replication is inhibited by 10^{-8} *M* (about 0.01 μg/mℓ) α-amanitin[160] suggest that RNA polymerase II might function in the replication or transcription of certain RNA as well as DNA templates in vivo.

XIII. QUANTITATIVE AND QUALITATIVE CHANGES DURING GROWTH AND DEVELOPMENT

The major classes of cellular RNA (e.g., 18 S and 28 S rRNA, tRNA, 5 S rRNA, and mRNA) are synthesized at different rates during many growth and developmental transitions in eukaryotes.[161] Numerous eukaryotic organisms have been studied to determine whether endogenous nuclear or solubilized RNA polymerase activities can be correlated with differential rates of synthesis of certain classes of cellular RNA during these transitions.

A. Changes During Development

Germination and subsequent growth of plant embryonic axes is a developmental transition where metabolically quiescent tissue undergoes a rapid activation which is initiated by imbibition. The emergence of the axis from the seed results primarily from elongation of cells and this rapid growth coincides with increased rates of RNA synthesis. With soybean axes, there is a 25-fold increase in RNA polymerase activity associated with isolated nuclei during the first 24 hr of germination and axis growth.[37,38] During this same period, there is no increase in the amount of RNA polymerases I or II as judged by chromatography of the solubilized enzymes on DEAE-Sephadex® and the purification of the class II enzymes to homogeneity. The subunit structures of RNA polymerase II purified from soybean axes germinated for 0, 12, 24, and 48 hr are identical on one and two dimensional polyacrylamide gels except for the molecular weight of the largest subunit which changes from 220,000 to 180,000 daltons during this period of axis growth. In germinating wheat, there is a similar alteration in the molecular weight of the largest subunit of RNA polymerase II, but in addition, the relative amounts of the 25,000 and 27,000 dalton subunits change during the first 36 hr postimbibition.[36] Although mixing experiments indicate that proteolysis of the largest subunit is not occurring during the purification of RNA polymerase II,[36,38] immuno-precipitation of the class II enzyme after partial purification suggests that at least a portion of the 180,000 dalton polypeptide may arise as a purification artifact[38] (See Section V.B).

RNA polymerase activities have also been monitored during developmental transitions in several fungi. Young and Whiteley[162] have reported changes in the specific activity and amounts of RNA polymerases I and II, but not III, during the transition from yeast-like to mycelial growth in *M. rouxii*. An enhanced level of RNA polymerase I which was solubilized and chromatographed on DEAE-Sephadex® could be correlated with enhanced synthesis of rRNA induced by aeration of the fungus. Solubilization and chromatography of RNA polymerases from ungerminated and germinated spores of *R. stolonifer* indicate that RNA polymerases I, II, and III are present in germinated spores, but that RNA polymerase II is absent in ungerminated spores.[125] The DEAE-cellulose chromatographic profiles reported in this study are, however, not consistant with data which indicates that RNA polymerase III elutes at low salt rather than high salt concentration on this resin (see Section II). Since RNA polymerase II as well as RNA polymerases I and III were relatively resistant to the α-amanitin concentration (10 μg/mℓ) tested in this study, it is possible that the DEAE-cellulose activity peak referred to as RNA polymerase III could, in fact, be a subform of RNA polymerase II.

RNA polymerases I, II, and III purified from the yeast and mycelial phases of *Histoplasma capsulatum,* a member of the fungi imperfecti, are reported to differ in α-amanitin sensitivity, mono- and divalent cation optima, temperature sensitivity, and subunit structure.[163] The relatively high levels of RNA polymerase III (about 80% of the total activity resolved on DEAE-Sephadex® chromatography) compared to RNA polymerases I and II as well as the interpretation of subunit structures from poor quality polyacrylamide gels makes this study somewhat questionable.

In the kingdom Protista, the RNA polymerases of *Acanthamoeba castellanii* have been studied in great detail during encystment of this soil amoeba. In the process of encystment, the metabolically active amoebae or trophozoites can be induced to form metabolically inactive cysts by transferring the trophozoites to a nutrient deficient medium. When RNA polymerases are solubilized from nuclei and chromatographed on DEAE-Sephadex®, it appears that the ratio of RNA polymerase I to II decreases during encystment, and this correlates with the decrease in in vitro rRNA synthesis.[128] More recent studies indicate that the total amount of RNA polymerases I, II, and III in whole cell extracts remains constant during encystment.[66] The difference between results obtained with whole cell extracts and isolated nuclei is suggested to arise from differential leaching of RNA polymerases from nuclei isolated from cysts compared to nuclei isolated from trophozoites.[66] Purification of *A. castellanii* RNA polymerases I and II to homogeneity from trophozoites and cysts indicates that the subunit structures of the enzymes are unaltered during the process of encystment.[65,68] In *Naegleria gruberi,* an amoeba-flagellate, no changes in the abundance or proportions of the nuclear RNA polymerases occur during the differentiation of the amoeba to the flagellate.[64] During this differentiation, proportionally less rRNA and more heterogeneous RNA (hnRNA) are synthesized in vivo, with no fluctuations in the amounts or properties of RNA polymerases. In *D. discoideum,* Pong and Loomis[67] have reported that during differentiation of vegetative amoeba into fruiting bodies, a reduction of both RNA polymerases I and II occurs with no change in the relative proportions of the enzymes. The subunit structure of RNA polymerase II is unaltered during this developmental process.

B. Changes During Proliferation

Plant RNA polymerases I and II have been studied in a number of organs where cell division is induced by the plant growth substance, auxin. Synthetic auxins such as 2,4-dichlorophenoxyacetic acid (2,4-D) induce the normally quiescent regions of stem and root tissues to proliferate.[164,165] This proliferation is preceded by the massive production of ribosomes and rRNA,[164] the swelling of nucleoli,[29] and a five- to tenfold enhancement of RNA polymerase I activity, but no increase in RNA polymerase II activity, assayed in chromatin or isolated nuclei.[28,72,166,167] Solubilization and DEAE-cellulose or DEAE-Sephadex® chromatography of the RNA polymerases from isolated nuclei or chromatin indicate that the level of RNA polymerase I increases in the nucleus after auxin treatment or that the specific activity of the class I enzyme is altered.[28,168,169] Increased RNA polymerase I activity has also been observed with auxin-treated lentil root.[170] Based on purification data where RNA polymerases I and II were purified to homogeneity from soybean hypocotyl induced to proliferate by 2,4-D, it has been shown that the amounts of RNA polymerases I and II increase by about ten- to twentyfold and sixfold, respectively, on a fresh weight tissue basis and about three- to sixfold and twofold, respectively, on a tissue DNA basis.[37,44] This latter study also showed that the increase in enzyme levels following auxin treatment resulted from *de novo* synthesis of the enzymes, that the relative incorporation of labeled amino acids into each RNA polymerase subunit was unaltered in slowly and rapidly proliferating hypocotyls, and that the subunit structures of RNA polymerases I and II were identical in quiescent and proliferating hypocotyls.

In yeast, Sebastian et al.[171] have reported a correlation between growth rate, cellular RNA content, and the amount of RNA polymerase I. It was suggested that RNA polymerases I and II are independently regulated in yeast cells and that the amount of RNA polymerase I may regulate the synthesis of rRNA. DEAE-Sephadex® fractionation of RNA polymerases I and II indicated about a threefold increase in RNA polymerase I, but a slight decrease in RNA polymerase II when yeast cells undergo a transition from slow to fast growth rates.

C. Changes During the Cell Cycle

The enzymatic activities of yeast RNA polymerases I and II have been investigated during various stages of the cell cycle.[172,173] The relative levels of RNA polymerases I and II, as judged by DEAE-Sephadex® profiles of solubilized RNA polymerases from whole cell extracts, indicate that the ratio of RNA polymerases I and II differs during G_1, S, early G_2, and late G_2 stages of the cell cycle. It was suggested that RNA polymerases I and II are regulated independently during the cell cycle in yeast and that RNA polymerase I is continuously synthesized while RNA polymerase II is synthesized in a step-wise manner during various stages of the cell cycle.[173] In contrast to the yeast studies, the levels of RNA polymerases I and II are reported to remain unchanged during the mitotic cycle of synchronously dividing *Physarum polycephalum* plasmodia.[174]

D. Summary

From the studies reported on the quantitative and qualitative changes in RNA polymerases during various developmental and growth transitions in plants and lower eukaryotes (as summarized in the above paragraphs) as well as animals,[15,18] several generalizations can be made. First, although there may be good correlation between in vivo rates of RNA synthesis and the levels of RNA polymerase activity expressed in isolated nuclei or chromatin, there is not necessarily a similar correlation with the total amounts of RNA polymerase I, II, and III in the tissue or cell. Second, RNA polymerases I, II, and III are undoubtedly present in both quiescent and dividing cells, and other factors or chromatin structure must play a large role in regulating the overall rates of transcription. Third, there is little evidence for qualitative changes (i.e., alteration in subunit structure) of RNA polymerases during transitions in growth and development. When changes in subunit structure are noted, they can generally be explained by purification artifacts (i.e., proteolysis of susceptible subunits); however, the presence of subforms of RNA polymerases that are present in different ratios during specific growth and developmental stages should not be entirely discounted at this time.

XIV. STIMULATORY AND INHIBITORY FACTORS

A variety of stimulatory and inhibitory factors that modify in vitro transcription have been described. This subject has been reviewed[3] and the author will only summarize some recent results where these factors have been characterized with respect to molecular structure and mode of action.

Link and Richter[175] have described the purification of a factor or group of factors that modify the in vitro activity of parsley RNA polymerase II. This factor copurifies with RNA polymerase II through DEAE-Sephadex® chromatography, but elutes from phosphocellulose at higher salt concentrations than those required to elute RNA polymerase II. When the purified factor is added back to pure parsley RNA polymerase II or to the enzyme that has been purified through phosphocellulose, it reduces the salt concentration optimum from 0.12 to 0.05 *M* ammonium sulfate, changes the preference of divalent cation from Mn^{++} to Mg^{++}, and increases transcriptional activity on double stranded parsley DNA by about threefold. The characteristics of purified parsley RNA polymerase II with added factor are similar to characteristics of the enzyme purified through DEAE-Sephadex® but not subjected to chromatography on phosphocellulose. The purified factor contains polypeptides of 30,000; 26,000; 25,000; and 14,000 daltons and some of these polypeptides have molecular weights identical to parsley RNA polymerase II subunits. Stimulation of RNA polymerase II by the factor on double stranded DNA templates appears to result from an enhanced elongation rate of RNA chains in vitro. The factor specifically stimulates RNA polymerase II transcription since no effect is observed when the factor is added to in vitro transcription reactions containing parsley RNA polymerase I or *E. coli* RNA polymerase.

A 37,000 dalton basic protein referred to as P_{37} has been purified from yeast and this protein stimulates in vitro RNA synthesis of both RNA polymerases I and II on double stranded DNA templates.[176-178] By mixing equimolar amounts of P_{37} and purified yeast RNA polymerase II, a complex of the factor and class II enzyme can be isolated by sedimentation on glycerol gradients. The complex has a dissociation constant of about 5×10^{-8} *M* and is unstable at high ionic strength (i.e., 0.3 *M* ammonium sulfate). Polyacrylamide gel analysis indicates that P_{37} preferentially associates with the subform of yeast RNA polymerase II containing the 220,000 dalton subunit (see Section V.B); however, stimulation of in vitro transcription on double stranded DNA is obtained with both subforms of yeast RNA polymerase II (e.g., subform A contains a 220,000 dalton subunit and subform B contains a 180,000 dalton subunit). Stimulation of in vitro transcription in the presence of the factor appears to be due to increased propagation of RNA chains with no effect on RNA chain initiation. Additional studies with P_{37} indicate that yeast RNA polymerase I in vitro transcription is also stimulated when the factor is added, but stimulation does not occur with *E. coli* holo or core RNA polymerase, wheat germ RNA polymerase II, or calf thymus RNA polymerase II. Antibodies raised against the individual subunits of yeast RNA polymerase I have been used to probe the subunits of yeast RNA polymerases I and II that interact with P_{37}. Results indicate that a subunit of 23,000 daltons which is common to both classes of RNA polymerase is primarily involved in the interaction with P_{37}. Some additional interaction must also occur, however, since one subform of yeast RNA polymerase II interacts preferentially with P_{37} and the dissociation constant of the RNA polymerase I-P_{37} complex is some two to five times greater than the RNA polymerase II complex.

A phosphoprotein from *P. polycephalum* has been purified and shown to stimulate transcription of ribosomal DNA cistrons of minichromosomes associated with isolated nucleoli.[179] This protein occurs as a dimer with a subunit molecular weight of 70,000. The phosphorylated protein binds to specific restriction fragments of ribosomal DNA, but this binding does not occur after dephosphorylation of the protein. When the ribosomal DNA minichromosomes are isolated from *Physarum,* the phosphorylated protein is found associated with them and is reported to have a binding constant of 10^{-10} *M.* Whether stimulation of rRNA synthesis on the minichromosomes is the result of increased initiation or increased RNA chain elongation has not been determined. In addition to the phosphoprotein stimulatory factor, a nuclear elongation factor has also been reported from *P. polycephalum.*[180] The elongation factor stimulates both RNA polymerases I and II purified from the slime mold when assayed on double stranded DNA templates, but no stimulation is obtained with *E. coli* RNA polymerase.

Small nonprotein molecules have also been implicated in the control of nuclear transcription. Polyphosphate guanosine derivatives isolated from the water mold, *Achlya,* inhibit in vitro transcription catalyzed by RNA polymerases I, II, or III from *Achlya* and *Blastocladiella*.[181,182] The synthesis of these inhibitors during development is thought to play some role in regulating in vivo transcription. A nucleolar inhibitor of RNA polymerase in *P. polycephalum* is a small molecule containing a glycerol-like moiety and phosphate.[183] The inhibitor binds to purified RNA polymerase but not to RNA polymerase engaged on the DNA template. Inhibition of in vitro transcription with the inhibitor is greater for the slime mold class I enzyme than the class II RNA polymerase. That the inhibitor may play some role in transcriptional regulation is suggested by the changes in concentration of this molecule during development and the cell cycle. Plant RNA polymerase I displays allosteric behavior with respect to nucleotides.[60,184-186] In the absence of divalent cations, nucleoside triphosphates act as allosteric inhibitors in contrast to the activation observed in the presence of divalent cations. It has been suggested that nucleoside triphosphates and Mg^{++} might act as low molecular weight regulators of RNA polymerase I transcription.

XV. PHOSPHORYLATION

Although there is little evidence for alteration in RNA polymerase subunit structure as a means of regulating RNA polymerase activities during growth and development, it is possible that more subtle modification of subunits might occur. One modification that has been studied in some detail is the phosphorylation of RNA polymerase subunits. When yeast cells are grown in a medium containing ^{32}P-phosphate, the RNA polymerases purified from these cells possess phosphorylated subunits.[9,187-189] Yeast RNA polymerase I contains phosphorylated subunits of 185,000; 44,000; 36,000; 24,000; and 20,000 daltons; RNA polymerase II has a phosphorylated polypeptide of 24,000 daltons; and RNA polymerase III contains phosphorylated polypeptides of 24,000 and 20,000 daltons. In addition, Buhler et al.[189] and Sentenac et al.[9] have reported the phosphorylation of a 220,000 dalton subunit of RNA polymerase II while the proteolyzed polypeptide of 185,000 daltons is unphosphorylated. One of the phosphorylated polypeptides, the 24,000 dalton subunit, is common to RNA polymerases I, II, and III, and a second phosphorylated subunit of 20,000 daltons is common to RNA polymerases I and III. When purified yeast RNA polymerases I, II, and III are incubated with a protein kinase purified from yeast and ^{32}P-ATP, a pattern of subunit phosphorylation similar to that found in vivo is obtained except that an additional polypeptide of 48,000 daltons is phosphorylated in RNA polymerase I and the 33,500 dalton subunit of RNA polymerase III is phosphorylated.[187,188]

The incorporation of ^{32}P-phosphate into RNA polymerase I subunits when yeast cell are incubated in the presence of cycloheximide suggest that phosphorylation can occur in the absence of protein synthesis and that an enzymatically regulated phosphorylation-dephosphorylation phenomenon can operate in yeast cells and could regulate RNA polymerase activity. Bell et al.[187] have suggested that phosphorylation of RNA polymerase subunits could affect the charge density of the enzyme and thus alter interaction among subunits of the enzyme or interaction with other molecules. Alternatively, they suggested that phosphorylation might affect enzyme turnover.

When RNA polymerases from yeast are phosphorylated in vitro with yeast protein kinase, there is no effect on enzyme activity assayed on native yeast DNA template.[187,188] Likewise, there is no change in enzyme activity when yeast RNA polymerase I is treated with acid or alkaline phosphatase prior to assay of the RNA polymerase activity.

Phosphorylation of plant RNA polymerase II has been reported in wheat[190] and soybean.[38,44] In each case, the 220,000 dalton subunit is found to be phosphorylated in vivo when tissues are incubated with ^{32}P-phosphate. No ^{32}P is incorporated into the

180,000 dalton subunit. This suggests that the phosphorylated moiety of the 220,000 dalton subunit of RNA polymerase II is susceptible to proteolytic cleavage. Mazus et al.[190] have additionally demonstrated that the 220,000 dalton subunit of RNA polymerase II from wheat germ can be phosphorylated in vitro when the enzyme is incubated with ^{32}P-ATP and a protein kinase purified from wheat germ.

XVI. CHLOROPLAST AND MITOCHONDRIAL RNA POLYMERASES

Chloroplast as well as mitochondrial RNA polymerases appear to be tightly associated with membrane-DNA complexes and solubilization of these complexes usually requires treatment with detergents such as Triton® X-100, digitonin, or *N*-laurylsarcosine, with metal chelators such as ethylenediaminetetraacetic acid (EDTA), with high salt concentrations, and in some cases with deoxyribonuclease.[191-195] Following solubilization, the chloroplast enzyme can be purified by conventional procedures such as chromatography on DEAE-cellulose or DEAE-Sephadex®, phosphocellulose, and DNA-cellulose as well as sedimentation through glycerol or sucrose density gradients. The size of maize chloroplast RNA polymerase is estimated to be 500,000 daltons and the enzyme is composed of two large subunits of 180,000 and 140,000 daltons[195] and two smaller subunits estimated to be 42,000 and 27,000 daltons.[196] Peptide mapping of maize nuclear RNA polymerase II subunits and chloroplast RNA polymerase subunits indicates that the 180,000 dalton polypeptides present in each enzyme are not related, nor are the 160,000; 43,000; and 28,000 dalton subunits of the nuclear class II enzyme related to the 140,000; 42,000; and 27,000 dalton subunits of chloroplast RNA polymerase.[196] These studies do not, however, dismiss the possibility that the RNA polymerase purified from chloroplasts is contaminated with nuclear RNA polymerase I or III. That nuclear RNA polymerases are not contaminants of the chloroplast enzyme preparation is important to demonstrate in view of the similarity in subunit structure of the presumptive chloroplast RNA polymerase and the nuclear RNA polymerases. In contrast to the maize enzyme, spinach chloroplast RNA polymerase is believed to consist of polypeptides of 69,000; 60,000; 55,000; 34,000; and 15,000 daltons.[197] Such a large difference in subunit structure of spinach and maize chloroplast RNA polymerases would hardly be expected in light of the strong conservation of subunit structure of the nuclear enzymes across several kingdoms of organisms. Whether, in fact, any of the polypeptides which have been reported to be associated with preparations containing chloroplast RNA polymerase activity are authentic subunits of the chloroplast enzyme remains questionable.

Upon illumination of etiolated maize seedlings, chloroplast RNA polymerase activity is reported to increase.[198] The fourfold increase in maize chloroplast RNA polymerase activity that results after 16 hr of illumination is apparently not due to an increase in the enzyme or to any subunit alteration in the enzyme and other factors are implicated in the regulation of chloroplast RNA synthesis.[199] This study, however, rests on the assumption that the polypeptides observed on SDS polyacrylamide gels are subunits of the chloroplast enzyme and this is far from clear as discussed in the preceding paragraph.

A protein, referred to as S factor, has been purified from maize chloroplasts that stimulates in vitro RNA synthesis catalyzed by maize chloroplast RNA polymerase.[200,201] Stimulation is greatest on supercoiled DNA templates and may be greater than fivefold the amount of transcription observed in absence of the factor. The S factor also promotes selective transcription of maize chloroplast DNA sequences on recombinant plasmids containing inserts of maize chloroplast DNA. The purified factor has a molecular weight of 27,500 and migrates as a single band on SDS polyacrylamide gels.

Transcriptionally active chromosomes have been isolated from *Euglena gracilis*[192,202,203] and spinach.[197,204] The chromosomes contain chloroplast DNA, DNA-dependent RNA polymerase, and some other proteins. The transcriptional complexes are isolated from chloroplasts by lysis of the organelles in buffers containing 4 m*M* EDTA and 1% Triton® X-100, and the complexes are purified by gel filtration. Both initiation and elongation of RNA chains occurs in vitro. Initiation in vitro takes place at selected loci and the most abundant in vitro transcripts hybridize to chloroplast DNA restriction fragments which code for 23 S, 16 S, and 5 S rRNA.[203,204] In addition, it appears that only the coding strand of the chloroplast DNA is transcribed in vitro.[203] Since rRNAs are the most abundant chloroplast transcripts in the cell and since the transcriptionally active chromosomes synthesize primarily rRNA, mechanisms that regulate selective transcription of these cistrons in vivo may also be operative in this in vitro system. The isolation of transcriptionally active chromosomes may allow the identification of regulatory factors that promote selective transcription of the chloroplast genome and may also allow the identification of factors that regulate transcription of the organellar DNA during development of the organelle and during environmentally (i.e., light) induced transitions.

In contrast to the complex subunit structures observed with chloroplast RNA polymerase preparations, the subunit structures reported for mitochondrial RNA polymerase are, in general, simple, consisting of single polypeptides with molecular weights in the range of 45,000 to 70,000 daltons. Mitochondrial RNA polymerase has been purified from two lower eukaryotes, yeast[205-213] and *Neurospora crassa*.[214] These enzymes like the chloroplast enzymes are resistant to inhibition by α-amanitin and generally the most pure preparations are relatively resistant to rifampicin.[213]

The most recent purification report on yeast mitochondrial RNA polymerase indicates that the native enzyme sediments on glycerol gradients at 6.3 S and has a presumptive molecular weight of 100,000 to 150,000.[213] This enzyme consists of a single polypeptide with a molecular weight of 45,000 as determined by SDS polyacrylamide gel electrophoresis. Antibodies raised against the 45,000 dalton polypeptide inhibit the activity of yeast mitochondrial RNA polymerase. The purified enzyme requires 10 m*M* Mg^{++} for optimal activity, and Mn^{++} will not substitute for Mg^{++}. The purified enzyme is tenfold more active on a poly (dAT) template than on yeast mitochondrial DNA, but is three- to fivefold more active on its homologous template than on other double stranded DNA templates. This yeast RNA polymerase shows some selectivity of transcription on its homologous template in that in vitro transcripts display much stronger hybridization to certain restriction fragments of the mitochondrial genome compared to other fragments of the genome.

A mitochondrial transcription complex has recently been purified from yeast.[215] Most of the RNA synthesized in vitro by this complex hybridizes to the 21 S and 14 S rRNA genes of yeast mitochondrial DNA. This is similar to the observation on the types of transcripts produced in chloroplast transcription complexes. Additional analysis of the in vitro transcription products synthesized by the mitochondrial transcription complex indicates that at least two promoters are utilized in the initiation of in vitro transcripts on the mitochondrial genome.

XVII. PROSPECTIVES

Despite the recent advances in the analysis of genome organization and genome structure in eukaryotes, there is very little known about the mechanisms involved in the regulation of eukaryotic gene expression. Recombinant DNA techniques have made it possible to isolate almost any gene, to determine the structure and sequence of that gene, and to mutate the sequence of that gene in vitro. Because of the current availability of eukaryotic genes and the increased numbers of genes that will surely be available in the near future, the templates required for studying transcriptional regulation in eukaryotes are not a limitation. Instead, the most obvious limitation in the study of plant and lower eukaryotic gene expression is the current lack of soluble transcription systems equivalent to those that have been developed with animal cells. Even if soluble transcription systems derived from animals are found to be generally capable of accurately transcribing plant and lower eukaryotic genes in vitro, the study of cell- and tissue-specific expression of these genes will most likely require the use of soluble systems derived from homologous cells and tissues. The development of plant and lower eukaryotic soluble transcription systems is certainly required for future advancement in the understanding of plant and lower eukaryotic RNA polymerases and transcriptional regulation.

ACKNOWLEDGMENTS

Research results reported from the author's laboratory were supported by U. S. Public Health Service Grant GM 24096 and USDA/SEA Competitive Research Grant 5901-0410-9-0312-0. The author is indebted to Dr. Gretchen Hagen for her critical comments on the manuscript, for her proof-reading of the manuscript, and for her help in preparing the figures and photographs.

REFERENCES

1. **Whittaker, R. H. and Margulis, L.,** Protist classification and the kingdom of organisms, *BioSystems,* 10, 3, 1978.
2. **Mans, R. J.,** RNA polymerases in higher plants, in *Nucleic Acid Biosynthesis,* Laskin, A. I. and Last, J. A., Eds., Marcel Dekker, New York, 1973, 93.
3. **Biswas, B. B., Ganguly, A., and Das, A.,** Eukaryotic RNA polymerases and factors that control them, *Prog. Nucl. Acids Res. Mol. Biol.,* 15, 145, 1975.
4. **Duda, C. T.,** Plant RNA polymerases, *Annu. Rev. Plant Physiol.,* 27, 119, 1976.
5. **Becker, W. M.,** RNA polymerases in plants, in *Nucleic Acids in Plants,* Vol. 1, Davies, J. W. and Hall, T. C., Eds., CRC Press, Boca Raton, Fla., 1978, 111.
6. **Guilfoyle, T. J.,** DNA and RNA polymerases, in *The Biochemistry of Plants,* Vol. 6, Marcus, A., Ed., Academic Press, New York, 1981, 207.
7. **Buhler, J. M., Dezelee, S., Huet, J., Iborra, F., and Sentenac, A.,** Structural properties and template requirements of yeast RNA polymerases, in *Biochemistry of the Cell Nucleus,* Hidvegi, E. J., Sumegi, J., and Venetianer, P., Eds., Elsevier/North Holland, Amsterdam, 1975, 289.
8. **Rutter, W. J., Valenzuela, P., Bell, G. I., Holland, M., Hager, G. L., Degennaro, L. J., and Bishop, R. J.,** The role of DNA-dependent RNA polymerases in transcriptive specificity, in *The Organization and Expression of the Eukaryotic Genome,* Bradbury, E. M. and Javaberian, K., Eds., Academic Press, New York, 1977, 279.
9. **Sentenac, A., Buhler, J. M., Ruet, A., Huet, J., Iborra, F., and Fromageot, P.,** Eukaryotic RNA polymerases, in *Gene Expression. Protein Synthesis and Control. RNA Synthesis and Control. Chromatin Structure and Function,* Clark, B. F. C., Klenow, H., and Zeuthen, J., Eds., Pergamon Press, Elmsford, N.Y., 1977, 187.
10. **Sebastian, J.,** Structure and function of the yeast RNA polymerases, *Trends Biochem. Sci.,* 2, 102, 1977.
11. **Seebeck, T. and Braun, R.,** Transcription in acellular slime moulds, *Adv. Microb. Physiol.,* 21, 1, 1980.
12. **Jacob, S. T.,** Mammalian RNA polymerases, *Prog. Nucl. Acids Res. Mol. Biol.,* 13, 93, 1973.
13. **Chambon, P.,** Animal RNA polymerases, in *The Enzymes,* Vol. 10, 3rd ed., Boyer, P. D., Ed., Academic Press, New York, 1974, 261.
14. **Chambon, P.,** Eukaryotic nuclear RNA polymerases, *Annu. Rev. Biochem.,* 44, 613, 1975.
15. **Roeder, R. G.,** Eukaryotic nuclear RNA polymerases, in *RNA Polymerase,* Losick, R. and Chamberlin, M., Eds., Cold Spring Harbor Laboratory, New York, 1976, 285.

16. **Beebee, T. J. and Butterworth, P. H. W.,** Eukaryotic deoxyribonucleic acid-dependent RNA polymerases, *Biochem. Soc. Symp.*, 42, 75, 1977.
17. **Jacob, S. T. and Rose, K. M.,** Basic enzymology of transcription in prokaryotes and eukaryotes, in *Cell Biology. A Comprehensive Treatise,* Vol. 3, Goldstein, L. and Prescott, D. M., Eds., Academic Press, New York, 1980, 113.
18. **Beebee, T. J. C. and Butterworth, P. H. W.,** Eukaryotic DNA-dependent RNA polymerases: an evaluation of their role in the regulation of gene expression, in *Eukaryotic Gene Regulation,* Vol. 2, Kolodney, G. M., Ed., CRC Press, Boca Raton, Fla., 1980, 1.
19. **Roeder, R. G. and Rutter, W. J.,** Multiple forms of DNA-dependent RNA polymerase in eukaryotic organisms, *Nature (London),* 224, 234, 1969.
20. **Guilfoyle, T. J.,** Purification and characterization of DNA-dependent RNA polymerases from cauliflower nuclei, *Plant Physiol.*, 58, 453, 1976.
21. **Schultz, L. D. and Hall, B. D.,** Transcription in yeast: α-amanitin sensitivity and other properties which distinguishes between RNA polymerases I and III, *Proc. Natl. Acad. Sci. U.S.A.*, 73, 1029, 1976.
22. **Hildebrandt, A. and Sauer, H. W.,** DNA-dependent RNA polymerases from *Physarum polycephalum, FEBS Lett.*, 35, 41, 1973.
23. **Gornicki, S. Z., Vuturo, S. B., West, T. V., and Weaver, R. F.,** Purification and properties of deoxyribonucleic acid-dependent ribonucleic acid polymerases from the slime mold *Physarum polycephalum, J. Biol. Chem.*, 249, 1792, 1974.
24. **Burgess, A. B. and Burgess, R. R.,** Purification and properties of two RNA polymerases from *Physarum polycephalum, Proc. Natl. Acad. Sci. U.S.A.*, 71, 1174, 1974.
25. **Weaver, R. F.,** Structural comparison of deoxyribonucleic acid-dependent ribonucleic acid polymerases I and II from the slime mold *Physarum polycephalum, Arch. Biochem. Biophys.*, 172, 470, 1976.
26. **Valenzuela, P., Hager, G., Weinberg, F., and Rutter, W. J.,** The molecular structure of yeast RNA polymerase III. Demonstration of the tripartite transcriptive system in lower eukaryotes, *Proc. Natl. Acad. Sci. U.S.A.*, 73, 1024, 1976.
27. **Mildvan, A. S. and Loeb, L. A.,** The role of metal ions in the mechanisms of DNA and RNA polymerases, *CRC Crit. Rev. Biochem.*, 6, 219, 1979.
28. **Guilfoyle, T. J., Lin, C. Y., Chen, Y. M., Nagao, R. T., and Key, J. L.,** Enhancement of soybean RNA polymerase I by auxin, *Proc. Natl. Acad. Sci. U.S.A.*, 72, 69, 1975.
29. **Chen, Y. M., Lin, C. Y., Chang, H., Guilfoyle, T. J., and Key, J. L.,** Isolation and properties of nuclei from control and auxin-treated soybean hypocotyl, *Plant Physiol.*, 56, 78, 1975.
30. **Lin, C. Y., Guilfoyle, T. J., Chen, Y. M., and Key, J. L.,** Isolation of nucleoli and localization of ribonucleic acid polymerase I from soybean hypocotyl, *Plant Physiol.*, 56, 850, 1975.
31. **Guilfoyle, T. J. and Key, J. L.,** Purification and characterization of soybean DNA-dependent RNA polymerases and the modulation of their activities during development, in *Nucleic Acids and Protein Synthesis in Plants,* Bogorad, L. and Weil, J., Eds., Plenum Press, New York, 1977, 37.
32. **Guilfoyle, T. J.,** Transcription of the cauliflower mosaic virus genome in isolated nuclei from turnip leaves, *Virology,* 107, 71, 1980.
33. **Adman, R., Schultz, L. D., and Hall, B. D.,** Transcription in yeast: separation and properties of multiple RNA polymerases, *Proc. Natl. Acad. Sci. U.S.A.*, 69, 1702, 1972.
34. **Gissinger, F., Kedinger, C., and Chambon, P.,** Animal DNA-dependent RNA polymerases. X. General enzymatic properties of purified calf thymus RNA polymerases AI and B, *Biochimie,* 56, 319, 1972.
35. **Dynan, W. S., Jendrisak, J. J., and Burgess, R. R.,** Templates for eukaryotic RNA polymerase II: artefacts can produce an apparent preference for denatured DNA over native DNA, *Anal. Biochem.*, 79, 181, 1977.
36. **Jendrisak, J. J.,** Purification, subunit structures, and functions of the nuclear RNA polymerases from higher plants, in *Genome Organization and Expression in Plants,* Leaver, C. J., Ed., Plenum Press, New York, 1980, 77.
37. **Guilfoyle, T. J., Olszewski, N., and Zurfluh, L. L.,** RNA polymerases and transcription during developmental transitions in soybean, in *Genome Organization and Expression in Plants,* Leaver, C. J., Ed., Plenum Press, New York, 1980, 93.
38. **Guilfoyle, T. J. and Malcolm, S.,** The amounts, subunit structures, and template-engaged activities of RNA polymerases in germinating soybean axes, *Dev. Biol.*, 78, 113, 1980.
39. **Jendrisak, J. J. and Burgess, R. R.,** A new method for the large-scale purification of wheat germ DNA-dependent RNA polymerase II, *Biochemistry,* 14, 4639, 1975.
40. **Guilfoyle, T. J. and Jendrisak, J. J.,** Plant DNA-dependent RNA polymerases: subunit structures and enzymatic properties of the class II enzymes from quiescent and proliferating tissues, *Biochemistry,* 17, 1860, 1978.

41. **Goto, H., Sasaki, Y., and Kamikubo, T.**, Large-scale purification and subunit structure of DNA-dependent RNA polymerase II from cauliflower inflorescence, *Biochim. Biophys. Acta*, 517, 195, 1978.
42. **Jendrisak, J. J.**, Purification and subunit structure of DNA-dependent RNA polymerase III from wheat germ, *Plant Physiol.*, 67, 438, 1981.
43. **Guilfoyle, T. J.**, Purification, subunit structure, and immunological properties of chromatin-bound ribonucleic acid polymerase I from cauliflower inflorescence, *Biochemistry*, 19, 5966, 1980.
44. **Guilfoyle, T. J.**, Auxin-induced deoxyribonucleic acid dependent ribonucleic acid polymerase activities in mature soybean hypocotyl, *Biochemistry*, 19, 6112, 1980.
45. **Hager, G. L., Holland, M. J., and Rutter, W. J.**, Isolation of RNA polymerases I, II, and III from *Saccharomyces cerevisiae*, *Biochemistry*, 16, 1, 1977.
46. **Valenzuela, P., Weinberg, F., Bell, G., and Rutter, W. J.**, Yeast DNA-dependent RNA polymerase I. A rapid procedure for the large scale purification of homogeneous enzyme, *J. Biol. Chem.*, 251, 1464, 1976.
47. **Hager, G., Holland, M., Valenzuela, P., Weinberg, F., and Rutter, W. J.**, RNA polymerases and transcriptive specificity in *Saccharomyces cerevisiae*, in *RNA Polymerase*, Losick, R. and Chamberlin, M., Eds., Cold Spring Harbor Laboratory, Cold Spring Harbor, N.Y., 1976, 745.
48. **Valenzuela, P., Bell, G. I., Weinberg, F., and Rutter, W. J.**, Isolation and assay of eukaryotic DNA-dependent RNA polymerases, in *Methods in Cell Biology*, Vol. 19, Stein, G., Stein, J., and Kleinsmith, L. J., Eds., Academic Press, New York, 1978, 1.
49. **Stunnenberg, H. G., Wennekes, L. M. J., and van den Broek, H. W. J.**, RNA polymerase from the fungus *Aspergillus nidulans*. Large-scale purification of DNA-dependent RNA polymerase I (or A), *Eur. J. Biochem.*, 98, 107, 1979.
50. **Stunnenberg, H. G., Wennekes, L. M. J., Spierings, T., and van den Broek, H. W. J.**, An α-amanitin-resistant DNA-dependent RNA polymerase II from the fungus *Aspergillus nidulans*, *Eur. J. Biochem.*, 117, 121, 1981.
51. **Smith, S. S. and Braun, R.**, A new method for the purification of RNA polymerase II (or B) from the lower eukaryote *Physarum polycephalum*. The presence of subforms, *Eur. J. Biochem.*, 82, 309, 1978.
52. **Spindler, S. R., Duester, G. L., D'Alessio, J. M., and Paule, M. R.**, A rapid and facile procedure for the preparation of RNA polymerase I from *Acantamoeba castellanii*. Purification and subunit structure, *J. Biol. Chem.*, 253, 4669, 1978.
53. **D'Alessio, J. M., Spindler, S. R., and Paule, M. R.**, DNA-dependent RNA polymerase II from *Acanthamoeba castellanii*. Large scale preparation and subunit composition, *J. Biol. Chem.*, 254, 4085, 1979.
54. **Spindler, S. R., D'Alessio, J. M. D., Duester, G. L., and Paule, M. R.**, DNA-dependent RNA polymerase III from *Acanthamoeba castellanii*. A rapid procedure for the large scale preparation of homogeneous enzyme, *J. Biol. Chem.*, 253, 6242, 1978.
55. **Horgen, P. A. and Key, J. L.**, The DNA-directed RNA polymerases of soybean, *Biochim. Biophys. Acta*, 294, 227, 1973.
56. **Jendrisak, J. J. and Becker, W. M.**, Purification and subunit analysis of wheat germ ribonucleic acid polymerase II, *Biochem. J.*, 139, 771, 1974.
57. **Fabisz-Kijowska, A., Dullin, P., and Walerych, W.**, Isolation and purification of RNA polymerases from rye embryos, *Biochim. Biophys. Acta*, 390, 105, 1975.
58. **Guilfoyle, T. J., Lin, C. Y., Chen, Y. M., and Key, J. L.**, Purification and characterization of RNA polymerase I from a higher plant, *Biochim. Biophys. Acta*, 418, 344, 1976.
59. **Hahn, H. and Servos, D.**, The isolation of multiple forms of DNA-dependent RNA polymerases from nuclei of quiescent wheat embryos by affinity chromatography, *Z. Pflanzenphysiol.*, 97, 43, 1980.
60. **Grossmann, K., Friedrich, H., and Seitz, U.**, Purification and characterization of chromatin-bound DNA-dependent RNA polymerase I from parsley *(Petroselinum crispum)*, *Biochem. J.*, 191, 165, 1980.
61. **Horgen, P. A. and Griffin, D. H.**, Specific inhibitors of the three RNA polymerases from the aquatic fungus *Blastocladiella emersonii*, *Proc. Natl. Acad. Sci. U.S.A.*, 68, 338, 1971.
62. **Buhler, J. M., Sentenac, A., and Fromageot, P.**, Isolation, structure, and general properties of yeast ribonucleic acid polymerase A (or I), *J. Biol. Chem.*, 249, 5963, 1974.
63. **Tellez de Inon, M. T., Leoni, P. D., and Torres, H. N.**, RNA polymerase activities in *Neurospora crassa*, *FEBS Lett.*, 39, 91, 1974.
64. **Soll, D. R. and Fulton, C.**, The constancy of RNA polymerase activities during a transcriptionally dependent differentiation in *Naegleria*, *Dev. Biol.*, 36, 236, 1974.
65. **Detke, S. and Paule, M. R**, DNA-dependent RNA polymerase I from *Acanthamoeba castellanii:* comparison of the catalytic properties and subunit architectures of the trophozoite and cyst enzymes, *Arch. Biochem. Biophys.*, 185, 333, 1978.

66. **Detke, S. and Paule, M. R.,** DNA-dependent RNA polymerases from *Acanthamoeba castellanii.* Multiple forms of the class III enzyme and levels of activity of the polymerase classes during encystment, *Biochim. Biophys. Acta,* 520, 131, 1978.
67. **Pong, S. S. and Loomis, W. F.,** Multiple nuclear ribonucleic acid polymerases during development of *Dictyostelium discoideum, J. Biol. Chem.,* 248, 3933, 1973.
68. **Detke, S. and Paule, M. R.,** DNA-dependent RNA polymerase II from *Acanthamoeba castellanii.* Comparison of the catalytic properties and subunit architectures of the trophozoite and cyst enzymes, *Biochim. Biophys. Acta,* 520, 376, 1978.
69. **Dezelee, S. and Sentenac, A.,** Role of DNA-RNA hybrids in eukaryotes. Purification and properties of yeast RNA polymerase B, *Eur. J. Biochem.,* 34, 41, 1973.
70. **Mans, R. J. and Novelli, G. D.,** Ribonucleotide incorporation by a soluble enzyme from maize, *Biochim. Biophys. Acta,* 91, 186, 1964.
71. **Stout, E. R. and Mans, R. J.,** Partial purification and properties of RNA polymerase from maize, *Biochim. Biophys. Acta,* 134, 327, 1967.
72. **Lin, C. Y., Guilfoyle, T. J., Chen, Y. M., Nagao, R. T., and Key, J. L.,** The separation of RNA polymerases I and II achieved by fractionation of plant chromatin, *Biochem. Biophys. Res. Commun.,* 60, 498, 1974.
73. **Guilfoyle, T. J. and Key, J. L.,** The subunit structures of soluble and chromatin-bound RNA polymerase II from soybean, *Biochem. Biophys. Res. Commun.,* 74, 308, 1977.
74. **Jendrisak, J. J. and Guilfoyle, T. J.,** Eukaryotic RNA polymerases: comparative subunit structures, immunological properties, and α-amanitin sensitivities of the class II enzymes from higher plants, *Biochemistry,* 17, 1322, 1978.
75. **Link, G. and Richter, G.,** Properties and subunit composition of RNA polymerase II from plant cell cultures, *Biochim. Biophys. Acta,* 395, 337, 1975.
76. **Hodo, H. G. and Blatti, S. P.,** Purification using polyethylenimine precipitation and low molecular weight subunit analysis of calf thymus and wheat germ DNA-dependent RNA polymerase II, *Biochemistry,* 16, 2334, 1977.
77. **Sasaki, Y., Goto, H., Tomi, H., and Kamikubo, T.,** DNA-dependent RNA polymerase III from cauliflower. Characterization and template specificity, *Biochim. Biophys. Acta,* 517, 205, 1978.
78. **Sasaki, Y., Ishiye, M., Goto, H., and Kamikubo, T.,** Purification and subunit structure of RNA polymerase II from pea, *Biochim. Biophys. Acta,* 564, 437, 1979.
79. **Vaisius, A. C. and Horgen, P. A.,** Purification and characterization of RNA polymerase II resistant to α-amanitin from the mushroom *Agaricus bisporus, Biochemistry,* 18, 795, 1979.
80. **Teissere, M., Penon, P., Azou, Y., and Ricard, J.,** RNA polymerase III from wheat embryos. Purification by affinity and hydrophobic chromatographies. Characterization and molecular properties, *FEBS Lett.,* 82, 77, 1977.
81. **Young, H. A. and Whiteley, H. R.,** Deoxyribonucleic acid-dependent ribonucleic acid polymerases in the dimorphic fungus *Mucor rouxii, J. Biol. Chem.,* 250, 479, 1975.
82. **Mullinix, K. P., Strain, G. C., and Bogorad, L.,** RNA polymerases of maize. Purification and molecular structure of DNA-dependent RNA polymerase II, *Proc. Natl. Acad. Sci. U.S.A.,* 70, 2386, 1973.
83. **Jendrisak, J. J. and Burgess, R. R.,** Studies on the subunit structure of wheat germ ribonucleic acid polymerase II, *Biochemistry,* 16, 1959, 1977.
84. **Van Keulen, H., Planta, R. J., and Retel, J.,** Structure and transcription specificity of yeast RNA polymerase A, *Biochim. Biophys. Acta,* 395, 179, 1975.
85. **Buhler, J. M., Iborra, F., Sentenac, A., and Fromageot, P.,** Structural studies on yeast RNA polymerases. Existence of common subunits in RNA polymerases A (I) and B (II), *J. Biol. Chem.,* 251, 1712, 1976.
86. **Gong, C. S. and Van Etten, J. L.,** Purification and properties of RNA polymerases I and II from germinated spores of *Rhizopus stolonifer, Can. J. Microbiol.,* 20, 1267, 1974.
87. **Buhler, J. M., Huet, J., Davies, K. E., Sentenac, A., and Fromageot, P.,** Immunological studies of yeast nuclear RNA polymerases at the subunit level, *J. Biol. Chem.,* 255, 9949, 1980.
88. **Huet, J., Buhler, J., Sentenac, A., and Fromageot, P.,** Dissociation of two polypeptide chains from yeast RNA polymerase A, *Proc. Natl. Acad. Sci. U.S.A.,* 72, 3034, 1975.
89. **Huet, J., Dezelee, S., Iborra, F., Buhler, J. M., Sentenac, A., and Fromageot, P.,** Further characterization of yeast RNA polymerases. Effect of subunits removal, *Biochimie,* 58, 71, 1976.
90. **Bull, P., MacDonald, H., and Valenzuela, P.,** The interaction of yeast RNA polymerase I and cibacron blue F3GA, *Biochim. Biophys. Acta,* 653, 368, 1981.
91. **Link, G., Kidd, G. H., Richter, G., and Bogorad, L.,** Structural relationships among the multiple forms of DNA-dependent RNA polymerase II from cultured parsley cells, *Eur. J. Biochem.,* 91, 363, 1978.
92. **Smith, S. S. and Braun, R.,** New polypeptide chains associated with highly purified RNA polymerase II or B from *Physarum polycephalum, FEBS Lett.,* 125, 107, 1981.

93. **Dezelee, S., Wyers, F., Sentenac, A., and Fromageot, P.,** Two forms of RNA polymerase B in yeast. Proteolytic conversion in vitro of enzyme BI into BII, *Eur. J. Biochem.,* 65, 543, 1976.
94. **Guilfoyle, T. J.,** unpublished data, 1981.
95. **Lescure, B., Williamson, V., and Sentenac, A.,** Efficient and selective initiation by yeast RNA polymerase B in a dinucleotide-primed reaction, *Nucl. Acids Res.,* 9, 31, 1981.
96. **Paule, M. R.,** Comparative subunit composition of the eukaryotic nuclear RNA polymerases, *Trends Biochem. Sci.,* 6, 128, 1981.
97. **Thonart, P., Bechet, J., Hilger, F., and Burny, A.,** Thermosensitive mutations affecting ribonucleic acid polymerases in *Saccharomyces cerevisiae, J. Bacteriol.,* 125, 25, 1976.
98. **Valenzuela, P., Bell, G. I., Weinberg, F., and Rutter, W. J.,** Yeast DNA-dependent RNA polymerase I, II, and III. The existence of subunits common to the three enzymes, *Biochem. Biophys. Res. Commun.,* 71, 1319, 1976.
99. **Sentenac, A., Dezelee, S., Iborra, F., Buhler, J. M., Huet, J., Wyers, F., Ruet, A., and Fromageot, P.,** Yeast RNA polymerases, in *RNA Polymerase,* Losick, R. and Chamberlin, M., Eds., Cold Spring Harbor Laboratory, Cold Spring Harbor, N.Y., 1976, 763.
100. **D'Alessio, J. M., Perna, P. J., and Paule, M. R.,** DNA-dependent RNA polymerases from *Acanthamoeba castellanii.* Comparative subunit structures of the homogeneous enzymes, *J. Biol. Chem.,* 254, 11282, 1979.
101. **Hildebrandt, A., Sebastian, J., and Halvorson, H. O.,** Yeast nuclear RNA polymerases I and II are immunologically related, *Nature (London), New Biol.,* 246, 73, 1973.
102. **Kidd, G. H., Link, G., and Bogorad, L.,** Comparison of large subunits of type II DNA-dependent RNA polymerases from higher plants, *Plant Physiol.,* 64, 671, 1979.
103. **Lattke, H. and Weser, U.,** Functional aspects of zinc in yeast RNA polymerase B, *FEBS Lett.,* 83, 297, 1977.
104. **Falchuk, K. H., Mazus, B., Ulpino, L., and Vallee, B. L.,** *Euglena gracilis* DNA dependent RNA polymerase II: a zinc metalloenzyme, *Biochemistry,* 15, 4468, 1976.
105. **Petranyi, P., Jendrisak, J. J., and Burgess, R. R.,** RNA polymerase II from wheat germ contains tightly bound zinc, *Biochem. Biophys. Res. Commun.,* 74, 1031, 1977.
106. **Falchuk, F. H., Ulpino, L., Mazus, B., and Vallee, B. L.,** *E. gracilis* RNA polymerase I: a zinc metalloenzyme, *Biochem. Biophys. Res. Commun.,* 74, 1206, 1977.
107. **Auld, D. S., Atsuya, I., Campino, C., and Valenzuela, P.,** Yeast RNA polymerase I: a zinc metalloenzyme, *Biochem. Biophys. Res. Commun.,* 69, 548, 1976.
108. **Lattke, H. and Weser, U.,** Yeast RNA polymerase B: a zinc protein, *FEBS Lett.,* 65, 288, 1976.
109. **Wandzilak, T. M. and Benson, R. W,** Yeast RNA polymerase III: a zinc metalloenzyme, *Biochem. Biophys. Res. Commun.,* 76, 247, 1977.
110. **Wandzilak, T. M. and Benson, R. W.,** *Saccharomyces cerevisiae* DNA-dependent RNA polymerase III: a zinc metalloenzyme, *Biochemistry,* 17, 426, 1978.
111. **Martial, J., Zaldivar, J., Bull, P., Venegas, A., and Valenzuela, P.,** Inactivation of rat liver RNA polymerases I and II and yeast RNA polymerase I by pyridoxal 5′-phosphate. Evidence for the participation of lysyl residues at the active site, *Biochemistry,* 14, 4907, 1975.
112. **Valenzuela, P., Bull, P., Zaldivar, J., Venegas, A., and Martial, J.,** Subunits of yeast RNA polymerase I involved in interactions with DNA and nucleotides, *Biochem. Biophys. Res. Commun.,* 81, 662, 1978.
113. **Cooper, C. S. and Quincey, R. V.,** The role of subunits in yeast DNA-dependent ribonucleic acid polymerase A, *Biochem. J.,* 181, 301, 1979.
114. **Winsor, B., Lacroute, F., Ruet, A., and Sentenac, A.,** Isolation and characterization of a strain of *Saccharomyces cerevisiae* deficient in in vitro RNA polymerase B (II) activity, *Mol. Gen. Genet.,* 73, 145, 1979.
115. **Ruet, A., Sentenac, A., and Fromageot, P.,** A specific assay for yeast RNA polymerases in crude cell extracts, *Eur. J. Biochem.,* 90, 325, 1978.
116. **Ruet, A., Sentenac, A., Fromageot, P., Winsor, B., and Lacroute, F.,** A mutation of the B_{220} subunit gene affects the structural and functional properties of yeast RNA polymerase B in vitro, *J. Biol. Chem.,* 255, 6450, 1980.
117. **Brodner, O. G. and Wieland, T.,** Identification of the amatoxin-binding subunit of RNA polymerase B by affinity labeling experiments. Subunit B_3 — the true amatoxin receptor protein of multiple RNA polymerase B, *Biochemistry,* 15, 3480, 1976.
118. **Cochet-Meilhac, M. and Chambon, P.,** Animal DNA-dependent RNA polymerases. II. Mechanism of the inhibition of RNA polymerase B by amatoxins, *Biochim. Biophys. Acta,* 353, 160, 1974.
119. **Sklar, V. E. F., Yamamoto, M., and Roeder, R. G.,** Molecular structures of eukaryotic class III RNA polymerases, in *RNA Polymerase,* Losick, R. and Chamberlin, M., Eds., Cold Spring Harbor Laboratory, Cold Spring Harbor, N.Y., 1976, 803.
120. **Renart, J. and Sebastian, J.,** Characterization and levels of the RNA polymerases during the embryogenesis of *Artemia salina, Cell Differ.,* 5, 96, 1976.

121. **Horgen, P. A., Vaisius, A. C., and Ammirati, J. F.**, The insensitivity of mushroom nuclear RNA polymerase activity to inhibition by amatoxins, *Arch. Microbiol.*, 118, 317, 1978.
122. **Cain, A. K. and Nester, E. W.**, Ribonucleic acid polymerase in *Allomyces arbuscula, J. Bacteriol.*, 115, 769, 1973.
123. **Johnson B. C. and Preston, J. F.**, Unique amanitin resistance of RNA synthesis in isolated nuclei from *Amanita* species accumulating amanitins, *Arch. Microbiol.*, 122, 161, 1979.
124. **Johnson, B. C. and Preston, J. F.**, α-Amanitin-resistant RNA polymerase II from carpophores of *Amanita* species accumulating amatoxins, *Biochim. Biophys. Acta*, 607, 102, 1980.
125. **Gong, G. S. and van Etten, J. L.**, Changes in soluble ribonucleic acid polymerases associated with germination of *Rhizopus stolonifer* spores, *Biochim. Biophys. Acta*, 272, 44, 1972.
126. **Christian, B. and Bequeret, J.**, Isolation and partial characterization of the multiple forms of deoxyribonucleic acid-dependent ribonucleic acid polymerase in the fungus *Podospora anserina, J. Biol. Chem.*, 254, 11566, 1979.
127. **Seebeck, T., Stalder, J., and Braun, R.**, Isolation of a minichromosome containing the ribosomal genes from *Physarum polycephalum, Biochemistry*, 18, 484, 1979.
128. **Detke, S. and Paule, M. R.**, DNA-dependent RNA polymerases from *Acanthamoeba castellanii:* properties and levels of activity during encystment, *Biochim. Biophys. Acta*, 383, 67, 1975.
129. **Zieve, G. W.**, Two groups of small stable RNAs, *Cell*, 25, 296, 1981.
130. **Gurley, W. B., Lin, C. Y., Guilfoyle, T. J., Nagao, R. T., and Key, J. L.**, Analysis of plant RNA polymerase I transcript in chromatin and nuclei, *Biochim. Biophys. Acta*, 425, 168, 1976.
131. **Willmitzer, L., Schmalenbach, W., and Schell, J.**, Transcription of T-DNA in octopine and nopaline crown gall tumours is inhibited by low concentrations of α-amanitin, *Nucl. Acids Res.*, 9, 4801, 1981.
132. **Jendrisak, J. J.**, α-Amanitin inhibition of germination and RNA synthesis in wheat embryos, *J. Biol. Chem.*, 256, 5860, 1981.
133. **Schultz, L. D.**, Transcriptional role of yeast deoxyribonucleic acid dependent ribonucleic acid polymerase III, *Biochemistry*, 17, 750, 1978.
134. **Ide, G. J.**, Nucleoside 5′-(γ-S) triphosphates will initiate transcription in isolated yeast nuclei, *Biochemistry*, 20, 2633, 1981.
135. **Tekamp, P. A., Valenzuela, P., Maynard, T., Bell, G. I., and Rutter, W. J.**, Specific gene transcription in yeast nuclei and chromatin by added homologous RNA polymerases I and III, *J. Biol. Chem.*, 254, 955, 1979.
136. **Tekamp, P. A., Garcea, R. L., and Rutter, W. J.**, Transcription and in vitro processing of yeast 5 S rRNA, *J. Biol. Chem.*, 255, 9501, 1980.
137. **Chamberlin, M.**, RNA polymerase — an overview, in *RNA Polymerase*, Losick, R. and Chamberlin, M., Cold Spring Harbor Laboratory, Cold Spring Harbor, N.Y., 1976, 17.
138. **Dezelee, S., Sentenac, A., and Fromageot, P.**, Role of deoxyribonucleic acid-ribonucleic acid hybrids in eukaryotes. Study of the template requirements of yeast ribonucleic acid polymerases and nature of the ribonucleic acid product, *J. Biol. Chem.*, 249, 5971, 1974.
139. **Dynan, W. S. and Burgess, R. R.**, In vitro transcription by wheat germ ribonucleic acid polymerase II: effects of heparin and role of template integrity, *Biochemistry*, 18, 4581, 1979.
140. **Chandler, D. W. and Gralla, J.**, Specific binding and protection of form II SV-40 deoxyribonucleic acid by ribonucleic acid polymerase II from wheat germ, *Biochemistry*, 19, 1604, 1980.
141. **Lewis, M. K. and Burgess, R. R.**, Transcription of simian virus 40 DNA by wheat germ RNA polymerase II. Priming of RNA synthesis by the 3′-hydroxyl of DNA at single stranded nicks, *J. Biol. Chem.*, 255, 4928, 1980.
142. **Cooke, R., Durand, R., Teissere, M., Penon, P., and Ricard, J.**, Characterization of heparin-resistant complex formation and RNA synthesis by wheat germ RNA polymerases I, II, and III in vitro on cauliflower mosaic virus DNA, *Biochem. Biophys. Res. Commun.*, 98, 36, 1981.
143. **Seidman, S., Surzycki, S. J., DeLorbe, W., and Gussin, G. N.**, Interaction between wheat germ RNA polymerase II and adenovirus 2 DNA: evidence for two types of stable binary complexes, *Biochemistry*, 18, 3363, 1979.
144. **Witney, F. R., Surzycki, J. A., Seidman, S., Dodds, J. R., Gussin, G. N., and Surzycki, S. J.**, Location of binding sites for RNA polymerase II from wheat germ and from human placenta on adenovirus DNA, *Mol. Gen. Genet.*, 179, 627, 1980.
145. **Sargosti, S., Lescure, B., and Yaniv, M.**, Comparative study of calf thymus and wheat germ RNA polymerase II: stability of initiation complexes and elongation rates, *Biochem. Biophys. Res. Commun.*, 88, 1077, 1979.
146. **Sargosti, S., Croissant, O., and Yaniv, M.**, Localization of the binding sites of prokaryotic and eukaryotic RNA polymerases on simian virus 40 DNA, *Eur. J. Biochem.*, 106, 25, 1980.
147. **Grellet, F., Cooke, R., Teissere, M., Delseny, M., Xech, J., and Penon, P.**, Electron microscopic mapping of wheat germ RNA polymerase II binding sites on cloned CaMV DNA, *Nucl. Acids Res.*, 9, 3927, 1981.

148. **Lilley, D. M. J. and Houghton, M.**, The interaction of RNA polymerase II from wheat with supercoiled and linear plasmid templates, *Nucl. Acids Res.*, 6, 507, 1979.
149. **Hollenberg, C. P.**, Ribosomal ribonucleic acid synthesis by isolated yeast ribonucleic acid polymerases, *Biochemistry*, 12, 5320, 1973.
150. **Cramer, J. H., Sebastian, J., Rownd, R. H., and Halvorson, H.**, Transcription of *Saccharomyces cerevisiae* ribosomal DNA in vivo and in vitro, *Proc. Natl. Acad. Sci. U.S.A.*, 71, 2188, 1974.
151. **Van Keulen, H. and Retel, J.**, Transcription specificity of yeast RNA polymerase A. Highly specific transcription in vitro of the homologous ribosomal transcription units, *Eur. J. Biochem.*, 79, 579, 1977.
152. **Holland, M. J., Hager, G. L., and Rutter, W. J.**, Transcription of yeast DNA by homologous RNA polymerase I and II: selective transcription of ribosomal genes by RNA polymerase I, *Biochemistry*, 16, 16, 1977.
153. **Sawadogo, M., Sentenac, A., and Fromageot, P.**, In vitro transcription of cloned yeast ribosomal DNA by yeast RNA polymerase A, *Biochem. Biophys. Res. Commun.*, 101, 250, 1981.
154. **Ballario, P., Buongiorno-Nardelli, M., Carnerali, F., DiMauro, E., and Pedone, F.**, Selective in vitro transcription by purified yeast RNA polymerase II on cloned 2 μm DNA, *Nucl. Acids Res.*, 9, 3959, 1981.
155. **Matsui, T., Segall, J., Weil, P. A., and Roeder, R. G.**, Multiple factors required for accurate initiation of transcription by purified RNA polymerase II, *J. Biol. Chem.*, 255, 11992, 1980.
156. **Weil, P. A., Luse, D. S., Segall, J., and Roeder, R. G.**, Selective and accurate initiation of transcription at the Ad 2 major late promoter in a soluble system dependent on purified RNA polymerase II and DNA, *Cell*, 18, 469, 1979.
157. **Huet, J., Wyers, F., Buhler, J. M., Sentenac, A., and Fromageot, P.**, Association of RNase H activity with yeast RNA polymerase A, *Nature (London)*, 261, 431, 1976.
158. **Iborra, F., Huet, J., Breant, B., Sentenac, A., and Fromageot, P.**, Identification of two different RNase H activities associated with yeast RNA polymerase A, *J. Biol. Chem.*, 254, 10920, 1979.
159. **Rackwitz, H. R., Rohde, W., and Sanger, H. L.**, DNA-dependent RNA polymerase II of plant origin transcribes viroid RNA into full-length copies, *Nature (London)*, 291, 297, 1981.
160. **Muhlbach, H. P. and Sanger, H. L.**, Viroid replication is inhibited by α-amanitin, *Nature (London)*, 278, 185, 1979.
161. **Lewin, B.**, *Gene Expression 2. Eukaryotic Chromosomes*, 2nd ed., John Wiley & Sons, New York, 1980, 641.
162. **Young, H. A. and Whiteley, H. R.**, Changes in the levels of DNA-dependent RNA polymerases during the transition of the dimorphic fungus *Mucor rouxii* from yeast-like to mycelial growth, *Exp. Cell Res.*, 91, 216, 1975.
163. **Kumar, B. V., McMillian, R. A., Medoff, G., Gutwein, M., and Kobayashi, G.**, Comparison of the ribonucleic acid polymerases from both phases of *Histoplasma capsulatum*, *Biochemistry*, 19, 1080, 1980.
164. **Key, J. L.**, Hormones and nucleic acid metabolism, *Annu. Rev. Plant Physiol.*, 20, 449, 1969.
165. **Zurfluh, L. L. and Guilfoyle, T. J.**, Auxin-induced nucleic acid and protein synthesis in the soybean hypocotyl, in *Levels of Genetic Control in Development*, Subtelny, S. and Abbott, U. K., Eds., Alan R. Liss, New York, 1981, 99.
166. **O'Brien, T. J., Jarvis, B. C., Cherry, J. H., and Hanson, J. B.**, Enhancement by 2,4-D of chromatin RNA polymerase in soybean hypocotyl tissue, *Biochim. Biophys. Acta*, 169, 35, 1968.
167. **Olszewski, N. and Guilfoyle, T. J.**, A new method for determining the number of RNA polymerases active in chromatin transcription, *Biochem. Biophys. Res. Commun.*, 94, 553, 1980.
168. **Hardin, J. W. and Cherry, J. H.**, Solubilization and partial characterization of soybean chromatin-bound RNA polymerase, *Biochem. Biophys. Res. Commun.*, 48, 299, 1972.
169. **Lin, C. Y., Chen, Y. M., Guilfoyle, T. J., and Key, J. L.**, Selective modulation of RNA polymerase I activity during growth transitions in the soybean seedlings, *Plant Physiol.*, 58, 614, 1976.
170. **Teissere, M., Penon, P., and Ricard, J.**, Hormonal control of chromatin availability and of the activity of purified RNA polymerases in higher plants, *FEBS Lett.*, 30, 65, 1973.
171. **Sebastian, J., Mian, F., and Halvorson, H. O.**, Effect of growth rate on the level of the DNA-dependent RNA polymerases in *Saccharomyces cerevisiae*, *FEBS Lett.*, 34, 159, 1973.
172. **Sebastian, J., Takano, I., and Halvorson, H. O.**, Independent regulation of nuclear RNA polymerases I and II during the yeast cell cycle, *Proc. Natl. Acad. Sci. U.S.A.*, 71, 769, 1974.
173. **Carter, B. L. A. and Dawes, I. W.**, Synthesis of two DNA-dependent RNA polymerases in yeast, *Exp. Cell Res.*, 92, 253, 1975.
174. **Hildebrandt, A. and Sauer, H. W.**, Levels of RNA polymerases during the mitotic cycle of *Physarum polycephalum*, *Biochim. Biophys. Acta*, 425, 316, 1976.
175. **Link, G. and Richter, G.**, Characterization of a protein factor stimulating RNA synthesis by DNA-dependent RNA polymerase II from plant cell cultures, *Eur. J. Biochem.*, 76, 119, 1977.

176. **Sawadogo, M., Sentenac, A., and Fromageot, P.**, Interaction of a new polypeptide with yeast RNA polymerase B, *J. Biol. Chem.*, 255, 12, 1980.
177. **Sawadogo, M., Huet, J., and Fromageot, P.**, Similar binding site for P_{37} factor on yeast RNA polymerases A and B, *Biochem. Biophys. Res. Commun.*, 96, 258, 1980.
178. **Sawadogo, M., Lescure, B., Sentenac, A., and Fromageot, P.**, Native deoxyribonucleic acid transcription by yeast RNA polymerase-P_{37} complex, *Biochemistry*, 20, 3542, 1981
179. **Kuehn, G. D., Affolter, H. U., Atmar, V. J., Seebeck, T., Gubler, U., and Braun, R.**, Polyamine-mediated phosphorylation of a nucleolar protein from *Physarum polycephalum* that stimulates rRNA synthesis, *Proc. Natl. Acad. Sci. U.S.A.*, 76, 2541, 1979.
180. **Ernst, G. H. and Sauer, H. W.**, A nuclear elongation factor of transcription from *Physarum polycephalum* in vitro, *Eur. J. Biochem.*, 74, 253, 1977.
181. **McNaughton, D. R., Klassen, G. R., and LeJohn, H. B.**, Phosphorylated guanosine derivatives of eukaryotes: regulation of DNA-dependent RNA polymerases I, II, and III in fungal development, *Biochem. Biophys. Res. Commun.*, 66, 468, 1975.
182. **LeJohn, H. B., Klassen, G. R., McNaughton, D. R., Cameron, L. E., Goh, S. H., and Meuser, R. V.**, Unusual phosphorylated compounds and transcriptional control in *Achlya* and other aquatic molds, in *Eukaryotic Microbes as Model Developmental Systems*, O'Day, D. H. and Horgen, P. A., Eds., Marcel Dekker, New York, 1977, 69.
183. **Hildebrandt, A. and Mengel, R.**, Characterization of an endogenous transcription inhibitor from *Physarum polycephalum*, *Z. Naturforsch.*, 34c, 76, 1979.
184. **Grossmann, K. and Seitz, U.**, RNA polymerase I from higher plants. Evidence for allosteric regulation and interaction with a nuclear phosphatase activity controlled NTP pool, *Nucl. Acids Res.*, 7, 2015, 1979.
185. **Grossmann, K. and Seitz, U.**, Cooperative effects of RNA polymerase from higher plant cells and *Escherichia coli:* a comparison, *FEBS Lett.*, 116, 193, 1980.
186. **Grossmann, K., Haschke, H. P., and Seitz, H. U.**, Regulation of RNA synthesis in higher plant cells by the action of a nucleoside triphosphatase, *Planta*, 152, 457, 1981.
187. **Bell, G. I., Valenzuela, P., and Rutter, W. J.**, Phosphorylation of yeast RNA polymerases, *Nature (London)*, 261, 429, 1976.
188. **Bell, G. I., Valenzuela, P., and Rutter, W. J.**, Phosphorylation of yeast DNA-dependent RNA polymerases in vivo and in vitro, *J. Biol. Chem.*, 252, 3082, 1977.
189. **Buhler, J. M., Iborra, F., Sentenac, A., and Fromageot, P.**, The presence of phosphorylated subunits in yeast RNA polymerases A and B, *FEBS Lett.*, 71, 37, 1976.
190. **Mazus, B., Szurmak, B., and Buchowicz, J.**, Phosphorylation in vitro and in vivo of the wheat embryo RNA polymerase II, *Acta Biochim. Pol.*, 27, 9, 1980.
191. **Bottomley, W., Smith, H. J., and Bogorad, L.**, RNA polymerases of maize: partial purification and properties of the chloroplast enzyme, *Proc. Natl. Acad. Sci. U.S.A.*, 68, 2412, 1971.
192. **Hallick, R. B., Lipper, C., Richards, O. C., and Rutter, W. J.**, Isolation of a transcriptionally active chromosome from chloroplasts of *Euglena gracilis*, *Biochemistry*, 15, 3039, 1976.
193. **Schiemann, J., Wollgiehn, R., and Parthier, B.**, DNA-dependent RNA polymerase in *Euglena gracilis* broken chloroplasts, *Biochem. Physiol. Pflanzen.*, 172, 507, 1978.
194. **Polya, G. M. and Jagendorf, A. T.**, Wheat leaf RNA polymerases. 1. Partial purification and characterization of nuclear, chloroplast, and soluble DNA-dependent enzymes, *Arch. Biochem. Biophys.*, 146, 635, 1971.
195. **Smith, H. J. and Bogorad, L.**, The polypeptide subunit structure of DNA-dependent RNA polymerase of *Zea mays* chloroplasts, *Proc. Natl. Acad. Sci. U.S.A.*, 71, 4839, 1974.
196. **Kidd, G. H. and Bogorad, L.**, Peptide maps comparing subunits of maize chloroplast and type II nuclear DNA-dependent RNA polymerase, *Proc. Natl. Acad. Sci. U.S.A.*, 76, 4890, 1979.
197. **Briat, J. F. and Mache, R.**, Properties and characterization of a spinach chloroplast RNA polymerase isolated from a transcriptionally active DNA-protein complex, *Eur. J. Biochem.*, 111, 503, 1980.
198. **Hardin, J. W., Apel, K., Smith, J., and Bogorad, L.**, RNA polymerases of the nucleus and chloroplasts of maize, in *Isozymes*, Vol. 1, Markert, C. L., Ed., Academic Press., New York, 1975, 55.
199. **Apel, K. and Bogorad, L.**, Light-induced increase in the activity of maize plastid DNA-dependent RNA polymerase, *Eur. J. Biochem.*, 67, 615, 1976.
200. **Bogorad, L., Jolly, S. O., Kidd, G., Link, G., and McIntosh, L.**, Organization and transcription of maize chloroplast genes, in *Genome Organization and Expression in Plants*, Leaver, C. J., Ed., Plenum Press, New York, 1980, 291.
201. **Jolly, S. O. and Bogorad, L.**, Preferential transcription of cloned maize chloroplast DNA sequences by maize chloroplast RNA polymerase, *Proc. Natl. Acad. Sci. U.S.A.*, 77, 822, 1980.
202. **Schiemann, J., Wollgiehn, R., and Parthier, B.**, Isolation of a transcription-active RNA polymerase-DNA complex from *Euglena* chloroplasts, *Biochem. Physiol. Pflanzen.*, 171, 474, 1977.

203. **Rushlow, K. E., Orozco, E. M., Lipper, C., and Hallick, R. B.,** Selective in vitro transcription of *Euglena* chloroplast ribosomal RNA genes by a transcriptionally active chromosome, *J. Biol. Chem.,* 255, 3786, 1980.
204. **Briat, J. F., Laulhere, J. P., and Mache, R.,** Transcription activity of a DNA-protein complex isolated from spinach plastids, *Eur. J. Biochem.,* 98, 285, 1979.
205. **Scragg, A. H.,** Mitochondrial DNA-directed RNA polymerase from *Saccharomyces cerevisiae* mitochondria, *Biochem. Biophys. Res. Commun.,* 45, 701, 1971.
206. **Scragg, A. H.,** A mitochondrial DNA-directed RNA polymerase from yeast mitochondria, in *The Biogenesis of Mitochondria. Transcriptional, Translational, and Genetic Aspects,* Kroon, A. M. and Saccone, C., Eds., Academic Press, New York, 1974, 47.
207. **Scragg, A. H.,** The isolation and properties of a DNA-directed RNA polymerase from yeast mitochondria, *Biochim. Biophys. Acta,* 442, 331, 1976.
208. **Scragg, A. H.,** Origin of the mitochondrial RNA polymerase of yeast, *FEBS Lett.,* 65, 148, 1976.
209. **Rogall, G. and Wintersberger, E.,** Low molecular weight subunit of a rifampicin-resistant mitochondrial RNA polymerase from yeast, *FEBS Lett.,* 46, 333, 1974.
210. **Eccleshall, T. R. and Criddle, R. S.,** The DNA-dependent RNA polymerases from yeast mitochondria, in *The Biogenesis of Mitochondria. Transcriptional, Translational, and Genetic Aspects,* Kroon, A. M. and Saccone, C., Eds., Academic Press, New York, 1974, 31.
211. **Eccleshall, R. and Criddle, R. S.,** DNA-dependent RNA polymerases isolated from yeast mitochondria, *Arch. Biochem. Biophys.,* 164, 602, 1974.
212. **Tsai, M. J., Michaelis, G., and Criddle, R. S.,** DNA-dependent RNA polymerase from yeast mitochondria, *Proc. Natl. Acad. Sci. U.S.A.,* 68, 473, 1971.
213. **Levens, D., Lustig, A., and Rabinowitz, M.,** Purification of mitochondrial RNA polymerase from *Saccharomyces cerevisiae, J. Biol. Chem.,* 256, 1474, 1981.
214. **Kuntzel, H. and Schafer, K. P.,** Mitochondrial RNA polymerase from *Neurospora crassa, Nature (London), New Biol.,* 231, 265, 1971.
215. **Levens, D., Morimoto, R., and Rabinowitz, M.,** Mitochondrial transcription complex from *Saccharomyces cerevisiae, J. Biol. Chem.,* 256, 1466, 1981.

Chapter 2

RNA POLYMERASES FROM HIGHER EUKARYOTES

Kathleen M. Rose, Dean A. Stetler, and Samson T. Jacob

TABLE OF CONTENTS

I. INTRODUCTION

The existence of multiple DNA-dependent RNA polymerases (E.C. 2.7.7.6) in eukaryotic cells has now been well-established. The three nuclear enzymes, designated RNA polymerases I, II, and III, are involved in the synthesis of ribosomal, messenger, and low molecular weight RNAs, respectively. In addition to the nuclear enzymes, distinct RNA polymerases are found in mitochondria and chloroplasts. Over the past decade, the subcellular localization, subunit composition, and reaction properties of these complex enzymes have been elucidated. The activity of the DNA dependent RNA polymerases of higher eukaryotes in response to a variety of stimuli has also been the subject of a number of investigations. Despite these advances, it is becoming increasingly clear that the availability of the appropriate polymerase is not sufficient for accurate, selective transcription of individual genes. Rather, regulation of gene transcription in eukaryotic cells appears to be a multifaceted phenomenon which requires interactions between RNA polymerases, auxiliary factors, and DNA.

Numerous reviews have appeared describing the general properties of eukaryotic RNA polymerases (see References 1 to 6). Chapter 1 of this volume addresses the DNA-dependent RNA polymerases of plants and lower eukaryotes. This chapter will focus on the enzymes from higher eukaryotes, particularly those of the animal kingdom. Due to space limitations it is not possible to present a comprehensive review of RNA polymerases in the present treatise. Rather, we have selected examples which serve to illustrate general concepts. We extend apologies to those investigators whose work may not have been cited in this chapter.

II. RNA POLYMERASE I

A. Function

The ribosomal genes of eukaryotes are highly reiterated (50 — 2×10^6 gene copies per cell). These multiple gene copies are linked in tandem array and located at the nucleolar organizer. The transcription unit of each gene is separated from that of the next one by a nontranscribed spacer. While the size of the transcription unit is relatively constant (M_r 4×10^6 to 5×10^6), the nontranscribed spacer varies in length from organism to organism. Large (40S to 45S) ribosomal RNA (rRNA) precursor molecules are transcribed from the individual genes and subsequently processed to yield the mature 18S, 28S, and 5S rRNAs (see Reference 7). The association of RNA polymerase I with nucleoli[8-10] first suggested a role for this enzyme in transcription of the rDNA. Coupled with the observation that isolated nucleoli are capable of synthesizing rRNA precursor molecules,[11] these data make a strong argument for involvement of RNA polymerase I in the synthesis of rRNA. Recently, we have obtained anti-RNA polymerase I antibodies.[12] Using this antibody preparation, we have been able to establish the localization of RNA polymerase I in intact cells by indirect immunofluorescence.[12a] As shown in Figure 1, the enzyme is located almost exclusively in nucleoli of rat liver. Preliminary studies indicate that rRNA synthesis is inhibited by microinjection of anti-RNA polymerase I antibodies into intact cells.[12b] Hence, there seems little doubt that RNA polymerase I is responsible for the synthesis of rRNA in vivo.

Whether the ribosomal cistrons are the only genes transcribed by polymerase I in vivo has not yet been established. Thus far, no other gene products arising from transcription by this enzyme have been identified. However, at least one other function, namely, the synthesis of the RNA primers for DNA synthesis, has been ascribed to RNA polymerase I. The observations that production of the RNA primer requisite for initiation of DNA replication in isolated nuclei is inhibited by RNA polymerase I antibodies[13] and that extensively purified yeast RNA polymerase I is capable of producing

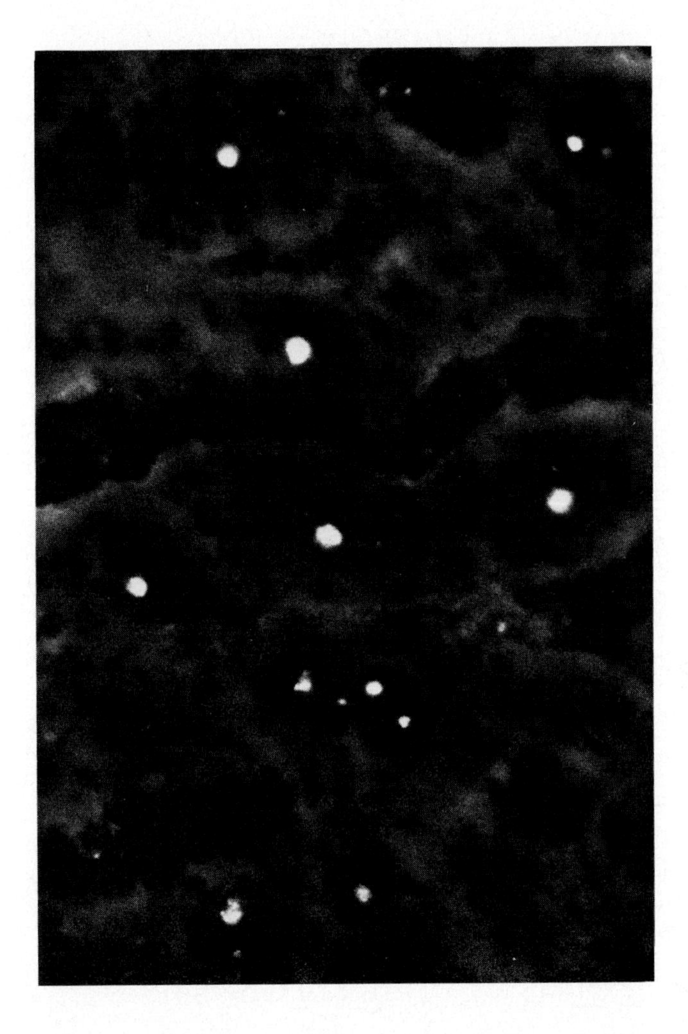

FIGURE 1. Subcellular localization of RNA polymerase I. Frozen sections of rat liver were rinsed with acetone and incubated with immuno-globulins[12] obtained from a rabbit injected with RNA polymerase I purified from rat hepatoma 3924A.[32] Antigen-antibody complexes were detected by fluorescence microscopy following incubation with FITC-conjugated goat anti-rabbit immunoglobulins. Accompanying phase contrast microscopy confirmed the nucleolar localization of the fluorescent areas. Tissue sections incubated with immunoglobulins from a control rabbit did not fluoresce (not shown). Photograph courtesy of Dr. Ulrich Scheer.

an RNA which becomes linked to DNA synthesized in vitro[14] led to this hypothesis. The involvement of polymerase I in synthesis of these RNA primers in vivo remains to be elucidated.

B. Regulation

By the late 1960s it was evident that rRNA synthesis was coordinated with a variety of growth stimuli as well as with neoplasia. Initial studies with nuclear, nucleolar, and chromatin preparations indicated that RNA polymerase I activity on the endogenous template often correlated with rRNA synthesis in vivo.

To differentiate between alterations in the enzyme and template, RNA polymerase I was solubilized and its activity ascertained on exogenous, deproteinized DNA. The first example of activation of solubilized RNA polymerase I in response to a physiological stimulus was reported by Sajdel and Jacob.[15] These investigators found that administration of glucocorticoids to adrenalectomized rats resulted in enhanced activity of RNA polymerase I extracted from liver. This observation has been confirmed in other laboratories.[16,17] Alterations in the activity of solubilized polymerase I have been detected in response to other hormonal and growth stimuli[18,19] (also see Reference 4). In many instances, the change in polymerase activity correlates with the in vivo modification of rRNA synthesis. Perhaps the best examples of increased RNA polymerase I activity are those found in neoplastic cells. As early as 1972, Chesterton et al.[20] reported that the activity of RNA polymerase I solubilized from a rat hepatoma was ninefold greater than that from normal rat liver. In the subsequent decade, at least 20 other laboratories have reported increases in this enzyme in tumor cells (reviewed in References 3, 4, and 6). Duceman and Jacob[21] did a systematic study which concluded that the elevation of polymerase I in a series of rat hepatomas reflects not only the increased growth rate, but also the degree of differentiation. That is, even though RNA polymerase I activity increases in normal growth, the enzyme activity from poorly differentiated tumors is markedly greater than that from normally growing tissue. Increases in RNA polymerase I activity have also been observed after administration of chemical carcinogens to animals.[6,22]

RNA polymerase I apparently exists as two populations of enzyme, one which is loosely associated with template ("free" population) and one which is tightly bound to the nucleolar DNA ("bound" or "engaged").[23-25] Since only the "bound" enzyme is actively transcribing rDNA, measurements of RNA synthesis by the total solubilized enzyme do not always reflect the amount of transcriptionally active polymerase. For example, Matsui et al.[24] have examined the activity of the two forms of liver RNA polymerase I after partial hepatectomy. Ribosomal RNA synthesis and the activity of extracted "bound" enzyme (RNA polymerase IB) increased at the same time (4h) after surgery. The activity of the "free" enzyme (RNA polymerase IA) was not altered during regeneration. Another illustration of the correlation between activity of the "bound" polymerase and rRNA synthesis is found in lymphocytes stimulated to proliferate by the mitogen, phytohemaglutinin. In 1975, Jaehning et al.[26] reported that total RNA polymerase I activity increased linearly after culture in the presence of mitogen; a 17-fold increase was observed after 4 days. Subsequent studies by Tillyer and Butterworth[27] suggested that this increase in RNA polymerase I activity was primarily due to an elevation of "free" enzyme, rather than the chromatin-engaged enzyme. The activity of the latter enzyme reached a plateau after one day and closely paralleled rRNA production. Interestingly, a recent study[28] indicated that the number of RNA polymerase I molecules actively transcribing rDNA (measured by number of nascent 3′OH termini insensitive to α-amanitin) is unchanged after stimulation with phytohemaglutinin. Rather, elevated rRNA synthesis is a result of an increased rate of rRNA chain elongation after mitogen treatment. Although there seems to be good correlation between rRNA synthesis and the activity of the "bound" polymerase after solubilization, it should be borne in mind that measurements made on deproteinized templates do not always provide an accurate assessment of in vivo transcription rates. Perhaps the most striking example of such a phenomenon is RNA polymerase I activity during the cell cycle. It is well-established that little or no rRNA is synthesized during mitosis. Transcriptional activity in isolated nuclei and chromosomes is also minimal during this stage of the cell cycle. On the other hand, enzyme solubilized from metaphase chromosomes is as active as that from cells in other stages of the cell cycle.[29] Clearly, factors other than polymerase I itself contribute to regulation of rDNA transcription during the cell cycle.

Thus far, it has been difficult to distinguish between RNA polymerase I activity and amount of enzyme. This distinction must be considered in quantitating alterations in response to biological stimuli. For example, a rapid, transient increase in liver polymerase I activity occurs in response to glucocorticoids.[15-17] The enhanced activity is observed in the absence of protein synthesis[30] and most likely reflects activation of existing enzyme rather than *de novo* synthesis. In contrast, the amount of enzyme found in tumors (estimated from quantity of pure protein) appears to be increased[6,31] relative to that of normal cells. We have recently observed both activation[6,12] and increased amount[6] of RNA polymerase I from a rapidly growing hepatoma. Activation of RNA polymerase I is most likely a result of enzyme phosphorylation. Several lines of evidence suggest that rat hepatoma polymerase I is more highly phosphorylated in vivo than liver enzyme. First, a nuclear protein kinase[32] (NII in the nomenclature of Desjardins et al.[33]) can phosphorylate and activate RNA polymerase I in vitro;[34] under these conditions, partially purified preparations of liver, but not of tumor, RNA polymerase I are activated by addition of excess kinase.[34a] Second, the specific activity of pure hepatoma enzyme is higher than that of liver enzyme,[6,35] even when measured under conditions in which endogenous protein kinase NII is not functional. Third, the hepatoma contains several-fold more protein kinase NII than liver.[32] Protein kinase NII is a cyclic nucleotide-independent casein kinase which uses both ATP and GTP as phosphoryl donor. This kinase is inhibited by heparin and stimulated by polyamines.[32] The possibility that protein kinase NII is a regulatory factor in polymerase I-catalyzed transcription in vivo is evidenced by the tight association of these enzymes in vitro. Specifically, this kinase is present in polymerase I preparations which have been subjected to sucrose gradient centrifugation in the presence of high salt. Under these conditions, the bulk of the NII protein kinase sediments at 7S whereas polymerase I migrates at 16S. In spite of the clear separation of the two enzyme activities on the gradient, some protein kinase activity is still present in the fractions containing the polymerase. (The kinase activity is more evident when the salt is removed by dialysis). The polymerase I from these gradients appears pure on polyacrylamide gels run under nondenaturing conditions. On such gels, protein kinase activity and RNA polymerase activity comigrate.[12] Using immunological criteria, it was ascertained that the two subunits of protein kinase NII, M_r 42,000 and 24,600, at least in part, account for the hepatoma polymerase I polypeptides of the same molecular weight.[12] Protein kinase NII phosphorylates the 120,000; 65,000; and 24,600 M_r subunits of RNA polymerase I. Stimulation of activity occurs even in the absence of the 65,000 M_r polypeptide.[34] Since the 24,600-dalton phosphopeptide is a component of the kinase and activation can be obtained without the 65,000 M_r polypeptide, modification of the 120,000 M_r subunit is most likely responsible for the altered RNA synthesis. The observation that phosphorylation results in an increased rate of RNA chain elongation implicates the 120,000-dalton subunit in this step of RNA synthesis. Protein kinase NI,[36] another cyclic nucleotide-independent nuclear casein kinase, does not activate or phosphorylate RNA polymerase I.[34] Protein kinase NI presumably plays a role in regulation of gene expression by modifying other nuclear proteins. One enzyme known to be phosphorylated and activated by protein kinase NI is nuclear poly(A) polymerase[36,37] (see also Chapter 5 of this volume), the enzyme responsible for the posttranscriptional addition of poly(A) tracts to heterogeneous nuclear RNA[38] (mRNA precursors). Although the role of protein kinase NII in phosphorylation of RNA polymerase I in vivo has not yet been established, preliminary experiments indicate that polymerase I is phosphorylated in vivo and that this modification occurs on the same polypeptides which are phosphorylated in vitro.[38a] Since only the NII kinase is capable of modifying polymerase I in vitro, protein kinase NII may also phosphorylate the 120,000; 65,000; and 24,600 M_r subunits in vivo.

C. Ribosomal Gene Transcription In Vitro

Isolated nuclei and nucleoli seem to be capable of synthesizing rRNA precursor molecules. Much of this transcription results from the elongation of rRNA chains which have been initiated in vivo. Attempts to reconstitute in vitro systems using RNA polymerase I partially purified from higher eukaryotes and purified ribosomal genes have met with little success. Significant transcription of ribosomal cistrons using deproteinized DNA has been observed only in the yeast system.

Cloned ribosomal DNA is accurately transcribed by polymerase I when microinjected into frog oocytes.[39,40] Using this technique, Bakken et al.[40] have mapped the initiation and termination sequences of the *Xenopus laevis* ribosomal transcription units. Homogenates of frog oocyte nuclei also accurately transcribe the rDNA.[41] In this system, the endogenous RNA polymerase I can reinitiate on the nuclear rDNA but not on exogenously added cloned rDNA. The observation that intact oocytes but not nuclear preparations, can form the transcription complexes with exogenous rDNA suggests that ribosomal gene transcription in vivo requires continuous synthesis of a factor which becomes bound to rDNA. This factor is apparently not shuttled among ribosomal cistrons, but remains associated with its initial rDNA host. Interestingly, Crampton and Woodland[42] have identified a protein factor in oocytes which stimulates rRNA synthesis in frog nuclear preparations. Although more precise experimental protocol is required, it appears as if this factor may affect initiation of rRNA chains.

Cloned mouse rDNA can be transcribed in vitro in the presence of cell extracts which contain the polymerase I as well as auxiliary factors.[43-46] In some of these systems, accurate transcription of the cloned gene as well as initial processing of the 5′ terminus of the nascent rRNA can take place. Apparently, some species specificity is exerted in vitro since homologous extracts are required for appropriate transcription.[46a] Cloned rDNA of *Drosophila melanogaster* can also be accurately transcribed in vitro by homologous cell extracts.[46b] Because these extracts do not initiate accurately on cloned rDNA from *D. virilis*, they appear to retain some of the in vivo control mechanisms. Clearly, much work remains before the nature of the factors which mediate accurate transcription of rDNA in vitro is elucidated.

Several investigators have identified factors which alter polymerase I-catalyzed transcription of deproteinized, undefined templates. Froehner and Bonner[47] identified a heat labile factor(s) which stimulates RNA polymerase I, but not polymerase II. These authors suggested that the factor may be a subunit of the polymerase itself. James et al.[48] have purified a nucleolar protein (C-14) of M_r 70,000 which stimulates RNA polymerase I-catalyzed transcription. Since protein C-14 also enhances RNA synthesis by *Escherichia coli* enzyme, the authors postulated that its effect was mediated via DNA binding. Protein C-14 appears to be related to a 70,000 M_r polypeptide from *Physarum polycephalum*.[49] Polyamine-mediated phosphorylation of the *Physarum* protein is involved in regulation of rRNA synthesis in vivo. Goldberg et al.[50] purified a protein factor from rat liver which stimulates RNA polymerase I and, to a lesser extent, polymerase II. The fraction contains two heat-stable proteins of M_r 11,000 and 12,000. These proteins appeared to bind to both DNA and rRNA. Interestingly, these workers also reported that polymerase I lost the ability to transcribe double-stranded DNA upon sucrose gradient centrifugation. Rose et al.[6,12,34] have also purified a factor from the sucrose gradient centrifugation step of the RNA polymerase I purification which stimulates transcription of double stranded DNA. The protein which regulates polymerase activity in the latter example has been identified as a cyclic AMP-independent nuclear protein kinase (NII).[6,12,32,34] The activation of polymerase I is mediated via polyamine-dependent phosphorylation of polymerase polypeptides of M_r 120,000; 65,000; and 24,600.[51] Protein kinase NII has been found associated with pure polymerase I and appears to be regulatory polypeptides of M_r 42,000 and 24,600 (see

Section II.B). It is not yet known whether any of the factors which stimulate polymerase I activity on calf thymus DNA also play a role in rDNA transcription.

III. RNA POLYMERASE II

A. Function

The structural genes for eukaryote proteins are large, complex DNA molecules, which, in many cases, contain a number of intervening sequences not found in the mature mRNA (see Reference 52). Transcription of these DNAs produces long mRNA precursor molecules, commonly referred to as heterogeneous nuclear RNA (HnRNA). The sensitivity of RNA polymerase II to α-amanitin,[53-55] an inhibitor which prevents mRNA production in vivo,[56,57] and the presence of this polymerase on transcriptionally active nucleoplasmic genes[58] suggests that this enzyme is responsible for the synthesis of mRNA precursors. That in vitro synthesis of specific mRNAs can be inhibited by concentrations of α-amanitin which affect only polymerases II was first observed in 1976.[59] These data directly implicate RNA polymerase II in mRNA production. Recently, anti-RNA polymerase II antibodies microinjected into nuclei of *X. laevis* have been shown to inhibit extranucleolar chromosomal transcription.[60] Since the antibodies were specific for polymerase II and did not cross-react with RNA polymerases I and III, these data provide direct evidence of the involvement of polymerase II in HnRNA synthesis in vivo.

B. Regulation

Regulation of the production of many eukaryote proteins occurs at the level of transcription. In some instances, in vitro transcription of chromatin or nuclear preparations by endogenous RNA polymerase II correlates with the alterations in gene expression observed in vivo (e.g., as in response to testosterone,[61] estradiol,[62] medroxyprogesterone,[63] partial hepatectomy,[64] and phytohemaglutinin[27]). Several laboratories have measured specific transcription products using chromatin and exogenous RNA polymerases. These studies indicate that specific genes become available for transcription after appropriate biological stimuli. These include transcription of histone genes during certain stages of the cell cycle,[65] the globin gene in erythroid cells,[66,67] the immunoglobulin gene in lymphoid cells,[68] and the ovalbumin gene in estrogen-stimulated oviducts.[69]

The actual number of RNA polymerase II molecules in response to physiological alterations has been quantitated by α-amanitin binding,[70] by immunological techniques,[71] and by estimation of the number of 3′OH termini of nascent RNA sensitive to α-amanitin in vitro.[72] Using *O*-[^{3}H]methyl-demethyl-γ-amanitin, Cochet-Meilhac et al.[70] estimated that transcriptionally active cells, e.g., rat liver, testis, and brain, contain 14,000 to 38,000 molecules of RNA polymerase II per haploid genome. Interestingly, quiescent tissue, such as the uterus from ovarectomized rats, had less polymerase II (~4700 molecules per haploid genome) than metabolically active tissue, i.e., uterus from mature rats, which had 16,000 molecules per haploid genome. These measurements, performed on whole tissue homogenates, suggest that the amount of RNA polymerase II correlates with the overall transcriptional state of the cell. Kastern et al.[73] recently measured the number of RNA polymerase II molecules in rooster liver by [^{3}H]amanitin binding and compared the amount of polymerase II with the enzyme activity. These investigators observed that administration of estrogen increased the amount of RNA polymerase to the same extent as the enzyme activity (measured either in intact nuclei or after solubilization). Transcription of deproteinized DNA by solubilized RNA polymerase II is also enhanced in response to other biological stimuli.[21,22,26,34,63,64] It is not yet clear whether the augmented polymerase II activity in the latter examples represents an increase in the number of transcriptionally active molecules or activation of preexisting enzyme or both.

Cox[72] estimated the amount of transcriptionally active polymerase II by determining the number of nascent RNA transcripts sensitive to α-amanitin in vitro. In contrast to the amanitin binding studies, the number of transcriptionally active enzyme molecules (in oviduct nuclei) was not altered by estrogen. However, it should be noted that little stimulation of RNA polymerase II-catalyzed RNA synthesis was obtained in this particular system.

Guialis et al.[71] have successfully used antibodies to RNA polymerase II to quantitate the enzyme concentration. These investigators observed that the activity of α-amanitin-insensitive RNA polymerase II increased in α-amanitin-resistant Chinese hamster ovary cell hybrids grown in the presence of the toxin. Concurrently, the amount of [^{3}H]amanitin bound to enzyme decreased and correlated with the loss of sensitivity to the inhibitor. The total mass of polymerase II, measured by radioimmunoassay, remained relatively constant throughout this period. Although this technique estimated only the total mass of the enzyme rather than the quantity of individual subunits, recent work from the same laboratory[74] indicates that the increase in amount of α-amanitin-resistant enzyme is due to a coordinate elevation in the synthesis of at least three of the polymerase II polypeptides. These data imply that fluctuations in enzyme concentration are responsible for differences in polymerase II activity.

Apparently, alterations in the concentration of RNA polymerase II do not always correlate with transcriptional activity. As observed for RNA polymerase I, there is apparently no decrease in the amount of RNA polymerase II during metaphase,[29] even though RNA synthesis is minimal during this period. Another example of interest is the observation that similar numbers of RNA polymerase II molecules initiate globin mRNA synthesis in mature and immature hen erythrocytes.[75] However, only the immature red cell produces full length globin mRNA. These results indicate that RNA polymerase II is clustered near the initiation point for globin RNA synthesis in the mature cell but that other factors control its traversal of the entire globin gene.

C. In Vitro Transcription

1. Factors Affecting General Transcription

Although RNA polymerase II possesses a complex subunit structure, this enzyme is not capable of specific transcription of deproteinized templates in vitro.[3] Rather, the DNA is transcribed symmetrically and at random (see References 2 to 5). This lack of specific initiation in vitro suggests that RNA polymerase II requires interaction with other cellular molecules to synthesize biologically relevant RNA. To investigate whether cellular molecules can regulate RNA synthesis, a number of investigators have isolated factors which alter RNA polymerase II-catalyzed transcription in vitro. Although several nonprotein factors, e.g., polyamines[76] and heparin,[77] influence RNA synthesis by eukaryote polymerase, these molecules are unlikely to mediate transcription of specific genes directly. Since many investigations have focused on the role of polypeptides in regulating RNA synthesis only the protein factors will be considered here. Initially, the effect of proteins on transcription of relatively undefined templates was investigated. The first factors capable of altering RNA polymerase II-catalyzed transcription were reported in 1970.[78-80] Description of other such proteins (see Tables 1 and 2) followed in the subsequent decade. In general, depending on their mode of action, these proteins can be classified as elongation or initiation factors.

a. Elongation Factors

The protein factors known to affect elongation of RNA chains by RNA polymerase II are summarized in Table 1. Two types of elongation factors have been described,

Table 1
FACTORS AFFECTING THE ELONGATION OF RNA CHAINS BY RNA POLYMERASE II

Source	Factor	Preparation	Properties	Mechanism	Ref.
Rat liver		Whole cell extract ssbjected to DEAE-cellulose chromatography and the effluent precipitated with $(NH_4)_2SO_4$ (30—70% saturation). Stimulatory activity purified on CM-cellulose, hydroxylapatite, Sephadex® G-100, and DEAE-cellulose, followed by preparative isoelectric focusing in a sucrose gradient.	M_r 30,000; pI 9—6; heat stable (90°C, 5 min)	Stimulates elongation rate and average product size but not initiation on a double-stranded DNA template. Has no effect on rat liver polymerases I and III	78, 81
Lamb thymus	HSF	Whole cell extract precipitated with $(NH_4)_2SO_4$ (50—75% saturation). HSF is then purified by Sp-Sephadex® chromatography, followed by heating at 90°C for 5 min, and hydroxylapatite chromatography	M_r 25,000; pI 8.0; heat stable (90°C, 5 min)	Binds to RNA polymerase II and increases rate of RNA chain elongation on double-stranded lamb thymus DNA, but not on denatured template. Has no effect on RNA polymerase I or *Escherichia coli* enzyme. Not involved in initiation	82
Mouse myeloma		Whole cell extract adsorbed batchwise to DEAE-cellulose and sedimented by centrifugation. The supernatant precipitated with $(NH_4)_2SO_4$ (45—60% saturation) and stimulatory activity purified on Sephadex® G-25, Sephadex® G-75, CM-Sephadex®, and DEAE-cellulose	M_r 20,000—30,000; heat labile (50°C, 5 min)	Stimulates the elongation reaction of RNA polymerase II. Has no effect on initiation. Stimulation is 20-fold with double-stranded (calf thymus or mouse myeloma) but only 2-fold with denatured DNA	83
Calf thymus	SF-2	Whole cell extract is precipitated with $(NH_4)_2SO_4$ (40% saturation). The pellet fraction is subjected to chromatography on DEAE-Sephadex® and CM-cellulose. Gel filtration results in the separation of SF-1 and SF-2. SF-2 is further purified by SP-Sephadex® chromatography	M_r 34,000 and 19,000; heat labile (98°C, 40 min)	Enhances elongation rate on form 1 SV40 template. Has no effect on initiation reaction or on polymerases I or III. Polypept de of M_r 19,000 is active in unwinding poly [d(A-T)]	84, 85

those which are heat-stable and those which are temperature-sensitive. The heat-stable proteins from rat liver[78,81] and lamb thymus[82] have molecular weights of 30,000 and 25,000, respectively. The M_rs of the heat-labile protein factors[83-85] range from 19,000 to 34,000. All of the elongation factors are specific for RNA polymerase II and have little or no effect on other eukaryotic or bacterial RNA polymerases. These proteins enhance polymerase II activity when double-stranded DNA, rather than denatured DNA, is the template. Other protein factors also stimulate transcription of double-stranded DNA[86-89] but their mechanism of action is unknown.

b. Initiation Factors

Polypeptides which increase the rate or extent of RNA chain initiation by RNA polymerase II have also been described. As observed for the elongation factors, these

Table 2
FACTORS AFFECTING THE INITIATION OF RNA CHAINS BY RNA POLYMERASE II

Source	Factor	Preparation	Properties	Mechanism	Ref.
Calf thymus	SF-1	Cell extracts precipitated with $(NH_4)_2SO_4$ and precipitates purified by chromatography on DEAE-Sephadex® and CM-Sephadex®. SF-1 separated from SF-2 by chromatography on Sephadex® G-75	Heat stable (98°C, 40 min). M_r 65,000; dimer of 27,000 and 36,000	Stimulates initiation on form 1 SV40 template. No detectable effect on elongation reaction. Does not affect RNA polymerases I or III	84, 85
Chicken myeloblastosis cells		Nuclear extracts subjected to DEAE-SePhadex® chromatography as for RNA polymerase purification. The fractions eluting at 0.1 *M* $(NH_4)_2SO_4$ were heated at 100°C for 15 min and the denatured protein removed by centrifugation	Heat stable (100°C, 15 min)	Causes increased initiation on denatured DNA. RNA polymerase IIb is enhanced more than IIa. No change in the size of RNA is observed	90, 91
Human KB3 cells	SF-A	Whole cell extract precipitated with $(NH_4)_2SO_4$ (10% saturation). The pellet fraction subjected to DEAE-cellulose chromatography, and the flow-through fractions applied to CM-cellulose. Two peaks of stimulatory activity were obtained (SF-A and SF-B eluting and 0.05 and 0.01 *M*KC1, respectively). SF-A further purified on Sephadex® G-150 and DNA-cellulose	Heat labile (70°C, 5 min). M_r 20,000—30,000	Interacts with RNA polymerase II and stimulates initiation on adenovirus 2 DNA. Has no effect on RNA polymerase I or on the *Escherichia coli* enzyme. In the absence of SF-A, the size of the RNA product is heterogeneous (4-29S). In the presence of SF-A, there is a definite peak between 4 and 18S on the gradient	92
	SF-B		Heat labile (70°C, 5 min). M_r 20,000—30,000	Amount varies greatly from preparation to preparation. Has not been characterized	
Ehrlich ascites tumor	SII	Whole cell extract is precipitated with $(NH_4)_2SO_4$ (50—85% saturation), the pelleted material applied to DEAE-cellulose, and the effluent subjected to phosphocellulose chromatography. Two peaks of stimulatory activity are obtained (SI and SII elute at 0.45 and 0.55 *M* KC1, respectively). CM-cellulose chromatography is also used to resolve SII into two separate components (SII and SII′). SII and SII′ are then purified to apparent homogeneity on a second DEAE-cellulose column	Heat labile (60°C, 10 min). M_r 40,500	SII (includes SII′) stimulates formation of initiation complexes by RNA polymerase II and causes marked increase in the size of RNA chains synthesized. Has no effect on RNA polymerase I. Ehrlich escites tumor DNA is a much better template than calf thymus DNA in the presence of SII, but about the same in its absence	88, 93

Table 2 (continued)
FACTORS AFFECTING THE INITIATION OF RNA CHAINS BY RNA POLYMERASE II

Source	Factor	Preparation	Properties	Mechanism	Ref.
Brine shrimp (*Artemia salina*)	S Protein	Nuclear extracts are subjected to DEAE-cellulose chromatography. The flow-through proteins containing S protein are then separated in a sucrose gradient	Heat labile (100°C, 10 min). M_r 400,000—500,000. (65% of complex is composed of 2 proteins M_r 183,000 and 67,000)	Stimulates RNA polymerase II from *Artemia* on either native or denatured *Artemia* or calf thymus template. Has no effect on *Artemia* RNA polymerase I or on RNA polymerase II from *Drosophila,* rat liver, mouse liver, or calf thymus, or on the *E. coli* enzyme. Results in transcription of a greater number of RNA molecules	94, 95

proteins include heat-stable[84,85,90,91] and heat-labile proteins[88,92,95] (see Table 2). In general, these proteins are specific for RNA polymerase II and have little or no effect on other DNA-dependent RNA polymerases. In contrast to the elongation factors, the effect of the initiation proteins does not appear to be dependent on a double-stranded template. Stimulation has been observed with heat-denatured DNA and poly[d(A-T)] as well as with double-stranded DNA.

The SII initiation factor from Ehrlich ascites tumor cells has been well-characterized. In a reconstituted system including homologous DNA template, SII stimulates the formation of initiation complexes by polymerase II.[96] Although the factor stimulates random initiation on a noncellular template (cloned adenovirus DNA),[97] it enhances cellular (globin) mRNA synthesis in isolated spleen cell nuclei.[98] Since anti-SII antibodies are capable of inhibiting RNA polymerase II-directed transcription in isolated nuclei,[99] it is likely that the polypeptide has a role in control of gene expression in vivo. The contention that SII functions in vivo was strengthened by the observation that SII is located in the cell nucleus during cell interphase,[100] but not during metaphase when there is minimal transcriptional activity. In the metaphase cells, SII is present throughout the cytoplasm.[100] Nuclei of a variety of cells including mouse 3T3 cells, HeLa cells, and salivary gland cells of flesh fly larvae contain SII (detected by indirect immunofluorescence).[100] This observation suggests that the RNA polymerase II-specific initiation factor, SII, has been conserved during evolution.

c. Protein Kinases as RNA Polymerase II Activators

RNA polymerase II exists as a phosphoprotein in vivo.[101,102] In vitro phosphorylation of RNA polymerase II by cyclic AMP-dependent and -independent protein kinases has been reported (Table 3). As early as 1974, Jungmann et al.[103] observed that cyclic nucleotide-dependent protein kinase is able to stimulate a partially purified preparation of RNA polymerase II. Subsequent work from the same laboratory indicates that a polypeptide of M_r 25,000 is the predominant phosphate acceptor of RNA polymerase.[105] In some instances, the 180,000 M_r polypeptide can also be phosphorylated. Although phosphorylation of polymerase II by cyclic AMP-dependent kinase correlates with enhanced activity it is not known what stage of transcription is affected.

Table 3
EFFECT OF PROTEIN KINASES ON RNA POLYMERASE II

Protein kinase			RNA Polymerase			
Protein kinase source	Designation	Cyclic nucleotide dependence	Source	Polypeptides phosphorylated (M_r)	Effect on activity	Ref.
Calf ovary cytosol	—	Dependent	Calf ovary nuclei	Not determined	Stimulates transcription of double-stranded calf thymus DNA	103
Novikoff ascites tumor cells	KI (previously HLF_2)	Independent	Novikoff ascites tumor nuclei	Not determined	Stimulates transcription of double-stranded Novikoff ascites tumor DNA	87
	KII	Independent	Novikoff ascites tumor nuclei	Not determined	None	104
Calf thymus nuclei	—	Dependent	Calf thymus nuclei	25,000 (180,000 in some experiments	Stimulates transcription of heat-denatured calf thymus DNA	105
	—	Independent	Calf thymus nuclei	25,000	Stimulates transcription of heat-denatured calf thymus DNA	106
Calf thymus cells	KI	Independent	Calf thymus nuclei	240,000; 214,000; 20,500	None	102, 107
	KII	Independent	Calf thymus nuclei	214,000; 20,500	None	102, 107
Morris hepatoma 3924A nuclei	NI	Independent	Morris hepatoma 3924A nuclei	None	None	108
Morris hepatoma 3924A nuclei	NII	Independent	Morris hepatoma 3924A nuclei	214,000; 140,000; 21,000	Stimulates transcription of double-stranded and heat-denatured calf thymus DNA and poly d(A-T). Results in an increased number of RNA chains initiated with no significant change in length of product	108

RNA polymerase II from calf thymus is also phosphorylated in vitro by homologous cyclic AMP-independent nuclear kinase.[106] As in the case of phosphorylation by the cyclic AMP-dependent kinase, this kinase phosphorylates a 25,000 M_r polypeptide n the polymerase II preparation and phosphorylation is accompanied by activation of the polymerase. Polymerase II from Novikoff ascites tumor cells is also activated by a homologous cyclic AMP-independent protein kinase, KI in the authors nomenclature (previously described as HLF_2).[87,104] This protein kinase has been shown to be located primarily in the nucleus of the Novikoff cells.[87] Two cellular casein kinases of thymus can phosphorylate homolooous RNA polymerase II.[102,107] These kinases, designated KI and KII, although structurally distinct from each other, modify similar polymerase II polypeptides (M_r 214,000 and 20,500). Protein kinase KI also phosphorylates the 240,000 M_r polypeptide of polymerase II. Unlike the Novikoff kinase, calf thymus KI does not activate polymerase II. Whether the lack of activation reflects differences in the kinese or polymerase preparations is not known.

Recently, Rose et al.[32,36] purified two major nuclear cyclic AMP-independent protein kinases from a rat hepatoma (Morris hepatoma 3924A). One of these kinases, NII,[32] activates and phosphorylates both RNA polymerase I[6,12,34] (see Section II.B) and polymerase II.[108] The other nuclear kinase, NI,[36] does not modify the RNA polymerases, but phosphorylates and stimulates another nuclear enzyme, poly(A) polymerase.[36,37] The polymerase II polypeptides of M_r 214,000; 140,000; and 21,000 are phosphorylated by the NII protein kinase. Interestingly, while phosphorylation of polymerase I enhances the rate of RNA chain elongation, phosphorylation of polymerase II by the same kinase stimulates RNA chain initiation.[108] As observed for other initiation factors (see Table 2), activation of polymerase II by protein kinase NII was observed using denatured DNA as well as double-stranded DNA as template.

Although protein kinase NII apparently plays a role in regulation of RNA polymerase II, protein phosphorylation is not involved in activation of the polymerase by several of the initiation factors listed in Table 2. For example, the SF-1 factor preparation from calf-thymus contains a very active kinase activity but this activity is completely inactivated by heat treatment whereas the stimulatory activity is heat stable.[84] Protein kinase does not copurify with the initiation factors from chicken myeloblastosis[91] or Ehrlich ascites tumor cells.[88,109] However, the SII initiation factor from Ehrlich ascites tumor is phosphorylated in vivo[110] and two forms of this protein (designated SII and SII′) differing only in their extent of phosphorylation exist. Although the biological significance of SII phosphorylation is not known, these data suggest that protein kinases also have a secondary role in the control of gene expression in vivo.

2. Transcription of Defined Templates

a. Accuracy of In Vitro Transcription

Although transcription of relatively undefined templates has provided understanding of the fundamental properties of RNA polymerase II, this approach has contributed little to our knowledge of what regslates transcription of specific genes. The availability of large quantites of specific genes (see Reference 111) and the technology to measure unique RNA transcripts have led to the development of in vitro test systems which faithfully synthesize distinct RNAs. The first report of accurate in vitro transcription of deproteinized DNA was published in 1978 by Guang-Jer Wu.[112] Wu observed that cell-free extracts of KB cells could transcribe the VA RNA_I genes of adenovirus 2 (Ad 2; see Section IV.C.1). Subsequently, a number of laboratories have utilized in vitro transcription systems to synthesize an array of specific RNAs. In general, two types of systems have been employed, those that utilize endogenous RNA polymerases and those that need to be supplemented with specific RNA polymerases. Many of the assays which have relied on endogenous enzyme are based on that described by Manley et al.[113] Basically, this technique involves incubation of defined genes and cofactors for RNA synthesis in the presence of whole cell extracts prepared by $(NH_4)_2SO_4$ precipitation of cell lysates (prepared as originally described by Sugden and Keller[92]). The in vitro systems which require exogenous polymerase II are often based on the protocol described by Weil et al.[114] The general method involves incubation of the DNA and cofactors for RNA synthesis in the presence of appropriate RNA polymerase II and the 100,000 × g supernatant (S-100) of a cell extract (prepared by a modification of the procedure of Wu and Zubay[115]). In many of the reconstituted transcription assays, the specific RNA products are detected as the appropriate "run-off" transcripts, i.e., RNAs of the same length as the distance from the site of appropriate RNA initiation to the end of a truncated DNA.

One of the genes which has been studied extensively in reconstituted transcription assays is the major late gene of adenovirus 2 (Table 4). In vivo, this gene is actively transcribed by RNA polymerase II.[116-118] Weil et al.[114] first demonstrated that specific transcription of the major late viral transcriptional unit occurs when Ad 2 DNA is incubated with purified RNA polymerase II in the presence of an S-100 preparation. Ad 2 DNA is accurately transcribed when S-100 from either virus-infected or uninfected cells is used. However, in the absence of the S-100, transcription is random.[114] Although appropriate initiation was observed with RNA polymerase II prepared from a variety of higher eukaryotic cells, it was not obtained using enzyme from wheat germ. This observation was confirmed by Wasylyk et al.[119] These investigators also found that neither yeast nor *Drosophila* enzyme is able to transcribe the virus (or cellular) genes accurately.

Manley et al.[113] demonstrated specific transcription of Ad 2 DNA by endogenous RNA polymerase II using a HeLa whole-cell extract. In this system the endogenous polymerase II activity initiates RNA synthesis at the early region 1B and polypeptide IX promoters as well as at the major late promoter of the Ad 2 DNA; transcription of the late gene predominates over that of the other regions. Since the Ad 2 major late transcription unit is generally not expressed in vivo until after the onset of viral DNA replication (see Reference 120), the ability of whole-cell[113] and S-100[114] extracts from uninfected cells to direct the specific transcription of this region suggests that not all components required for biological control are present in these reconstituted systems. In contrast to transcription of Ad 2 DNA, HeLa extracts initiate rna synthesis on the early regions of Herpes simplex virus DNA, but not on the late genes.[121]

Interestingly, extracts from HeLa cells that have been infected with polio virus are incapable of directing the transcription of the Ad 2 major late transcriptional unit.[122] Since RNA polymerase III-directed transcription is only slightly decreased in these extracts, this effect appears to be specific for RNA polymerase II-mediated RNA synthesis. When the poliovirus-infected cell extract is supplemented with S-100 from uninfected cells, transcription by RNA polymerase II is restored to normal levels. This observation suggests that one or more of the factors required for transcription by polymerase II are deficient in the infected cell extract.

b. Required DNA Sequences

Following development of cell-free systems capable of supporting accurate transcription of the major late Ad 2 gene in vitro,[113,114] several viral[119,121,123-130] and cellular[119,123,131-138] genes were found to be selectively transcribed in similar whole cell or S-100 extract systems (Table 4). By using the in vitro transcription systems and cloned deletion mutants, attempts have been made to define the promotors of these genes. DNA sequence analyses have revealed three highly conserved regions upstream from the cap site of eucaryotic mRNA. One (PyCATTCPu), called the capping box, is located close to the 5′ end of the mRNA.[139] The second, an A + T rich region,

Table 4
IN VITRO TRANSCRIPTION OF DEFINED TEMPLATES BYSRA POLYMERASE II

Gene	Source of extract	Source of RNA polymerase II	Sequences required	Ref.
Adenovirus 2 major late	KBScell S-100	*Xenopus laevis* KB cell Mouse plasmacytoma Calf thymus	Not determined	114
	Bombyx mori whole cell	Endogenous	Not determ ned	137
	HeLaswhole cell	Endogenous	−47 to −12 (−32)eTATAAAA	128
Adefovirus 2 major late, early IB, and polypeptide 1X	HeLa whole cell	Endogenous	Not determined	113
Adenovirus 2 early Ia	HeLa cell S-100	Calf thymus	Not determined	119
Adenovirus 2 early EIIa	HeLa cell S-100	Calf thymus	EIIaE1: non-TETA, sequence not defined EIIaE2: (−60)TTAAATT EIIaL: TACAAAT	130
Adenovirus 2-associated virus type 2 major viral mRNA	KB cell S-100	KB cell	Not determined	124
Simian virus 40 early and late	HeLa whole cell	Endogenous	Not determined	127
Simian virus 40 early	HeLa cell S-100	Calf thymus	(−28)TATTTAT	129
Moloney murine leukemia virus long terminal repeat	HeLa whole cell	Endogenous	Not determined	126
Herpes simplex virus type I early	HeLaswhole cell	Endogenous	Not determined	121
Rous sarcoma virus	KBE whole cell	Endogenous	Not determined	125
Mouse β-globin	KB cell S-100	KB cell	Not determined	131
	Bombyx mori whole cell	Endogenous	Not determined	137
Human globins	HeLa whole cell	Endogenous	Not determined	132
Rabbit β-globin	HeLa whole cell	Endogenous	−34 to −20	134
Ovalbumin	HeLae cell S-100	Calf thymus	Not determined	135
	HeLa whole cell	Endogenous	−61sto −26 (−33)TATATAT	
Conalbumin	HeLa cell S-100	Calf thymus	−44 to −8 (−32)ZATAAAA substitution of T at position −29 to A or G decrease efficiency	133,136
Bombyx mori	HeLa whole cell	Endogenous	−29 to +6	137, 138
Fibroin	*Bombyxsmori* whole cell	Endogenous	−29 to +6 upstream of −74	137

referred to as the Goldberg-Hogness or TATA box, is typically centered 25 base pairs upstream from the cap site (see Reference 140). Finally, a third region of homology, called the CCAAT box, is found 70 to 80 nucleotides upstream from the cap site.[141,142] Deletion of the cap site and CCAAT box apparently has no effect on in vitro transcription of β-globin,[134] conalbumin,[123] or Ad 2 major late[123,128] genes. However, the CCAAT region promotes high efficiency transcription of the β-globin gene in vivo.[142a] Local sequences around the initiation or cap site appear to be involved only in directing the selection of the exact starting point.[123,129,143,144] The importance of the TATA box in directing transcription of several genes has been demonstrated in vitro. Specifically:

1. Deletion of this region essentially abolishes accurate transcription of cloned ovalbumin, conalbumin, β-globin, Ad 2 major late, and SV40 early genes (Table 4). Although several of the other genes transcribed in vitro have a TATA-like sequence in an appropriate position,[121,126,130-132,137] the necessity for this region has not been directly demonstrated.
2. A transition[136] or transversion[133] of the third nucleotide in the conalbumin TATA box drastically decreases the efficiency by which polymerase II transcribes the gene.
3. Deletions of a given size downstream from the rabbit β-globin TATA box shift the initiation of transcription downstream approximately an equal distance.[134]
4. Cloning of the TATA region of the major late Ad 2 gene into pBR322 results in initiation of RNA synthesis 25 base pairs downstream from this site.[145]

These results indicate that the TATA region directs polymerase II to initiate transcription 25 to 30 base pairs downstream. Interestingly, the major late Ad 2 gene and the conalbumin gene are transcribed in vitro with about the same efficiency[119] and also have the same base sequences from −21 to −32 (Table 4). In contrast, the ovalbumin and early Ad 2 genes have different A + T rich 5′-flanking sequences and are transcribed with lower efficiency. Hence slight variations in the TATA box sequence apparently affect the efficiency of initiation.

Although the TATA box is both necessary and sufficient for accurate transcription of a number of genes in vitro (Table 4), the requirement for this region in vivo is less stringent. Deletion of the TATA regions of the sea urchin histone H2A gene,[143] the SV40 early region genes,[144,146] or the β-globin gene[142a] decreases, but does not abolish, transcription in vivo. Since transcripts with different 5′ ends, as well as normal transcripts are formed in the absence of TATA, other sequences can apparently substitute for this region. In addition, SV40 mutants with deletions far upstream from the TATA box, in a region which contains a tandom repeat of a 72 base pair sequence (−166 to −188 and −189 to −261), transform cells inefficiently and do not express significant T antigen (an early gene product) even though the TATA box is intact.[144] Since in vivo transcription of cloned β-globin genes is enhanced more than 200-fold when the 72 base pair repeat is inserted either upstream or downstream from the globin gene,[147] the 72 base pair tandom repeat is apparently not simply an upstream promotor. Finally, certain genes, including papovavirus late genes and Ad 2 EIIaEI and IVa2 genes, lack TATA-like sequences at the appropriate position.[148-150] Thus, other control regions probably function in vivo. The reconstituted transcription systems apparently contain the factor(s) necessary for recognition of these regions since the EIIa gene of Ad 2 is accurately transcr bed in the HeLa S-100 system.[130]

Although cell-free systems provide an important tool for the identification of transcriptional control mechanisms, in vitro studies should be accompanied by investigation of defined genes in intact cells. It should also be pointed out that extracts from some cell types may be better than others for studying expression of particular genes and that an extract from a homologous source is most likely the system of choice. For example, the silk fibroin gene from *Bombyx mori* contains a promoter region in the TATA box position (−29 to +6) which is required for selective transcription of the gene in the presence of extracts from either *Bombyx* or HeLa cells.[137,138] However, a sequence element upstream from position −74 enhances transcription of the fibroin gene in the homologous system but not in the heterologous system. This observation suggests that the HeLa extract lacks a component necessary for regulation by the far upstream region.

c. Active Components of Cell-Free Extracts

The KB cell S-100 extracts and the HeLa whole-cell extracts used in the reconstituted systems have been separated into multiple components. Four distinct fractions of the KB S-100 (in addition to RNA polymerase II) are required for efficient and selective transcription of Ad 2 DNA.[151] Three of these fractions contain factors which are different from those required for accurate transcription of cloned 5S rRNA genes by polymerase III. The HeLa whole-cell extract has been separated into several fractions which are essential for selective transcription of cloned ovalbumin genes.[152] Of particular interest is the observation that extracts from chick oviduct and HeLa cells contain similar factors.[152] The HeLa and oviduct factors may be analogous to those from KB cells. Whether these fractions contain general factors, necessary for all RNA polymerase II-mediated transcription, or whether they are specific for certain genes is not yet known. The possibility that some of the previously described general transcription factors (Tables 1 to 3) are related to those proteins involved in directing the specific transcription observed in the reconstituted systems remains to be elucidated. Although Wasylyk et al.[119] apparently failed to increase the transcriptional efficiency of an S-100 extract by addition of the RNA polymerase II-stimulatory factors of Stein and Hausen,[79] it is possible that these factors were already in excess in the S-100 preparation.

IV. RNA POLYMERASE III

A. Function

Chromatographic resolution of three types of nuclear RNA polymerases was first reported in 1969.[153] Subsequent structural analyses indicated that the three enzymes, RNA polymerases I, II, and III, in order of elution from DEAE-Sephadex®, are structurally distinct.[3,154,155] Failure to detect RNA polymerase III in many early investigations was a result of (1) its leakage from the nucleus due to initial homogenization in iso- or hypotonic buffers; (2) resolution of the polymerases on DEAE-cellulose rather than DEAE-Sephadex® (RNA polymerase III coelutes with polymerase I in the former case); or (3) incomplete separation from polymerase II on DEAE-Sephadex® due to use of steep salt gradients for elution (for more complete discussion see References 3 and 4). The observation that vertebrate RNA polymerase III was sensitive to high (> 100 μg/mℓ) concentrations of α-amanitin[156-158] permitted evaluation of this enzyme in the presence of the other polymerases and provided the impetus to study transcription by polymerase III in more detail.

The demonstration that the synthesis of 5S rRNA and 4.5S tRNA precursors in isolated nuclei is sensitive to high concentrations of α-amanitin[159,160] indicated that polymerase III is responsible for the synthesis of these cellular genes. Transcription of two small RNAs associated with adenovirus, VA RNA_I and VA RNA_{II}, is also catalyzed by RNA polymerase III.[161]

B. Regulation

Physiological stimuli have been reported to alter the activity of solubilized RNA polymerase III. For example, transcription of deproteinized DNA by this enzyme is enhanced in bovine lymphosarcoma,[162] a mouse myeloma,[163] several ascites tumors,[164] a poorly differentiated rat hepatoma,[21,35] and regenerating liver,[21,64] compared to control tissues. Solubilized RNA polymerase III activity is also increased in lymphocytes in the presence of phytohemaglutinin[26] and in adrenal glands of guinea pigs treated with adrenocorticotropic hormone.[165] In all cases, the elevation of solubilized enzyme correlates with increased 5S RNA and tRNA production in vivo. In contrast to those stimuli which call for elevated cellular polymerase III transcription products, adenovirus infection apparently does not alter the level of extracted RNA polymerase III.[166,167]

C. In Vitro Transcription of Defined Templates

1. 5S rRNA Genes

The ease of analysis of the RNA polymerase III transcription products and the facility with which this enzyme initiates RNA synthesis in vitro have permitted the study of polymerase III-specific transcription in nuclear preparations and reconstituted systems. In 1977, Parker and Roeder[168] reported that exogenous polymerase III (but not polymerases I and II) purified from *X. laevis* could accurately transcribe 5S rRNA genes using isolated chromatin. However, polymerase III was unable to synthesize the same RNA when cloned DNA fragments were used as templates. Although polymerase III does not specifically transcribe the cloned 5S genes, this mammalian polymerase preferentially transcribes the 5S regions of intact homologous DNA.[169] In the latter case, the investigators noted that intact DNA was required for high efficiency transcription of the 5S genes.

As mentioned previously (Section III.C.2.a), the first report of accurate transcription of a defined DNA template using a cell-free system was that of Wu[112] in 1978. It was demonstrated that the 20,000 × g supernatant prepared from KB cells, containing RNA polymerase III, could accurately transcribe KB cell DNA and Ad 2 DNA to yield 5S rRNA and VA RNA_I, respectively.

Following the initial success of Wu in achieving accurate transcription of 5S rDNA in cell free extracts,[112] a number of other investigators have used similar systems to study transcription of this gene in detail (Table 5). Birkenmeier et al.[171] reported similar findings using extracts from oocyte nuclei. Sakonju et al.[174] have also utilized the *Xenopus* oocyte system to determine which regions of the 5S rDNA are necessary for correct gene function. By using cloned *X. borealis* 5S rRNA genes it was found that 5S rRNA is synthesized even after enzymatic removal of the entire 5′-flanking region of the cloned gene. Thus, at least in the *Xenopus* system, the DNA located upstream from the 5S gene does not control transcription. In fact, efficient transcription of the 5S rDNA proceeds even after removal of the first 50 nucleotides of the 5′ coding region.[174] Similar experiments utilizing 3′-deletion mutants revealed that accurate initiation occurs in mutants containing 83 or more 5′ gene residues but not in those containing fewer than 80.[175] Further, insertion of nucleotides between positions +40 and +41 moves the point of initiation downstream from the normal site the same distance as the length of the insert.[174] These observations suggest that a control region exists within the 5S rRNA gene, extending from gene residue +50 to residue +80. This region directs RNA polymerase III to initiate transcription at a position approximately 50 base pairs upstream. Although the internal control region is important in influencing the general region for initiation, the exact site of initiation is determined by local sequences.

Termination of transcription by RNA polymerase III has also been studied in vitro. Bogenhagen and Brown[176] have reported that RNA polymerase III recognizes a short sequence at the end of the 5S rRNA gene (*X. laevis*) that signals the end of transcription. Unlike the *E. coli* models,[177,178] a hairpin loop does not seem to be required for

Table 5
IN VITRO TRANSCRIPTION OF DEFINED TEMPLATES BY RNA POLYMERASE III

Gene	Source of extract	Regions required	Ref.
KB cell DNA (5S rRNA)	KBE cells	Not determined	112,170
Xenopus laevis oocyte 5S rRNA	*X. laevis* oocytes	Not determined	171,172
X. borealis oocyte 5S rRNA	KB cells Mouse plasmacytoma cells *X. laevis* kidney cells	Not determined	173
	X. laevis oocytes	Not determined	172
X. boralis somatic 5S rRNA	*X. laevis* oocytes	+ 50 to + 83	174,175
Ad 2 VA RNA$_I$	KB cells	Not determined	112,170
	X. laevis oocytes	Not determined	171
Ad 2 VA RNA$_I$ JNA$_{II}$	KB cells or *X. laevis* oocytes KBScells only	Not determined	173
	HeLa cells	−33 to + 9; + 9 to + 72 Undefined coding sequences	188
Drosophila melanogaster tRNA^{Arg}, tRNA^{Asn}, tRNA^{Lys}	*X. laevis* oocytes	Not determined	207,208
D. melanogaster tRNA^{Lys}	*X. laevis* oocytes	Coding sequences only; certain 5′-flanking sequences capable of repressing transcription	191
D. melanogaster tRNA^{Arg}	*X. laevis* oocytes HeLa cells *D. melanogaster* Kc cells	+ 8 to + 25; + 50 to + 58	195
X. laevis tRNA$^{Met}_{1}$	*X laevis* oocytes	Both 5′ and 3′ half of coding sequences	192
	X. laevis oocytes	+ 8 to + 13; + 51 to + 72	194
X. laevis tRNA$^{Leu}_{CUG}$	*X. laevis* oocytes	+ 13 to + 20; + 51 to + 64	193
Bombyx mori tRNA$^{Ala}_{2}$	*X. laevis* oocytes	Not determined	209
	X. laevis oocytes	Coding sequences only	205
	X. laevis oocytes	Coding sequences only	197
	B. mori posterior silk glands or ovaries	Coding sequences *and* upstream from −11	
Schizosaccharomyces pombe tRNA	*X. laevis* oocytes	Not determined	207
Saccharomyces cerevisiae tRNA^{Trp}	*X. laevis* oocytes	Not determined	198
S. cerevisiae tRNA$^{Leu}_{3}$	*X. laevis* oocytes	21 base pair insertion into center of coding sequence has no effect	196
S. cerevisiae SUP4 tRNA^{Tyr}	*X. laevis* oocytes	+ 56	210
S. cerevisiae tRNA$^{Leu}_{3}$	*X. laevis* oocytes	3′ half of gene not required	211
	HeLa cells	Not determined	212
Chinese hamster ovary type 2 *Alu*-equivalent sequence	HeLa cells	Not determined	213

termination of the 5S genes. Similarly, the instability of the rU:dA duplex, thought to be a major force in procaryotic termination,[179] does not seem to have an important role in 5S rRNA gene termination.

Extracts from immature oocytes[172] or KB cells[173] devoid of RNA polymerase III activity are capable of directing the accurate transcription of 5S rRNA genes in the presence of exogenous polymerase III. This ability is sensitive to heat treatment and trypsin, and is inhibited by high DNA concentrations.[173] Since the purified enzyme alone does not accurately transcribe the cloned genes, these results indicate that the extracts contain at least one protein factor capable of stimulating accurate transcription by interacting with the DNA. Fractionation of the cell extracts has led to the identification of at least three protein factors required for the transcription of 5S rRNA genes by polymerase III.[180,181] One of the factors, termed TFIIIA, has been purified to homogeneity from *X. laevis* ovaries.[181] Unfertilized egg extracts contain all of the components necessary for transcription of 5S rRNA genes except for this factor, a basic protein with a molecular weight of 37,000.[181] Using the "footprinting" method of Galas and Schmitz[182] it has been determined[181,183] that TFIIIA binds to an intragenic region (positions +45 to +96) of both the somatic and oocyte-type 5S rRNA genes. This reaction is not mediated by RNA polymerase III or components in the unfertilized egg extracts. Because this binding is in the same approximate position as the intragenic control region described by Brown and co-workers,[174,175] it was postulated that TFIIIA binds to the 5S rRNA gene and directs subsequent initiation by RNA polymerase III. There are several lines of evidence in support of a central role for TFIIA in control of 5S rRNA synthesis in vivo. First, in addition to binding to 5S rDNA,[181,183] TFIIIA also binds to the 5S rRNA,[184,185] suggesting that TFIIIA is involved in a feedback mechanism for the control of 5S rRNA transcription. Second, in unfertilized, mature *Xenopus* oocytes, TFIIIA is depleted[181,185] and no 5S rRNA transcription is detectable.[186] Hence, the cellular level of TFIIIA appears to correlate with in vivo 5S rRNA synthesis. Third, somatic cells do not express the same set of 5S rRNA genes as immature oocytes[186] and contain a factor that is different from the oocyte TFIIIA.[187] Although the two factors may be responsible for transcription of distinct 5S rRNA genes in vivo, accurate transcription of both oocyte- and somatic-type 5S rRNA genes occurs in vitro in the presence of either factor. In addition, extracts from human and murine kidney cells[173] and HeLa[187a] can direct accurate transcription of cloned oocyte 5S rRNA genes even though the oocyte-type gene is not normally expressed in these cells. Thus, not all levels of control of polymerase III gene transcription appear to be present in the current cell-free systems. On the other hand, since TFIIIA is not required for accurate transcription of the $tRNA^{met}$ gene,[181] it appears to be specific for the 5S genes.

2. Adenovirus-Associated RNA

Several groups have reported accurate transcription of Ad 2 VA RNA genes in cell-free systems containing crude extracts from a variety of cell types[170,171,173,188] (Table 5). At least some of the factors required for specific transcription of these genes are different from those of the 5S rRNA gene. A cytoplasmic fraction (S-20) from human KB cells directs accurate transcription of both the VA RNA and KB cell 5S rRNA genes.[112] In contrast, when the S-20 extract is subjected to higher speed (100,00 × g) centrifugation, it loses most of its ability to transcribe the KB cell genes but retains the ability to direct specific transcription of the Ad 2 DNA.[170]

Extracts derived from various cell types differ in their ability to support transcription of the two VA RNA and the 5S rRNA genes.[173] Human cell extracts transcribe the VA RNA_I and VA RNA_{II} genes and the two distinct *Xenopus* 5S rRNA genes with similar efficiencies. In contrast, *Xenopus* extracts apparently transcribe only the VA RNA_I gene and the dominant oocyte 5S rRNA gene. Studies using restriction endonuclease fragments of Ad 2 DNA[173] clearly demonstrate that the VA RNA_I and RNA_{II} genes have separate control regions. However, the two genes apparently require at least one common factor since the presence of the VA RNA I gene reduces transcription of the VA RNA_{II} gene.[188]

The S-100 fraction can be separated into at least three components by phosphocellulose chromatography.[189] Two of these components (factors IIIB and IIIC) are required for selective transcription of Ad 2 VA RNA_I, tRNA, and 5S rRNA genes and are distinct from the factors required for RNA polymerase II-catalyzed transcription. The other component (factor IIIA) is required for specific transcription of the cloned 5S rRNA gene only. However, this activity has not been separated from RNA polymerase II factor IIA. Nevertheless, these data indicate that the S-100 fraction contains general transcription factors for polymerase III-mediated synthesis (IIIB and IIIC) as well as a possible gene-specific factor (IIIA).

Fowlkes and Shenk[188] have identified the control regions of the VA RNA_I gene. The first region is located in the 5′-flanking sequences (−33 to +9) and is required for accurate initiation and efficient transcription of the gene. The second region appears to be internal to the coding sequences since VA RNA_I genes deleted from the 5′ end to +24 are not transcribed but are capable of inhibiting transcription of the wild-type genes. These data suggest that the modified genes are able to sequester a factor, present in limiting amounts in the cell extracts, which is required for transcription of the VA RNA_I gene. Thus, the second region controlling transcription of the VA RNA_I gene is similar to the 5S rRNA gene control region in that it is located within the coding region. However, the 5′ boundary of the VA RNA_I gene control region is closer to the 5′ end of the transcript (+9) than that of the 5S rRNA gene control region (+50). It should also be noted that there is little sequence homology between the control regions of the two genes and that the TFIIIA protein factor which binds to the 5S rRNA gene[183] apparently is not required for VA RNA_I gene transcription (see Reference 188). On the other hand, the striking homologies between the intragenic control region of the VA RNA_I gene and similarly located regions of tRNA genes, (see Section IV.C.3) suggest that they may share specific transcription factors.

3. *Transfer RNA*

Extracts from *Xenopus* oocytes which direct accurate transcription of 5S rRNA genes are also capable of accurate synthesis of transfer RNA (tRNA) from a variety of cloned genes (Table 5). Like the 5S rRNA[174,175,182,184] and the Ad 2 VA RNA[188] genes, tRNA genes contain an intragenic control region. However, the TFIIIA factor that binds to this region of the 5S rRNA gene is not required for tRNA gene transcription.[181] So far, the VA RNA_I and tRNA genes do not appear to require different factors for transcription. This fact, the sequence homology found in the intragenic control region,[188] and the hairpin-like structure postulated for VA RNAs,[190] have led to the hypothesis that VA RNAs may have originated as inverted duplications of tRNA genes.[188]

The tRNA gene intragenic control region has been demonstrated to be split into two different regions.[191-195] One region is located near the 5′ end of the coding sequence (+8 to +25). The second region is located near the posterior of the gene (+50 to +72) (see Table 5 for exact position in particular genes). Substitution,[194] extension,[196] or small deletions[193,194] of the sequences in the area located between these two regions have no effect on in vitro or in vivo transcription. However, complete deletion of these sequences abolishes transcriptional activity.[194] Apparently, the middle region serves to keep the anterior and posterior control regions separated by a critical distance. This distance can be lengthened but not shortened. Competition studies[197] with truncated, inactive tRNA genes have provided evidence that the tRNA gene internal control regions act by binding factor(s) which influence RNA polymerase III transcription.

The requirement for 5′ flanking sequences of various tRNA genes has also been investigated. No obvious common sequences appear to exist in this region.[198-203] A *Xenopus* $tRNA^{met}$ gene containing only 22 5′-flanking base pairs has been shown to be transcribed when injected into *Xenopus* oocytes.[204] Subsequently, using *Xenopus* extracts, it was demonstrated that a *Bombyx* $tRNA_2^{gla}$ gene with only six 5′-flanking base pairs[205] and two *Drosophila* $tRNA^{lys}$ genes with no 5′-flanking sequence[191] can be transcribed as efficiently as the wildtype genes.

A comparison of the sequences of the 5′-flanking regions of four $tRNA_2^{lys}$ genes from *Drosophila* revealed a highly conserved sequence (GTCAGTTTTTA) located around position −20.[206] However, the significance of this conserved region is unknown since its deletion does not affect in vitro transcription.[191] Although they are not required for transcription, the 5′-flanking regions of the *Drosophila* $tRNA_2^{lys}$ genes apparently influence the efficiency of transcription in *Xenopus* extracts. Spenifically, even though the coding sequences of the four $tRNA^{lys}$ genes are identical, the number 4 $tRNA_2^{lys}$ gene is transcribed almost 14 times better than the number 2 gene.[191] Removal of the 5′-flanking sequence from gene 2 increases its transcriptional efficiency to a level similar to that of gene 4. Further, when the 5′-flanking sequences of genes 2 and 4 are exchanged, the relative transcriptional efficiencies of the two genes are reversed. These results indicate that the 5′-flanking sequences are not equivalent in all tRNA genes and that certain 5′-flanking sequefces are capable of repressung the transcription of the coding sequences. Whether the 5′-flanking sequefces exejt their effect by interacting directly with RNA polymerasesUII or whether the effekt is mediated via an extract factor is unknown.

Sprague et al.[197] have studied the function of the 5′-flanking sequence using a cloned $tRNA_2^{ala}$ gene from the silkworm *B. mori* and extracts from either *B. mori* or *X laevis.* As Garber and Gage[205] had previously reported, these investigators observed that removal of the 5′-flanking sequences of the *Bombyx* $tRNA_2^{ala}$ gene has no effect on its ability to serve as a template for RNA polymerase III in *Xenopus* extracts. However, when *Bombyx* extracts were used deletion of the 5′-flanking sequences abolished transcriptional activity.[197] The reason for the contrasting results with extracts from different sources (*Xenopus* and *Bombyx*) is unknown. It is possible that the differences in the RNA polymerase III molecules of silkworms and frogs are responsible. It is also feasible that *Bombyx* extracts may include a factor which binds to polymerase III resulting in the requirement for a particular 5′-flanking sequence for initiation of transcription. Alternatively, the *Xenopus* extract may contain a factor which can replace or overcome the need for a 5′-flanking sequence.

V. PERSPECTIVES

The role of the three RNA polymerases in the synthesis of distinct classes of RNA in vivo is now well-established. Accurate transcription of specific genes using highly purified RNA polymerase, deproteinized DNA, and purified regulatory factors is fast approaching reality. A major breakthrough in the field of eukaryotic transcription in recent years has been the construction of cell-free systems capable of accurate initiation and transcription of specific genes. Using these systems, considerable progress has been made in identifying the regions of the DNA which control transcription by RNA polymerases II and III. The advent of the cell-free system has also resulted in elucidation of some of the protein factors involved in regulation of transcription by these polymerases. Although accurate transcription of ribosomal genes has been obtained using cell extracts, the nature of the factors which mediate this transcription has not been established. Recent studies on the activation of purified RNA polymerase I by an endogenous protein kinase are of interest in this regard. Whether protein phosphorylation regulates rDNA transcription remains to be determined.

Although many appropriate controls exist in the cell-free transcription assays, not all levels of regulation are operative in vitro. For example, simultaneous initiation at the early and late genes of Ad 2 DNA occurs in vitro. This observation as well as the ability of extracts from both infected and uninfected cells to initiate transcription of adenovirus genes imply that biological control of viral RNA synthesis is absent in vitro. Likewise, synthesis of cellular RNAs is not always appropriately regulated in vitro. Specifically, *Xenopus* extracts, from different stages of development, cannot distinguish between oocyte and somatic 5S rRNA genes. *Xenopus* extracts also fail to recognize the 5′ flanking sequence of a *Bombyx* tRNA gene which influences transcription in the homologous system. Similarly, a region far upstream from the *B. mori* fibroin gene, which regulates transcription in the homologous system, does not influence initiation when a heterologous (HeLa) extract is used. Finally, the observation that factors for chick ovalbumin gene transcription are similar in HeLa cells and chick oviduct, in spite of the fact that ovalbumin is not a normal product of the transformed cells, indicates that gene-specific regulation is not exerted in vitro. Elucidation of the factors which regulate precise transcription of specific genes poses a challenging problem for the future.

Another aspect of in vitro transcription systems which must be considered is the source of RNA polymerase itself. Specifically, genes normally transcribed in mammalian cells are accurately copied in in vitro cell extracts (S-100 fraction) supplemented with RNA polymerase II from mammalian sources, but not with enzyme from insects, yeast, or wheat germ. The requirement for compatible RNA polymerase II suggests that some levels of control in gene selection are retained by the purified enzyme. In this regard it should be noted that RNA polymerases I (yeast)[214-216] and III (thymus)[169] preferentially transcribe appropriate genes from DNA molecules containing many different cistrons.

One area in which virtually no information is available is the role of individual enzyme polypeptides in RNA synthesis. Although preliminary data from the yeast system suggests that the larger polypeptide of polymerases I[217] and II[218] are involved in DNA binding and chain elongation, comparable studies have not been performed with enzymes from higher eukaryotes. One approach to studying the function of individual polypeptides would be to use subunit-specific inhibitors. Monoclonal antibodies to RNA polymerase II[219,220] and I[221] have recently been obtained. Use of these antibodies should prove valuable in identification of subunit-specific interactions and, ultimately, in elucidation of the function of the individual subunits of the enzymes.

ACKNOWLEDGMENT

We would like to thank Ms. Edna J-A. Mayeski for her invaluable assistance in preparation of the manuscript.

REFERENCES

1. **Jacob, S. T.**, Mammalian RNA polymerases, *Prog. Nucl. Acid Res. Mol. Biol.*, 13, 93, 1973.
2. **Chambon, P.**, Eucaryotic RNA polymerases, in *The Enzymes*, Vol. 10, 3rd ed., Boyer, P. D., Ed., Academic Press, New York, 1974, 261.
3. **Roeder, R. G.**, Eukaryotic nuclear RNA polymerases, in *RNA Polymerase*, Losick, R. and Chamberlin, M., Eds., Cold Spring Harbor Laboratory, Cold Spring Harbor, N.Y., 1976, 285.
4. **Jacob, S. T. and Rose, K. M.**, RNA polymerases and poly(A) polymerase from neoplastic tissues and cells, *Methods Cancer Res.*, 14, 191, 1978.
5. **Jacob, S. T. and Rose, K. M.**, Basic enzymology of transcription in prokaryotes and eukaryotes, in *Cell Biology A Comprehensive Treatise*, Vol. 3, Goldstein, L. and Prescott, D., Eds., Academic Press, New York, 1980, 113.
6. **Rose, K. M., Duceman, B. W., and Jacob, S. T.**, RNA polymerases in neoplasia, in *Isozymes*, Vol. 5, Ratazzi, M., Scandalios, J., and Whitt, G., Eds., Alan Liss, New York, 1981, 115.
7. **Hadjiolov, A. A. and Nikolaev, N.**, Maturation of ribosomal ribonucleic acids and the biogenesis of ribosomes, *Prog. Biophys. Mol. Biol.*, 31, 95, 1976.
8. **Jacob, S. T., Sajdel, E. M., and Munro, H. N.**, Presence of two RNA polymerase activities in liver nucleoli, *Biochim. Biophys. Acta*, 157, 421, 1968.

9. **Roeder, R. G. and Rutter, W. J.,** Specific nucleolar and nucleoplasmic RNA polymerases, *Proc. Natl. Acad. Sci. U.S.A.,* 65, 675, 1970.
10. **Jacob, S. T., Sajdel, E. M., and Munro, H. N.,** Mammalian RNA polymerases and their selective inhibition by α-amanitin, *Adv. Enzyme Regul.,* 9, 169, 1971.
11. **Ballal, N. R., Choi, Y. C., Mouche, R., and Busch, H.,** Fidelity of synthesis of preribosomal RNA in isolated nucleoli and nucleolar chromatin, *Proc. Natl. Acad. Sci. U.S.A.,* 74, 2446, 1977.
12. **Rose, K. M., Stetler, D. A., and Jacob, S. T.,** Protein kinase activity of RNA polymerase I purified from a rat hepatoma: probable function of M_r 42,000 and 24,600 polypeptides, *Proc. Natl. Acad. Sci., U.S.A.,* 78, 2833, 1981.

12a. **Scheer, U. and Rose, K.,** unpublished.

12b. **Schlegel, R., Miller, L., and Rose, K.,** unpublished.

13. **Brun, G. and Weissbach, A.,** Initiation of Hela cell DNA synthesis in a subnuclear system, *Proc. Natl. Acad. Sci. U.S.A.,* 75, 5931, 1978.
14. **Plevani, P. and Chang, L. M. S.,** Initiation of enzymatic DNA synthesis by yeast RNA polymerase I, *Biochemistry,* 17, 2530, 1978.
15. **Sajdel, E. M. and Jacob, S. T.,** Mechanism of early effect of hydrocortisone on the transcriptional process. Stimulation of the activities of purified rat liver nucleolar RNA polymerases, *Biochem. Biophys. Res. Commun.,* 45, 707, 1971.
16. **Schmid, W. and Sekeris, C. E.,** Nucleolar RNA synthesis in the liver of partially hepatectomized and cortisol-treated rats, *Biochim. Biophys. Acta,* 402, 244, 1975.
17. **Todhunter, J. A., Weissbach, H., and Brot, N.,** Modification of rat liver RNA polymerase I after *in vivo* stimulation by hydrocortisone or methylisobutylxanthine, *J. Biol. Chem.,* 253, 4514, 1978.
18. **Smuckler, E. M. and Tata, J. R.,** Changes in hepatic nuclear DNA-dependent RNA polymerase caused by growth hormone and triiodothyronine, *Nature (London),* 234, 37, 1971.
19. **Cooke, A. and Brown, M.,** Stimulation of the activities of solubilized pig lymphocyte RNA polymerases by phytohaemagglutinin, *Biochem. Biophys. Res. Commun.,* 51, 1042, 1973.
20. **Chesterton, C. J., Humphrey, S. M., and Butterworth, P. H. W.,** Comparison of the multiple deoxyribonucleic acid-dependent ribonucleic acid forms of whole rat liver and a minimal deviation rat hepatoma cell line, *Biochem. J.,* 126, 675, 1972.
21. **Duceman, B. W. and Jacob, S. T.,** Transcriptionally active RNA polymerases from Morris hepatoma and rat liver. Elucidation of the mechanism for the preferential increase in the tumor RNA polymerase I, *Biochem. J.,* 190, 781, 1980.
22. **Leonard, T. B. and Jacob, S. T.,** Alterations in DNA-dependent RNA polymerases I and II from rat liver by thioacetamide: preferential increase in the level of chromatin-associated nucleolar RNA polymerase IB, *Biochemistry,* 16, 4538, 1977.
23. **Yu, F. L.,** Two functional states of the RNA polymerases in the rat hepatic nuclear and nucleolar fractions, *Nature (London),* 251, 344, 1974.
24. **Matsui, T., Onishi, T. and Muramatsun, M.,** Nucleolar DNA-dependent RNA polymerase from rat liver. 2. Two forms and their physiological significance, *Eur. J. Biochem.,* 71, 361, 1976.
25. **Kellas, B. L., Austoker, J. L., Beebee, T. J. C., and Butterworth, P. H. W.,** Forms AI and AII DNA-dependent RNA polymerases as components of two defined pools of polymerase activity in mammalian cells, *Eur. J. Biochem.,* 72, 583, 1977.
26. **Jaehning, J. A., Stewart, C. C., and Roeder, R. G.,** DNA-dependent RNA polymerase levels during the response of human peripheral lymphocytes to phytohaemagglutinin, *Cell,* 4, 51, 1975.
27. **Tillyer, C. R. and Butterworth, P. H. W.,** Relationship between the activities of different pools of RNA polymerases I and II during stimulation of human lymphocytes, *Nucl. Acid Res.,* 5, 2099, 1978.
28. **Dauphinais, C.,** The control of ribosomal RNA transcription in lymphocytes. Evidence that the rate of chain elongation is the limiting factor, *Eur. J. Biochem.,* 114, 487, 1981.
29. **Matsui, S., Weinfeld, H., and Sandberg, A. A.,** Quantitative conservation of chromatin-bound RNA polymerases I and II in mitosis, *J. Cell Biol.,* 80, 451, 1979.
30. **Jacob, S. T., Jänne, O., and Sajdel-Sulkowska, E. M.,** Hormonal regulation of RNA polymerases in liver and kidney, in *Isozymes: Developmental Biology,* Vol. 3, Markert, C. L., Ed., Academic Press, New York, 1975, 11.
31. **Schwartz, L. B. and Roeder, R. G.,** Purification and subunit structure of deoxyribonucleic acid-dependent ribonucleic acid polymerase I from the mouse myeloma, MOPC 315, *J. Biol. Chem.,* 249, 5898, 1974.
32. **Rose, K. M., Bell, L. E., Siefken, D. A., and Jacob, S. T.,** A heparin-sensitive nuclear protein kinase. Purification properties and increased activity in rat hepatoma relative to liver, *J. Biol. Chem.,* 256, 7468, 1981.
33. **Desjardins, P. R., Lue, P. F., Liew, C. C., and Gornall, A. G.,** Purification and properties of rat liver nuclear protein kinases, *Can. J. Biochem.,* 50, 1250, 1972.
34. **Duceman, B. W., Rose, K. M., and Jacob, S. T.,** Activation of purified hepatoma RNA polymerase I by homologous protein kinase NII, *J. Biol. Chem.,* 256, 10755, 1981.

34a. **Duceman, B., Rose, K., and Jacob, S.** unpublished.
35. **Rose, K. M., Ruch, P. A., Morris, H. P., and Jacob, S. T.,** RNA polymerases from a rat hepatoma. Partial purification and comparison of properties with corresponding liver enzymes, *Biochim. Biophys. Acta,* 432, 60, 1976.
36. **Rose, K. M. and Jacob, S. T.,** Phosphorylation of nuclear poly(A) polymerase. Comparison of liver and hepatoma enzymes, *J. Biol. Chem.,* 254, 10256, 1979.
37. **Rose, K. M. and Jacob, S. T.,** Phosphorylation of nuclear poly(adenylic acid) polymerase by protein kinase. Mechanism of enhanced poly(adenylic acid) synthesis, *Biochemistry,* 19, 1472, 1980.
38. **Rose, K. M., Roe, F. J., and Jacob, S. T.,** Two functional states of poly(adenylic) acid polymerase in isolated nuclei, *Biochim. Biophys. Acta,* 478, 180, 1977.
38a. **Pizer, L. and Rose, K.,** unpublished.
39. **Trendelenburg, M. F. and Gurdon, J. B.,** Transcription of cloned *Xenopus* ribosomal genes visualised after injection into oocyte nuclei, *Nature (London),* 276, 292, 1978.
40. **Bakken, A., Morgan, G., Sollner-Webb, B., Roan, J., Busky, S., and Reeder, R. H.,** Mapping of transcription initiation and termination signals on *Xenopus laevis* ribosomal DNA, *Proc. Natl. Acad. Sci. U.S.A.,* 79, 56, 1982.
41. **Hipskind, R. A. and Reeder, R. H.,** Initiation of ribosomal RNA chains in homogenates of oocyte nuclei, *J. Biol. Chem.,* 255, 7896, 1980.
42. **Crampton, J. M. and Woodland, H. R.,** Isolation from *Xenopus laevis,* embryonic cells of a factor which stimulates ribosomal RNA synthesis by isolated nuclei, *Dev. Biol.,* 70, 467, 1979.
43. **Grummt, I.,** Specific transcription of mouse ribosomal DNA in a cell-free system that mimics control *in vivo, Proc. Natl. Acad. Sci. U.S.A.,* 78, 727, 1981.
44. **Miller, K. and Sollner-Webb, B.,** Transcription of mouse ribosomal genes by RNA polymerase I: *in vitro* and *in vivo* initiation and processing sites, *Cell,* 27, 165, 1981.
45. **Grummt, I.,** Mapping of a mouse ribosomal DNA promotor by in vitro transcription, *Nucl. Acid Res.,* 9, 6093, 1981.
46. **Mishima, Y., Yamamoto, O., Kominami, R., and Muramatsu, M.,** *In vitro* transcription of a cloned mouse ribosomal RNA gene, *Nucl. Acid Res.,* 9, 6773, 1981.
46a. **Grummt, I., Roth, E., and Paule, M. R.,** Ribosomal RNA transcription *in vitro* is species specific, *Nature (London),* 296, 173, 1982.
46b. **Kohorn, B. D. and Rae, P. M. M.,** Accurate transcription of truncated ribosomal DNA templates in a Drosophila cell-free system, *Proc. Natl. Acad. Sci. U.S.A.,* 79, 1501, 1982.
47. **Froehner, S. C. and Bonner, J.,** Ascites tumor ribonucleic acid polymerases. Isolation, purification and factor stimulation, *Biochemistry,* 12, 3064, 1973.
48. **James, G. T., Yeoman, L. C., Matsui, S., Goldberg, A., and Busch, H.,** Isolation and characterization of nonhistone chromosomal protein C-14 which stimulates RNA synthesis, *Biochemistry,* 16, 2384, 1977.
49. **Kuehn, G. D., Affolter, H., Atmar, V. J., Seebeck, T., Gubler, G., and Brown, R.,** Polyamine-mediated phosphorylation of a nucleolar protein from *Physarum polycephalum* that stimulates rRNA synthesis, *Proc. Natl. Acad. Sci. U.S.A.,* 76, 2541, 1979.
50. **Goldberg, M. I., Perriard, J.-C., and Rutter, W. J.,** A protein cofactor that stimulates the activity of DNA-dependent RNA polymerase I on double-stranded DNA, *Biochemistry,* 16, 1648, 1977.
51. **Jacob, S. T., Duceman, B. W., and Rose, K. M.,** Spermine-mediated phosphorylation of RNA polymerase I and its effect on transcription, *Med. Biol.,* 59, 381, 1981.
52. **Molloy, G. and Puckett, L.,** The metabolism of heterogeneous nuclear RNA and the formation of cytoplasmic messenger RNA in animal cells, *Prog. Biophys. Mol. Biol.,* 31, 1, 1976.
53. **Jacob, S. T., Sajdel, E. M., and Munro, H. N.,** Different responses of soluble whole nuclear RNA polymerase and soluble nucleolar RNA polymerase to divalent cations and to inhibition by α-amanitin, *Biochem. Biophys. Res. Commun.,* 38, 765, 1970.
54. **Kedinger, C., Gniadowski, M., Mandel, J. L., Gissinger, F., and Chambon, P.,** α-Amanitin: a specific inhibitor of one of two DNA-dependent RNA polymerase activities from calf thymus, *Biochem. Biophys. Res. Commun.,* 38, 165, 1970.
55. **Lindell, T. J., Weinberg, F., Morris, P. W., Roeder, R. G., and Rutter, W. J.,** Specific inhibition of nuclear RNA polymerase II by α-amanitin, *Science,* 170, 447, 1970.
56. **Tata, J. R., Hamilton, M. J., and Shields, D.,** Effects of alpha-amanitin *in vivo* on RNA polymerase and nuclear RNA synthesis, *Nature (London), New Biol.,* 238, 161, 1972.
57. **Jacob, S. T., Sajdel, E. M., Muecke, W., and Munro, H. N.,** Soluble RNA polymerases of rat liver nuclei: properties, template specificity and amanitin responses *in vitro* and *in vivo, Cold Spring Harbor Symp. Quant. Biol.,* 35, 681, 1970.
58. **Jamrich, M., Greenleaf, A. L., and Bautz, E. K. F.,** Localization of RNA polymerase in polytene chromosomes of *Drosophila melanogaster, Proc. Natl. Acad. Sci. U.S.A.,* 74, 2079, 1977.
59. **Suzuki, Y. and Giza, P. E.,** Accentuated expression of silk fibroin genes *in vivo* and *in vitro. J. Mol. Biol.,* 107, 183, 1976.

60. **Bona, M., Scheer, U., and Bautz, E. K. F.,** Antibodies to RNA polymeraseII(B) inhibit transcription in lampbrush chromosomes after microinjection into living amphibian oocytes, *J. Mol. Biol.,* 151, 81, 1981.
61. **Jänne, O., Bullock, L. P., Bardin, C. W., and Jacob, S. T.,** Early androgen action in kidney of normal and androgen-insensitive (Tfm/Y) mice. Changes in RNA polymerase and chromatin template activities, *Biochim. Biophys. Acta,* 418, 330, 1976.
62. **O'Malley, B. W., Schwartz, R. J., and Schrader, W. T.,** A review of regulation of gene expression by steroid hormone receptors, *J. Steroid Biochem.,* 7, 1151, 1976.
63. **Lin, Y. C., Bullock, L. P., Bardin, C. W., and Jacob, S. T.,** Effect of medroxyprogesterone acetate and testosterone on solubilized RNA polymerases and chromatin template activity in kidney from normal and androgen-insensitive (Tfm/Y) mice, *Biochemistry,* 17, 4833, 1978.
64. **Yu, F. L.,** Increased levels of rat hepatic nuclear free and engaged RNA polymerase activities during liver regeneration, *Biochem. Biophys. Res. Commun.,* 64, 1107, 1975.
65. **Stein, G., Park, W., Thrall, C., Mans, R., and Stein, J.,** Regulation of cell cycle stage-specific transcription of histone genes from chromatin by non-histone chromosomal proteins, *Nature (London),* 257, 764, 1975.
66. **Wilson, G. N., Steggles, A. W., and Nienhaus, A. W.,** Strand-selective transcription of globin genes in rabbit erythroid cells and chromatin, *Proc. Natl. Acad. Sci. U.S.A.,* 72, 4835, 1975.
67. **Crouse, G. F., Fodor, E. J. B., and Doty, P.,** *In vitro* transcription of chromatin in the presence of mercurated nucleotide, *Proc. Natl. Acad. Sci. U.S.A.,* 73, 1564, 1976.
68. **Smith, M. M. and Huang, R. C. C.,** Transcription *in vitro* of immunoglobulin kappa light chain genes in isolated mouse myeloma nuclei and chromatin, *Proc. Natl. Acad. Sci. U.S.A.,* 73, 775, 1976.
69. **Towle, H. C., Tsai, M. J., Tsai, W. Y., and O'Malley, B. W.,** Effect of estrogen on gene expression in the chick oviduct, *J. Biol. Chem.,* 252, 2396, 1977.
70. **Cochet-Meilhac, M., Nuret, P., Courvalin, J. C., and Chambon, P.,** Animal DNA-dependent RNA polymerases. 12. Determination of the cellular number of RNA polymerase B molecules, *Biochim. Biophys. Acta,* 353, 185, 1974.
71. **Guialis, A., Beatty, B., Ingles, C. J., and Crerar, M. M.,** Regulation of RNA polymerase II activity in α-amanitin-resistant CHO hybrid cells, *Cell,* 10, 53, 1977.
72. **Cox, R. F.,** Quantitation of elongating form A and B RNA polymerases in chick oviduct nuclei and effects of estradiol, *Cell,* 7, 455, 1976.
73. **Kastern, W. H., Christmann, J. L., Eldridge, J. D., and Mullinix, K. P.,** Estrogen regulates the number of RNA polymerase II molecules in rooster liver, *Biochem. Biophys. Acta,* 653, 259, 1981.
74. **Guialis, A., Morrison, K. E., and Ingles, J. E.,** Regulated synthesis of RNA polymerase II polypeptides in Chinese hamster ovary cell lines, *J. Biol. Chem.,* 254, 4171, 1979.
75. **Gariglio, P., Bellard, M., and Chambon, P.,** Clustering of RNA polymerase B molecules in the 5′ moiety of the adult β-globin gene of hen erythrocytes, *Nucl. Acid Res.,* 9, 2589, 1981.
76. **Jänne, O., Bardin, C. W., and Jacob, S. T.,** DNA-dependent RNA polymerases I and II from kidney. Effect of polyamines on the *in vitro* transcription of DNA and chromatin, *Biochemistry,* 14, 3589, 1975.
77. **Cox, R.,** Transcription of high molecular weight RNA from hen oviduct chromatin by bacterial and endogenous form B RNA polymerases, *Eur. J. Biochem.,* 39, 49, 1973.
78. **Siefart, K. H.,** A factor stimulating the transcription on double-stranded DNA by purified RNA polymerase from rat liver nuclei, *Cold Spring Harbor Symp. Quant. Biol.,* 35, 719, 1970.
79. **Stein, H. and Hausen, P,.** Factors influencing the activity of mammalian RNA polymerase, *Cold Spring Harbor Symp. Quant. Biol.,* 35, 709, 1970.
80. **Stein, H. and Hausen, P.,** A factor from calf thymus stimulating DNA-dependent RNA polymerase isolated from this tissue, *Eur. J. Biochem.,* 14, 270, 1970.
81. **Seifart, S. H., Juhasz, P. P., and Benecke, B. J.,** A protein factor from rat-liver tissue enhancing the transcription of native templates by homologous RNA polymerase B, *Eur. J. Biochem.,* 33, 181, 1973.
82. **Revie, D. and Dahmus, M. E.,** Purification and partial characterization of a stimulatory factor for lamb thymus RNA polymerase II, *Biochemistry,* 18, 1813, 1979.
83. **Lentfer, D. and Lezius, A. G.,** Mouse-myeloma RNA polymerase B. Template specificities and the role of a transcription-stimulating factor, *Eur. J. Biochem.,* 30, 278, 1972.
84. **Benson, R. H., Spindler, S. R., Hodo, H. G., and Blatti, S. P.,** DNA-dependent RNA polymerase II stimulatory factors from calf thymus: purification and structural studies, *Biochemistry,* 17, 1387, 1978.
85. **Spindler, S. R.,** Deoxyribonucleic acid dependent ribonucleic acid polymerase II specific initiation and elongation factors from calf thymus, *Biochemistry,* 18, 4042, 1979.
86. **Lee, S. C. and Dahmus, M. E.,** Stimulation of eucaryotic DNA-dependent RNA polymerase by protein factors, *Proc. Natl. Acad. Sci. U.S.A.,* 70, 1383, 1973.

87. **Dahmus, M. E.,** Stimulation of ascites tumor RNA polymerase II by protein kinase, *Biochemistry,* 15, 1821, 1976.
88. **Natori, S., Takeuchi, K., Takahashi, K., and Mizuno, D.,** DNA dependent RNA polymerase from Ehrlich ascites tumor cells. II. Factors stimulating the activity of RNA polymerase II, *J. Biochem.,* 73, 879, 1973.
89. **Nakanishi, Y., Sekimizu, K., Mizuno, D., and Natori, S.,** Apparent difference in the way of RNA synthesis stimulation by two stimulatory factors of RNA polymerase II, *FEBS Lett.,* 93, 357, 1978.
90. **Chuang, R., Chuang, L., and Laszlo, J.,** A new eukaryotic RNA polymerase factor: a factor from chicken myeloblastosis cells which stimulates transcription of denatured DNA, *Biochem. Biophys. Res. Commun.,* 57, 1231, 1974.
91. **Chuang, R. Y. and Chuang, L. F.,** Increased frequency of initiation of RNA synthesis due to a protein factor from chicken myeloblastosis nuclei, *Proc. Natl. Acad. Sci. U.S.A.,* 72, 2935, 1975.
92. **Sugden, B. and Keller, W.,** Mammalian deoxyribonucleic acid-dependent ribonucleic acid polymerases. I. Purification and properties of an α-amanitin-sensitive ribonucleic acid polymerase and stimulatory factors from Hela and KB cells, *J. Biol. Chem.,* 248, 3777, 1973.
93. **Sekimizu, K., Kobayashi, N., Mizuno, D., and Natori, S.,** Purification of a factor from Ehrlich ascites tumor cells specifically stimulating RNA polymerase II, *Biochemistry,* 15, 5064, 1976.
94. **D'Alessio, J. M. and Bagshaw, J. C.,** DNA-dependent RNA polymerases from *Artemia salina.* IV. Appearance of nuclear RNA polymerase activity during pre-emergence development of encysted embryos, *Differentiation,* 8, 53, 1977.
95. **D'Alessio, J. M. and Bagshaw, J. C.,** DNA-dependent RNA polymerases from *Artemia salina.* Characterization of a protein factor from developing embryos that stimulates artemia RNA polymerase II, *Dev. Biol.,* 70, 71, 1979.
96. **Sekimizu, K., Mizuno, D., and Natori, S.,** Enhancement of formation of the initiation complex by a factor stimulating RNA polymerase II from Ehrlich ascites tumor cells, *Biochim. Biophys. Acta,* 479, 180, 1977.
97. **Nakanishi, Y., Mitsuhashi, Y., Sekimizu, K., Yokoi, H., Tanaka, Y., Horikoshi, M., and Natori, S.,** Characterization of three proteins stimulating RNA polymerase II, *FEBS Lett.,* 130, 69, 1981.
98. **Ueno, K., Sekimizu, K., Obinata, M., Mizuno, D., and Natori, S.,** Stimulation of messenger ribonucleic acid synthesis in isolated nuclei by a protein that stimulates RNA polymerase II, *Biochemistry,* 20, 634, 1981.
99. **Ueno, K., Sekimizu, K., Mizuno, D., and Natori, S.,** Antibody against a stimulatory factor of RNA polymerase II inhibits nuclear RNA synthesis. *Nature (London),* 277, 145, 1979.
100. **Sekimizu, K., Mizuno, D., and Natori, S.,** Localization of a factor stimulating RNA polymerase II in the nucleoplasm, revealed by immunofluorescence, *Exp. Cell Res.,* 124, 63, 1979.
101. **Bell, G. I., Valenzuela, P., and Rutter, W.J.,** Phosphorylation of yeast DNA-dependent RNA polymerases *in vivo* and *in vitro, J. Biol. Chem.,* 252, 3082, 1977.
102. **Dahmus, M. E.,** Phosphorylation of eucaryotic DNA-dependent RNA polymerase, *J. Biol. Chem.,* 256, 3332, 1981.
103. **Jungmann, R. A., Hiestand, P. C., and Schweppe, J. S.,** Adenosine 3′:5′-monophosphate-dependent kinase and the stimulation of ovarian nuclear ribonucleic acid polymerase activities, *J. Biol. Chem.,* 249, 5444, 1974.
104. **Dahmus, M. E. and Natzle, J.,** Purification and characterization of Novikoff ascites tumor protein kinase, *Biochemistry,* 16, 1901, 1977.
105. **Kranias, E. G., Schweppe, J. S., and Jungmann, R. A.,** Phosphorylative and functional modifications of nucleoplasmic RNA polymerase by homologous adenosine 3′-5′-monophosphate-dependent protein kinase from calf thymus and by heterologous phosphatase, *J. Biol. Chem.,* 6750, 1977.
106. **Kranias, E. G. and Jungmann, R. A.,** Phosphorylation of calf thymus RNA polymerase by nuclear cyclic 3′,5′-AMP-independent protein kinase, *Biochim. Biophys. Acta,* 517, 439, 1978.
107. **Dahmus, M. E.,** Purification and properties of calf-thymus casein kinases I and II, *J. Biol. Chem.,* 256, 3319, 1981.
108. **Stetler, D. A. and Rose, K. M.,** Phosphorylation of DNA-dependent RNA polymerase II by nuclear protein kinase NII. Mechanism of enhanced RNA synthesis, *Biochemistry,* in press.
109. **Kuroiwa, A., Sekimizu, K., Ueno, K., Mizuno, D., and Natori, S.,** Separation of a stimulatory factor of RNA polymerase II from protein kinase activity of Ehrlich ascites tumor cells, *FEBS Lett.,* 75, 183, 1977.
110. **Jekimizu, K., Kubo, Y., Segawa, K., and Natori, S.,** Difference in phosphorylation of two factors stimulating RNA polymerase II of Ehrlich ascites tumor cells, *Biochemistry,* 20, 2286, 1981.
111. **Abelson, J. and Butz, E., Eds.,** Recombinant DNA, *Science,* 209, 1, 1980.
112. **Wu, G.,** Adenovirus DNA-directed transcription of 5.5S RNA *in vitro, Proc. Natl. Acad. Sci. U.S.A.,* 75, 2175, 1978.
113. **Manley, J. L., Fire, A., Cano, A., Sharp, P. A., and Gefter, M. L.,** DNA-dependent transcription of adenovirus genes in a soluble whole-cell extract, *Proc. Natl. Acad. Sci. U.S.A.,* 77, 3855, 1980.

114. **Weil, P. A., Luse, D. S., Segall, J., and Roeder, R. G.,** Selective and accurate initiation of transcription at the Ad2 major late promotor in a soluble system dependent on purified RNA polymerase II and DNA, *Cell,* 18, 469, 1979.
115. **Wu, G. J. and Zubay, G.,** Prolonged transcription in a cell-free system involving nuclei and cytoplasm, *Proc. Natl. Acad. Sci. U.S.A.,* 71, 1803, 1974.
116. **Price, R. and Pennann, S.,** Transcription of the adenovirus genome by an α-amanitin-sensitive ribonucleic acid polymerase in Hela cells, *J. Virol.,* 9, 621, 1972.
117. **Wallace, R. D. and Kates, J.,** State of adenovirus 2 deoxyribonucleic acid in the nucleus and its mode of transcription: studies with isolated viral deoxyribonucleic acid-protein complexes and isolated nuclei, *J. Virol.,* 9, 627, 1972.
118. **Weinmann, R., Raskas, H. J., and Roeder, R. G.,** Role of DNA-dependent RNA polymerases II and III in transcription of the adenovirus genome late in productive infection, *Proc. Natl. Acad. Sci. U.S.A.,* 71, 3426, 1974.
119. **Wasylyk, B., Kedinger, C., Corden, J., Brison, O., and Chambon, P.,** Specific *in vitro* initiation of transcription on conalbumin and ovalbumin genes and comparison with adenovirus-2 early and late genes, *Nature (London),* 285, 367, 1980.
120. **Flint, J.,** The topography and transcription of the adenovirus genome, *Cell,* 10, 153, 1977.
121. **Frink, R. J., Draper, K. G., and Wagner, E. K.,** Uninfected cell polymerase efficiently transcribes early but not late herpes simplex virus type 1 mRNA, *Proc. Natl. Acad. Sci. U.S.A.,* 78, 6139, 1981.
122. **Crawford, N., Fire, A., Samuels, M., Sharp, P. A., and Baltimore, D.,** Inhibition of transcription factor activity by poliovirus, *Cell,* 27, 555, 1981.
123. **Corden, J., Wasylyk, B., Buchwalder, A., Sassone-Corsi, P., Kedinger, C., and Chambon, P.,** Expression of cloned genes in new environment, *Science,* 209, 1406, 1980.
124. **Green, M. R. and Roeder, R. G.,** Definition of a novel promoter for the major adenovirus-associated virus mRNA, *Cell,* 22, 231, 1980.
125. **Yamamoto, T., deCrombugghe, B., and Pastan, I.,** Identification of a functional promoter in the long terminal repeat of Rous sarcoma virus, *Cell,* 22, 787, 1980.
126. **Fuhrman, S. A., VanBevereu, C., and Verma, I. M.,** Identification of a RNA polymerase II initiation site in the long terminal repeat of Moloney murine leukemia viral DNA, *Proc. Natl. Acad. Sci. U.S.A.,* 78, 5411, 1981.
127. **Handa, H., Kaufman, R. J., Manley, J., Gefter, M., and Sharp, P. A.,** Transcription of Simian Virus 40 DNA in a Hela whole cell extract, *J. Biol. Chem.,* 256, 478, 1981.
128. **Hu, S. L. and Manley, J. L.,** DNA sequence required for initiation of transcription *in vitro* from the major late promoter of adenovirus 2, *Proc. Natl. Acad. Sci. U.S.A.,* 78, 820, 1981.
129. **Mathis, D. J. and Chambon, P.,** The SV40 early region TATA box is required for accurate *in vitro* initiation of transcription, *Nature (London),* 290, 310, 1981.
130. **Mathis, D. J., Elkaim, R., Kedinger, C., Sassone-Corsi, P., and Chambon, P.,** Specific *in vitro* initiation of transcription on the adenovirus type 2 early and late EII transcription, *Proc. Natl. Acad. Sci. U.S.A.,* 78, 7383, 1981.
131. **Luse, D. and Roeder, R.,** Accurate transcription initiation on a purified mouse β-globin DNA fragment in a cell-free system, *Cell,* 20, 691, 1980.
132. **Proudfoot, N. J., Shander, M. H. M., Manley, J. L., Gefter, M. L., and Maniatis, T.,** Structure and *in vitro* transcription of human globin genes, *Science,* 209, 1331, 1980.
133. **Wasylyk, B., Derbyshire, R., Guy, A., Molko, D., Roget, A., Teoule, R., and Chambon, P.,** Specific *in vitro* transcription of conalbumin gene is drastically decreased by single point mutation in T-A-T-A box homology sequence, *Proc. Natl. Acad. Sci. U.S.A.,* 77, 7024, 1980.
134. **Grosveld, G. C., Shewmauaker, C. K., Jat, P., and Flavell, R. A.,** Localization of DNA sequences necessary for transcription of the rabbit β-globin gene *in vitro, Cell,* 25, 215, 1981.
135. **Tsai, S. Y., Tsai, M. J., and O'Malley, B. W.,** Specific 5′ flanking sequences are required for faithful initiation of *in vitro* transcription of the ovalbumin gene, *Proc. Natl. Acad. Sci. U.S.A.,* 78, 879, 1981.
136. **Wasylyk, B. and Chambon, P.,** A T to A base substitution and small deletions in the conalbumin TATA box drastically decrease specific *in vitro* transcription, *Nucl. Acids Res.,* 9, 1813, 1981.
137. **Tsuda, M. and Suzuki, Y.,** Faithful transcription initiation of fibroin gene in a homologous cell-free system reveals an enhancing effect of 5′-flanking sequence far upstream, *Cell,* 27, 175, 1981.
138. **Tsujimoto, Y., Hirose, S., Tsuda, M., and Suzuki, Y.,** Promoter sequence of fibroin gene assigned by *in vitro* transcription system, *Proc. Natl. Acad. Sci. U.S.A.,* 78, 4838, 1981.
139. **Busslinger, M., Portmann, R., Irminger, J. C., and Birnstiel, M. L.,** Ubiquitous and gene-specific regulatory 5′ sequences in a sea urchin histone DNA clone coding for histone protein variants, *Nucl. Acids Res.,* 8, 957, 1980.
140. **Gannon, F., O'Hare, K., Perrin, F., LePennec, J. P., Benoist, C., Cochet, M., Breathnach, R., Royal, A., Garapin, A., Cami, B., and Chambon, P.,** Organization and sequences at the 5′ end of a cloned complete ovalbumin gene, *Nature (London),* 278, 428, 1979.

141. **Benoist, C., O'Hare, K., Breathnach, R., and Chambon, P.,** The ovalbumin gene-sequence of putative control regions, *Nucl. Acids Res.,* 127, 1980.
142. **Efstratiadis, A., Posakoney, J. W., Maniatis, T., Lawn, R. M., O'Connell, C., Spritz, R. A., DeRiel, J. K., Forget, B. C., Weissman, S. M., Slightom, J. L., Blecht, A. E., Smithies, O., Barelle, F. E., Shoulders, C. C., and Proudfoot, N. J.,** The structure and evolution of the human β-globin gene family, *Cell,* 21, 653, 1980.
142a. **Grosveld, G. C., De Boer, E., Shewmaker, C. K., and Flavell, R. A.,** DNA sequences necessary for transcription of the rabbit β-globin gene *in vivo, Nature (London),* 295, 120, 1982.
143. **Grosschedl, R. and Birnstiel, M. L.,** Identification of regulatory sequences in the prelude sequences of an H2A histone gene by the study of specific deletion mutants *in vivo, Proc. Natl. Acad. Sci. U.S.A.,* 77, 1432, 1980.
144. **Benoist, C. and Chambon, P.,** *In vivo* sequence requirements of the SV40 early promoter region, *Nature (London),* 290, 304, 1981.
145. **Sassone-Corsi, P., Corden, J., Kedinger, C., and Chambon, P.,** Promotion of specific *in vitro* transcription by excised TATA box sequences inserted in a foreign nucleotide environment, *Nucl. Acids Res.,* 9, 3941, 1981.
146. **Benoist, C. and Chambon, P.,** Deletions covering the putative promoter region of early mRNAs of simian virus 40 do not abolish T-antigen expression, *Proc. Natl. Acad. Sci. U.S.A.,* 77, 3865, 1980.
147. **Banerji, J., Rusconi, S., and Schaffner, W.,** Expression of a β-globin gene is enhanced by remote SV40 DNA sequences, *Cell,* 27, 299, 1981.
148. **Baker, C. C., Herigge, J., Courteis, G., Galibert, F., and Ziff, E.,** Messenger RNA for the Ad2 DNA binding protein: DNA sequence encoding the first leader and heterogeneity at the 5′ end, *Cell,* 18, 569, 1979.
149. **Baker, C. C. and Ziff, E. B.,** Promoters and heterogeneous 5′ termini of the messenger-RNA's of adenovirus serotype-2, *J. Mol. Biol.,* 149, 189, 1981.
150. **Flavell, A. J., Cowie, A., Ahrand, J. R., and Kamen, R.,** Localization of three major capped 5′ ends of polyoma virus late mRNA's within a single tetranucleotide sequence in the viral genome, *J. Virol.,* 33, 902, 1980.
151. **Matsui, T., Segall, J., Weil, A., and Roeder, R. G.,** Multiple factors required for accurate initiation of transcription by purified RNA polymerase II, *J. Biol. Chem.,* 255, 11992, 1980.
152. **Tsai, S. Y., Tsai, M. J., Kops, L. E., Minghetti, P. P., and O'Malley, B. W.,** Transcription factors from oviduct and Hela cells are similar, *J. Biol. Chem.,* 256, 13055, 1980.
153. **Roeder, R. G. and Rutter, W. J.,** Multiple forms of DNA-dependent RNA polymerase in eukaryotic organisms, *Nature (London),* 224, 234, 1969.
154. **Sklar, V. E. F., Schwartz, L. B., and Roeder, R. G.,** Distinct molecular structures of nuclear class I, II and III DNA-dependent RNA polymerases, *Proc. Natl. Acad. Sci. U.S.A.,* 72, 348, 1975.
155. **Wittig, B. and Wittig, S.,** Purification of class A, B, and C DNA-dependent RNA polymerases from chick embryos, *Biochim. Biophys. Acta,* 520, 598, 1978.
156. **Seifart, K. H., Benecke, B. J., and Juhasz, P. P.,** Multiple RNA polymerase species from rat liver tissue: possible existence of a cytoplasmic enzyme, *Arch. Biochem. Biophys.,* 151, 519, 1972.
157. **Austoker, J. L., Trevor, J. C., Beebee, C., Chesterton, J., and Butterworth, P. H. W.,** DNA-dependent RNA polymerase activity of Chinese hamster kidney cells sensitive to high concentrations of α-amanitin, *Cell,* 3, 227, 1974.
158. **Wilhelm, J., Dina, D., and Crippa, M.,** A special form of deoxyribonucleic acid-dependent ribonucleic acid polymerase from oocyte of *Xenopus laevis.* Isolation and characterization, *Biochemistry,* 13, 1200, 1974.
159. **Weinman, R. and Roeder, R. G.,** Role of DNA-dependent RNA polymerase III in the transcription of the tRNA and 5S RNA genes, *Proc. Natl. Acad. Sci. U.S.A.,* 71, 1790, 1974.
160. **Weil, P. A. and Blatti, S. P.,** Hela cell deoxyribonucleic acid dependent RNA polymerases: function and properties of the class III enzymes, *Biochemistry,* 15, 1500, 1976.
161. **Weinman, R., Raskas, H., and Roeder, R. G.,** Role of DNA-dependent RNA polymerases II and III in the transcription of the adenovirus genome in KB cells, *Proc. Natl. Acad. Sci. U.S.A.,* 71, 3426, 1974.
162. **Furth, J. J., Nicholson, A., and Austin, G. G.,** The enzymatic synthesis of ribonucleic acid in animal tissue III. Further purification of soluble RNA polymerase from lymphoid tissue and some general properties of the enzyme, *Biochim. Biophys. Acta,* 213, 124, 1970.
163. **Schwartz, L. B., Sklar, V. E. F., Jaehning, J., Weinman, R., and Roeder, R. G.,** Isolation and partial characterization of the multiple forms of deoxyribonucleic acid-dependent ribonucleic acid polymerase — mouse myeloma MOPC 315, *J. Biol. Chem.,* 249, 5889, 1974.
164. **Blair, D. G. R.,** DNA-dependent RNA polymerases of Ehrlich carcinoma, other murine ascites tumors, and murine normal tissues, *J. Natl. Cancer Inst.,* 55, 397, 1975.
165. **Fuhrman, S. A. and Gill, G. N.,** Adrenocorticotropic hormone regulation of adrenal RNA polymerases. Stimulation of nuclear RNA polymerase III, *Biochemistry,* 15, 5520, 1976.

166. **Hosenlopp, P., Wells, D., and Chambon, P.,** Animal DNA-dependent RNA polymerases. Partial purification and properties of three classes of RNA polymerases from uninfected and adenovirus-infected Hela cells, *Eur. J. Biochem.*, 58, 237, 1975.
167. **Weinmann, R., Brendler, T., Raskas, H. J., and Roeder, R. G.,** Low molecular weight viral RNAs transcribed by RNA polymerase III during adenovirus infection, *Cell*, 7, 557, 1976.
168. **Parker, C. S. and Roeder, R. G.,** Selective and accurate transcription of the *Xenopus laevis* 5S RNA genes in isolated chromatin by purified RNA polymerase III, *Proc. Natl. Acad. Sci. U.S.A.*, 74, 44, 1977.
169. **Ackerman, S. and Furth, J. J.,** Selective *in vitro* transcription of the 5S RNA genes of a DNA template, *Biochemistry*, 18, 3243, 1979.
170. **Wu, G.,** Faithful transcription of adenovirus 5.5S RNA gene by RNA polymerase III in a human KB cell-free extract, *J. Biol. Chem.*, 255, 251, 1980.
171. **Birkenmeier, E. H., Brown, D. D., and Jordan, E.,** A nuclear extract of *Xenopus laevis* oocytes that accurately transcribes 5S RNA genes, *Cell*, 15, 1077, 1978.
172. **Ng, S. Y., Parker, C. S., and Roeder, R. G.,** Transcription of cloned *Xenopus* 5S RNA genes by *X. laevis* RNA polymerase III in reconstituted systems, *Proc. Natl. Acad. Sci. U.S.A.*, 76, 136, 1979.
173. **Weil, P. A., Segall, I., Harris, B., Ng, S. Y., and Roeder, R. G.,** Faithful transcription of eucaryotic genes by RNA polymerase III in systems reconstituted with purified templates, *J. Biol. Chem.*, 254, 6163, 1979.
174. **Sakonju, S., Bogenhagen, D. F., and Brown, D. D.,** A control region in the center of the 5S RNA gene directs specific initiation of transcription: I. The 5′ border of the region, *Cell*, 19, 13, 1980.
175. **Bogenhagen, D. F., Sakonju, S., and Brown, D. D.,** A control region in the center of the 5S RNA gene directs specific initiation of transcription: II. The 3′ border of the region, *Cell*, 19, 27, 1980.
176. **Bogenhagen, D. F. and Brown, D. D.,** Nucleotide sequences in *Xenopus* 5S DNA required for transcription termination, *Cell*, 24, 261, 1981.
177. **Gilbert, W.,** Starting and stopping sequences for the RNA polymerase, in *RNA Polymerase*, Losick, R. and Chamberlin, M., Eds., Cold Spring Harbor Laboratory, Cold Spring Harbor, N.Y., 1976, 193.
178. **Rosenberg, M. and Court, D.,** Regulatory sequences involved in promotion and termination of RNA transcription, *Annu. Rev. Genet.*, 13, 319, 1979.
179. **Martin, F. H. and Tinoco, I., Jr.,** DNA-RNA hybrid duplexes containing oligo (dA:rU) sequences are exceptionally unstable and may facilitate termination of transcription, *Nucl. Acids Res.*, 8, 2295, 1980.
180. **Roeder, R. G., Engelke, D. R., Harris, B., Ng, S. Y., Segall, J., Shastry, B. S., and Weil, P. A.,** Factors involved in the transcription of purified genes by RNA polymerase III, in *Eukaryotic Gene Regulation*, Axel, R., Maniatis T., and Fox, C. F., Eds., Academic Press, New York, 1979, 521.
181. **Engelke, D. R., Ng, S. Y., Shastry, B. S., and Roeder, R. G.,** Specific interaction of a purified transcription factor with an internal control region of 5S RNA genes, *Cell*, 19, 717, 1980.
182. **Galas, D. and Schmitz, A.,** DNase footprinting: a simple method for detection of protein-DNA binding specificity, *Nucl. Acids Res.*, 5, 3157, 1978.
183. **Sakonju, S., Brown, D. D., Engelke, D., Ng, S. Y., Shastry, B. S., and Roeder, R. G.,** The binding of a transcription factor to deletion mutants of a 5S ribosomal RNA gene, *Cell*, 23, 665, 1981.
184. **Pelham, H. R. B. and Brown, D. D.,** A specific transcription factor that can bind either the 5S RNA gene or 5S RNA, *Proc. Natl. Acad. Sci. U.S.A.*, 77, 4170, 1980.
185. **Honda, B. M. and Roeder, R. G.,** Association of a 5S gene transcription factor with 5S RNA and altered levels of the factor during cell differentiation, *Cell*, 22, 119, 1980.
186. **Ford, P. J. and Brown, R. D.,** Sequences of 5S ribosomal RNA from *Xenopus mulleri* and the evolution of 5S gene-coding sequences, *Cell*, 8, 485, 1976.
187. **Pelham, H. R. B., Wormington, W. M., and Brown, D. D.,** Related 5S RNA transcription factors in *Xenopus* oocytes and somatic cells, *Proc. Natl. Acad. Sci. U.S.A.*, 78, 1760, 1981.
187a. **Gruissem, W. and Seifart, K. H.,** Transcription of 5S RNA genes *in vitro* is feedback-inhibited by HeLa 5S RNA, *J. Biol. Chem.*, 257, 1468, 1982.
188. **Fowlkes, D. M. and Shenk, T.,** Transcriptional control regions of the adenovirus VA 1 RNA gene, *Cell*, 22, 405, 1980.
189. **Segall, J., Matsui, T., and Roeder, R.,** Multiple factors are required for the accurate transcription of purified genes by RNA polymerase III, *J. Biol. Chem.*, 255, 11986, 1980.
190. **Akusjarvi, G. and Pettersson, U.,** Sequence analysis of adenovirus DNA: complete nucleotide sequence of the spliced 5′ noncoding region of adenovirus 2 hexon messenger RNA, *Cell*, 16, 841, 1979.
191. **DeFranco, D., Schmidt, O., and Soll, D.,** Two control regions for eucaryotic tRNA gene transcription, *Proc. Natl. Acad. Sci. U.S.A.*, 77, 3365, 1980.
192. **Kressman, A., Hofstetter, H., DiCapua, E., Grosschedl, R., and Birnstiel, M. L.,** A tRNA gene of *Xenopus laevis* contains at least two sites promoting transcription, *Nucl. Acids Res.*, 7, 1749, 1979.

193. **Galli, G., Hofstetter, H., Birnstiel, M. L.**, Two conserved sequence blocks within eukaryotic tRNA genes are major promoter elements, *Nature (London)*, 294, 626, 1981.
194. **Hofstetter, H., Kressman, A., and Birnstiel, M. L.**, A split promoter for a eucaryotic tRNA gene, *Cell*, 24, 573, 1981.
195. **Sharp, S., DeFranco, D., Dingermann, T., Farrell, P., and Soll, D.**, Internal control regions for transcription of eukaryotic tRNA genes, *Proc. Natl. Acad. Sci. U.S.A.*, 78, 6657, 1981.
196. **Johnson, J. D., Ogden, R., Johnson, P., Abelson, J., Demmbeck, P., and Itakura, R.**, Transcription and processing of a yeast tRNA gene containing a modified intervening sequence, *Proc. Natl. Acad. Sci. U.S.A.*, 77, 2564, 1980.
197. **Sprague, K. U., Larson, D., and Morton, D.**, 5′ flanking sequences are required for activity of Silkworm alanine tRNA genes in homologus *in vitro* transcription system, *Cell*, 22, 171, 1980.
198. **Ogden, R. C., Beckman, J. S., Abelson, J., Kang, H. S., Soll, D., and Schmidt, O.**, *In vitro* transcription and processing of a yeast tRNA gene containing an intervening sequence, *Cell*, 17, 399, 1979.
199. **Goodman, H. M., Olson, M. V., and Hall, B. D.**, Nucleotide sequence of a mutant eukaryotic gene: The yeast tyrosine-inserting ochre suppressor SUP4-0, *Proc. Natl. Acad. Sci. U.S.A.*, 74, 5453, 1977.
200. **Valenzuela, P., Venegas, A., Weinburg, F., Bishop, R., and Rutter, W. J.**, Structure of yeast phenylalanine-tRNA genes: An intervening DNA segment within the region coding for the tRNA, *Proc. Natl. Acad. Sci. U.S.A.*, 75, 190, 1978.
201. **Venegas, A., Quiroga, M., Zaldivar, J., Rutter, W. J., and Valenzuela, P.**, Isolation of yeast $tRNA^{Leu}$ genes, *J. Biol. Chem.*, 254, 12306, 1979.
202. **Hovemann, B., Sharp, S. J., Yamada, H., and Soll, D.**, Analysis of *Drosophila* tRNA gene cluster, *Cell*, 19, 889, 1980.
203. **Mao, J., Schmidt, O., and Soll, D.**, Dimeric transfer RNA precursor in *S. pombe*, *Cell*, 21, 509, 1980.
204. **Telford, J., Kressman, A., Koski, R., Grosschedl, R., Muller, F., Clarkson, S. G., and Birnstiel, M. L.**, Delimination of a promoter for RNA polymerase III by means of a functional test, *Proc. Natl. Acad. Sci. U.S.A.*, 76, 2590, 1979.
205. **Garber, R. L. and Gage, L. P.**, Transcription of a cloned *Bombyx mori* $tRNA_2^{Ala}$ gene: nucleotide sequence of the tRNA precursor and its processing *in vitro*, *Cell*, 18, 817, 1979.
206. **Hovemann, B., Schmidt, O., Yamada, H., Silverman, S., Mao, J., DeFranco, D., and Soll, D.**, in *tRNA: Biological Aspects*, Soll, D., Abelsen, J., and Schimmel, P., Eds., Cold Spring Harbor Laboratory, Cold Spring Harbor, N.Y., 1980, pp. 325.
207. **Schmidt, O., Mao, J., Silverman, S., Hovemann, B., and Soll, D.**, Specific transcripton of eucaryotic tRNA genes in *Xenopus* germinal vesicle extracts, *Proc. Natl. Acad. Sci. U.S.A.*, 75, 4819, 1978.
208. **Silverman, S., Schmidt, O., Soll, D., and Hovemann, B.**, The nucleotide sequence of a cloned *Drosophila* arginine tRNA gene and its *in vitro* transcription in *Xenopus* germinal vesicle extracts, *J. Biol. Chem.*, 254, 10290, 1979.
209. **Hagenbuchle, O., Larson, D., Hall, G. I., and Sprague, K. U.**, The primary transcription product of a silkworm alanine tRNA gene: identification of *in vitro* sites of initiation, termination, and processing, *Cell*, 18, 1217, 1979.
210. **Koski, R. A., Clarkson, S. G., Kunjan, J., Hall, B. D., and Smith, M.**, Mutations of the yeast SUP4 tRNA locus: transcription of the mutant genes *in vitro*, *Cell*, 22, 415, 1980.
211. **Carrara, G., DiSegni, G., Otsuka, A., and Tocchini-Valentini, G. P.**, Deletion of the 3′ half of the yeast $tRNA's_3^{Leu}$ gene does not abolish promotor function *in vitro*, *Cell*, 27, 371, 1981.
212. **Standring, D. N., Venegas, A., and Rutter, W. J.**, Yeast $tRNA_3^{Leu}$ gene transcribed and spliced in a Hela cell extract, *proc. Natl. Acad. Sci. U.S.A.*, 78, 5963, 1981.
213. **Haynes, S. R. and Jelinek, W. R.**, Low molecular weight RNAs transcribed *in vitro* by RNA polymerase III from Alutype dispersed repeats in Chinese hamster DNA are also found *in vivo*, *Proc. Natl. Acad. Sci. U.S.A.*, 78, 6130, 1981.
214. **VanKeulen, H., Plonta, R. J., and Retel, J.**, Structure and transcription specificity of yeast RNA polymerase A, *Biochim. Biophys. Acta*, 395, 179, 1975.
215. **VanKeulen, H. and Retel, J.**, Transcription specificity of yeast RNA polymerase A. Highly specific transcription *in vitro* of the homologous ribosomal transcription units, *Eur. J. Biochem.*, 79, 579, 1977.
216. **Holland, M. J., Hager, G. L., and Rutter, W. J.**, Transcription of yeast DNA by homologous RNA polymerases I and II: selective transcription of ribosomal genes by RNA polymerase I, *Biochemistry*, 16, 16, 1977.
217. **Valenzuela, P., Bull, P., Zaldivar, J., Venegas, A., and Martial, J.**, Subunits of yeast RNA polymerase I involved in interactions with DNA and nucleotides, *Biochem. Biophys. Res. Commun.*, 81, 662, 1978.

218. **Ruet, A., Sentenac, A., Fromageot, P., Winsor, B., and Lacronte, F.,** A mutation of the B_{220} subunits gene affects the structural and functional properties of yeast RNA polymerase B *in vitro, J. Biol. Chem.,* 255, 6450, 1980.
219. **Kramer, A., Haars, R., Kabisch, R., Will, H., Bautz, F. A., and Bautz, E. K. F.,** Monoclonal antibody directed against RNA polymerase II of *Drosophila melanogaster, Mol. Gen. Genet.,* 180, 193, 1980.
220. **Christmann, J. L. and Dahmus, M. E.,** Monoclonal antibody specific for calf thymus RNA polymerases II_O and II_A, *J. Biol. Chem.,* 256, 11798, 1982.
221. **Rose, K.,** unpublished.

Chapter 3

RNA SPLICING IN VITRO

S. J. Flint

TABLE OF CONTENTS

I. INTRODUCTION

It is only 5 years since segments of noncoding DNA (termed insertions, introns, or intervening sequences) present in eukaryotic genes specifying tRNA, rRNA, or mRNA were first described. Since 1977, the catalogue of such discontinuous genes has greatly expanded and much has been learned about their structural organization and mode of expression. It is, for example, firmly established that such intervening sequences are transcribed colinearly with coding sequences to be removed posttranscriptionally in the step known as splicing. To remove transcripts of intervening sequences while preserving the integrity of the coding sequence, splicing necessarily comprises precise endonucleolytic cleavages at the boundaries of an intervening sequence and ligation of the ends of coding segments. The vast body of information describing the primary sequences of discontinuous genes and their mature RNA products has proved an enormous stimulus to considerations of the mechanisms of splicing. Nevertheless our actual knowledge of these mechanisms and the cellular machinery that mediates splicing reactions has accrued relatively slowly, apparently constrained by the rate of development of systems that will perform splicing reactions in vitro.

Splicing of yeast pre-tRNA species in soluble extracts was, for example, first reported in 1979 and we now possess a detailed, although not complete, picture of the mechanism of pre-tRNA splicing as well as some information about the relevant enzymes. Splicing of pre-rRNA or pre-mRNA species in cell extracts has been described only within the last year: our knowledge of these reactions is correspondingly less complete but, at least in the case of pre-rRNA splicing, advancing rapidly.

In this article, such in vitro systems and their contribution to our present understanding of splicing mechanisms and machines are discussed in the context of more general questions about splicing, such as those concerning the modes of substrate recognition and the specificities of splicing reactions.

II. SPLICING OF PRECURSORS TO tRNA

A. Occurrence and Structure of Eukaryotic tRNA Genes with Intervening Sequences

Transcripts of tRNA genes that contain intervening sequences (IVS) were first identified in yeast; sequencing of cloned yeast tRNA genes, including one encoding $tRNA^{tyr}$ and three specifying $tRNA^{phe}$ identified a short intervening sequence in the coding regions of these genes.[1,2] Yeast genes encoding $tRNA^{trp}$,[3] $tRNA_3^{leu}$,[4,5] and the minor speciese $tRNA_{UCG}^{ser}$[6] also contain an IVS, whereas genes specifying $tRNA^{ser}$, $tRNA^{asp}$, $tRNA_3^{glu}$, $tRNA_2^{ser}$, and $tRNA^{arg}$ do not.[4,6-8] The observation that only the precursors transcribed from the five TRNA genes containing an IVS listed above accumulate in certain yeast mutants[9,10] argues that all other yeast tRNA genes indeed possess no IVS. Discontinuous tRNA genes certainly seem to be the exception in higher eukaryotes: intervening sequences have been reported to be present in one *Xenopus laevis* $tRNA^{tyr}$ gene[11] and two $tRNA^{leu}$ genes of *Drosophila melanogaster*.[12] Of course, the total number of fully characterized tRNA genes of higher eukaryotes is not large, some 10 to 20, so it is difficult to make general statements about the distribution of those that possess an IMS. Similarly, it is not yet possible to assess the significance of the fact that the discontinuous *Xenopus* and *Drosophila* tRNA genes appear to possess counterparts in the yeast genome.

The first demonstration that some yeast tRNA genes possesseed intervening sequences included a result that ruled out any possibility that the presence of the IVS serves to inactivate the gene: one of the $tRNA^{tyr}$ genes sequenced by Goodman et al.[1] was that specifying the ochre suppressor SUP4-o, whose activity could be demonstrated readily. Sequencing of individual tRNA precursor (pre-tRNA) species that ac-

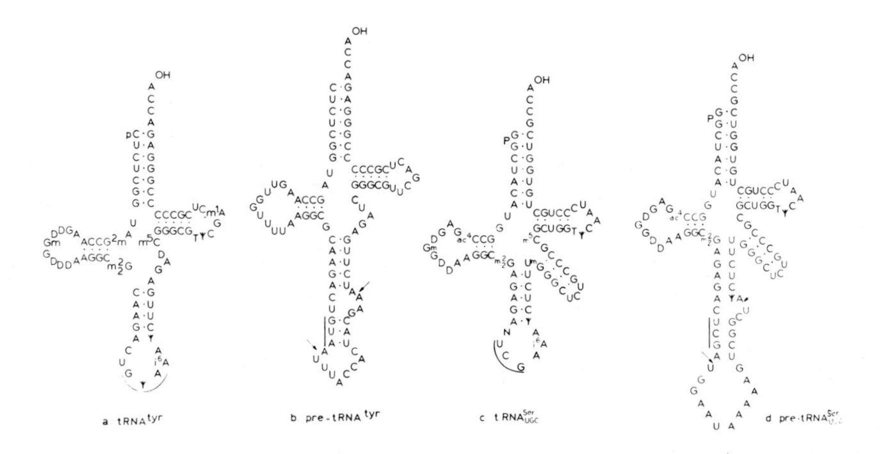

FIGURE 1. Structure of yeast tRNA species and their intervening-sequence containing precursors. The most favorable secondary structures of yeast $tRNA^{tyr}$, pre-$tRNA^{tyr}$, $tRNA^{ser}_{ucg}$, and pre-$tRNA^{ser}_{ucg}$ are shown in parts a, b, c and d, respectively. The structures of $tRNA^{tyr}$ and its precursor and of $tRNA^{ser}_{ucg}$ and its precursor are based on those described by O'Farrell et al.[14] and Etcheverry et al.,[6] respectively. In all parts of the figure, the anticodon is marked by a line. In parts b and d, likely sites of splicing cleavage are indicated by the arrows.

cumulate in those yeast mutants defective in processing mentioned previously, such as rna-1,[9] provided biochemical evidence for the transcription of genes that contain an IVS and established that the IVS itself is transcribed, to be removed posttranscriptionally.[13]

The intervening sequences present in tRNA genes are short, some 13 to 50 nucleotide pairs. The 14 base-pair (bp) insertions in the three yeast $tRNA^{tyr}$ genes that have been seqeuenced are identical to one another at all but one position.[1] Similarly, the 18 or 19 bp intervening sequences found in three yeast $tRNA^{phe}$ genes are nearly identical.[2] On the other hand, essentially no sequence homology can be detected among the intervening sequences present in yeast tRNA genes specifying different tRNA species although conservation of certain IVS sequences in yeast and *Drosophila* $tRNA^{leu}$ has been noted.[12] Despite such lack of direct sequence homology, discontinuous yeast tRNA genes and their transcripts do share a number of properties: the intervening sequences are, for example, all A + T-rich and, although varying from 14 bp in yeast $tRNA^{tyr}$ to 45 bp in *Drosophila* $tRNA^{leu}$,[1,12] always lie very close to the 3′ side of the anticodon triplet. Although the exact location of an IVS cannot be deduced from the primary sequences of the gene and its mature tRNA product, because of duplicated residues at the IVS-coding sequence boundaries, it is clear that the IVS always begins with 0 to 3 nucleotides of the 3′ end of the anticodon. Most significantly, perhaps, the primary sequences of transcripts of these tRNA genes that possess an IVS can be arranged to form a similar secondary structure: this preserves the overall clover-leaf form of a typical, mature tRNA but is distinguished from that structure by the formation of base-pairs between the anticodon and residues of the IVS. In this form, illustrated in Figure 1 for the precursors to yeast $tRNA^{tyr}$ and $tRNA^{ser}_{UCG}$, the sites at which the transcript must be cleaved to excise the IVS are contained within a single-stranded loop characteristic of the structure of the unspliced molecule. The results of limited nuclease digestion of yeast pre-$tRNA^{tyr}$ and $tRNA^{ser}_{UCG}$ provide evidence that the secondary structures illustrated in Figure 1 are more than a mere figment of model-building; precursor tRNA molecules containing an IVS are cut under such conditions within that part of the IVS that forms the loop of the extended anticodon arm (see Figure 1), whereas the anticodon itself is not susceptible, as it is the mature tRNA species.[6,14] It is therefore clear that the anticondon must be sequestered in a base-paired structure in a tRNA precursor possessing an IVS. As would be predicted from this model, such a tRNA precursor cannot be amino-acylated in vitro under conditions that permit efficient amino-acylation of mature tRNA.[14] The exposure of the sites at which the precursor must be cleaved during splicing in a structure that can be adopted by all precursors that contain an IVS suggests that the secondary structure of a molecule is important in splicing of tRNA transcripts. As we shall see, this view is strengthened by the results of characterization of the intermediates formed when splicing reactions take place in vitro.

B. Processing of tRNA Precursors

The transcription and/or subsequent processing of pre-tRNA molecules, which carry 5′-flanking and 3′-trailing sequences and may or may not possess an IVS, will take place in several in vitro systems. The best-characterized coupled system must be *X. laevis* oocytes or extracts prepared from their nuclei; when a cloned tRNA gene, of homologous or heterologous origin (for example yeast, nematode or silkworm) is supplied, not only is the tRNA gene transcribed correctly, but the pre-tRNA product is also processed efficiently.[15-22] Thus, *Xenopus* oocyte nuclei must contain all enzymatic activities necessary to trim pre-tRNA species, to perform the modifications that occur in vivo (although these may be done incorrectly in the case of heterologous pre-tRNA species[22]) and to remove intervening sequences, when present. At least one *X. laevis* tRNA gene is discontinuous,[11] so the presence of such splicing activity in *Xenopus* oocytes is not too surprising.

Sequencing of the primary products of transcription and processing intermediates made in these systems has established the scheme of tRNA processing illustrated in Figure 2. Maturation of a primary transcript includes removal of the 5′-flanking sequence, including the triphosphate of the initiating nucleotide, as well as removal of the precursor-specific nucleotides that comprise its 3′ trailer sequence. The 5′ leader sequences of tRNA gene products vary in length and sequence even between different copies of one kind of tRNA gene within one organism (see, for example, References 1 and 21). The cleavage that liberates the mature 5′ terminus may be performed by an RNase P-like enzyme and occur in several steps.[22,23] Maturation also includes base modification, in a sequential and orderly fashion as illustrated in Figure 2, and removal of an IVS, when present, apparently a late processing step that occurs when mature 5′ and 3′ termini have been created and the majority of base modifications performed.

C. Splicing in Soluble Yeast Extracts

Xenopus oocyte nuclei have proved extremely valuable in elucidating the pathways of pre-tRNA processing, but greatest progress towards understanding the splicing reaction itself, and the enzymes that mediate it, has been made using a homologous system. As we have seen, the yeast rna-1 mutation leads to an accumulation of those tRNA precursors that possess an IVS.[9,13] Such mutant cells thus provide a relatively convenient source of unspliced tRNA species, which can be resolved from one another and purified in two-dimensional acrylamide gels. When pre-tRNA^{tyr} or -tRNA^{phe} purified in this way are incubated with extracts of wild-type yeast cells, they are converted to 4S molecules the size of the corresponding mature tRNA species. Upon RNAase T1 digestion, the 4S products no longer yield oligonucleotides that are characteristic of the IVS but rather those diagnostic of the mature tRNA species.[13,14] This result establishes that the yeast extracts employed possess both an activity that will excise the IVS precisely and one that can perform the ligation necessary to create a mature tRNA molecule.

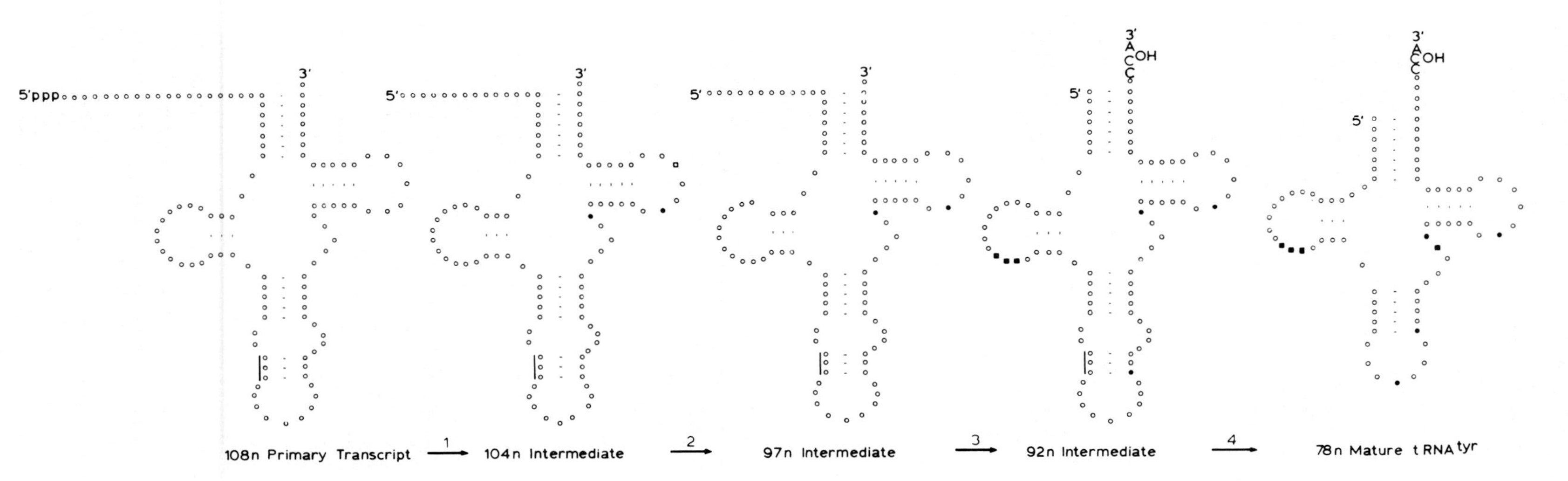

FIGURE 2. Processing of yeast pre-tRNA^{tyr}. The 108 nucleotide pre-tRNA^{tyr} includes 5′ flanking and 3′ trailing sequences and a 14 nucleotide IVS. The secondary structure in which this precursor is depicted is that given by DeRobertis and Olson.[21] In step 1, trimming of the 5′ end, including the initiating triphosphate, and some base modifications occur. Further 5′ end trimming (steps 2 and 3) and base modifications (step 3) then take place. In step 3, the 3′ trailing sequence is also removed and replaced by the CCA_{OH} terminus characteristic of mature tRNA. Finally, (step 4) the IVS is spliced out of the 92 nucleotide precursor, which is matured by additional base modifications. In all parts of the figure, the position of the anticodon is marked by a line and X, ●, ○, and ■ represent 5′-methylcytosine, pseudouridine, methyladenosine, and dihydrouridine, respectively. This scheme is based upon data reported by DeRobertis and Olson[21] and Melton et al.[22]

These experiments with crude extracts initially generated some confusion about the location within the cell of the enzymes that splice pre-tRNA, found by different authors to reside in a cytoplasmic, ribosomal wash fraction,[13] or in nuclear extracts.[14] The former result and the frequently observed partitioning of pre-tRNA species between nuclear and cytoplasmic fractions, probably reflect the difficulty of obtaining perfect fractionation and more especially of preventing leakage of nuclear components; it is now firmly established that splicing of pre-tRNA molecules is indeed a nuclear event. Maturation of a 92 nucleotide precursor to yeast tRNA^{tyr} that bears mature 5′ and 3′ termini but retains its IVS is observed only when the molecule is injected into the nuclei of *Xenopus* oocytes: no splicing is observed following injection into the cytoplasm or into enucleated oocytes.[22] Experiments employing extremely careful separations of the nuclear envelope and nuclear contents of oocyte nuclei have led to a similar conclusion.[24]

Despite the use of crude yeast extracts in the initial experiments, an absolute requirement of the splicing reaction for both divalent metal ions and ATP could be demonstrated.[13,14,25] In such systems, splicing is inhibited in the presence of mature tRNA, whether this be homologous or heterologous to the pre-tRNA substrate supplied. This latter observation suggests that one enzyme or enzyme-system mediates splicing of all yeast tRNA precursors that contain an intervening sequence.

Interestingly, the requirement of the in vitro splicing reaction for ATP provides a means to uncouple excision, cleavage at either end of the IVS, from ligation, rejoining of the novel termini of the precursor created by excision.[25] During a 60 min incubation of pre-tRNA^{phe} with the yeast ribosomal wash fraction in the presence of ATP, splicing goes essentially to completion. But at early times during the reaction, RNA products that are smaller than the mature tRNA, in fact the size of half-molecules, appear transiently and with kinetics that suggest they are intermediates in the splicing process. Such smaller species also accumulate when production of mature tRNA is inhibited by removal of ATP from, or addition of mature tRNA to the reaction mix.[25] Fingerprinting of such smaller molecules generated from pre-tRNA^{phe}, -tRNA^{tyr}, -tRNA$^{ser}_{UCG}$, -tRNA$^{leu}_{3}$, or -tRNA^{trp} under such conditions in vitro has established that they do indeed comprise 5′ and 3′ half-molecules of the respective tRNA and contain no sequences of the IVS.[26] The appearance of these half-molecules early in the in vitro reaction and their accumulation when ligation is inhibited suggests that the excision and ligation steps of pre-tRNA splicing are not of necessity coupled. Moreover, when purified half-molecules, for example, those cut from pre-tRNA^{phe}, are incubated with yeast extracts in the presence of ATP they are joined correctly, as judged by the oligonucleotides produced upon RNase T1 digestion, to yield mature tRNA^{phe}. Thus, it would appear that half-molecules are normal intermediates of the splicing process.

The ability to uncouple excision from ligation in vitro has permitted a detailed characterization of intermediates of the splicing reaction, the half-molecules described in the previous paragraph and the excised intervening sequence itself, which accumulates in parallel with half-molecules,[25,26] information from which several important features of the splicing reaction have been deduced. The intervening sequence is, for example, excised as in linear molecule, the result expected of the kind of two stage excision-ligation process described in the previous paragraph and illustrated in Figure 3. This is a significant observation, for it clearly eliminates models of pre-tRNA splicing in which the enzyme(s) switches phospho-diester bonds at the junctions between coding and intervening sequences to produce in one step the mature tRNA species and a circular intervening sequence.

The half molecules that are the products of the excision step carry, as illustrated in Figure 3, 3′-phosphate and 5′-hydroxyl termini. Thus, the splicing ligase appears to be unique, for all other known ligases require the converse termini, 3′-hydroxyl and 5′-

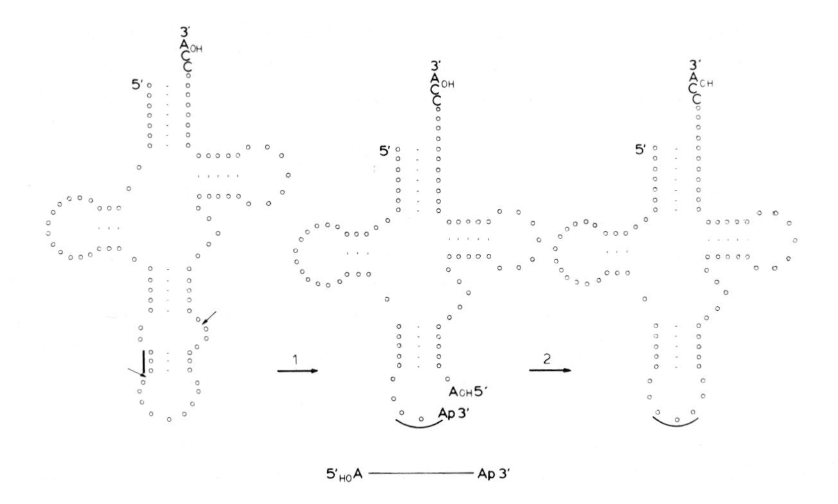

FIGURE 3. Two two-step splicing reaction of yeast pre-tRNA^{tyr}. The yeast tRNA^{tyr} splicing precursor of 92 nucleotides is depicted as in Figure 2. In the first step of splicing, two endonucleolytic cleavages occur at the sites indicated to liberate the IVS as a linear molecule and 5′ and 3′ half-molecules held-together by H-bonding. In the ATP-dependent step 2 the novel 3′ phosphate and 5′ hydroxyl termini of this splicing intermediate are ligated. This scheme is based upon the results of Peebles et al.[25] and Knapp et al.[26]

phosphate. The 5′-hydroxyl and 3′-phosphate termini of tRNA half-molecules have been shown to be essential for their religation.[26] Intermediates generated during splicing of *Tetrahymena* pre-RNA also carry such unusual termini (see Section C) and an enzyme that will ligate RNA molecules carrying 3′-phosphate and 5′-hydroxyl termini has recently been observed in wheat germ extracts.[28] It will be of considerable interest to learn whether the requirement for such termini is a general property of splicing ligases, and their creation that of splicing endonucleases; it could, for example, be imagined that such unusual termini serve to distinguish substrates of the splicing ligase from other RNA molecules.

The specific cleavages that are necessary to excise the IVS occur in regions that have been inferred to be single-stranded and in all five pre-tRNA species studied, a splice point is located directly adjacent to that nucleotide in the anticodon stem that is usually hypermodified in mature tRNA, such that the substrates of the ligation reaction are structurally very similar. Moreover, the structure of these intermediates would seem to be such as to facilitate the close proximity of the nucleotides to be joined (see Figure 3). It is hardly surprising that the nature of sequence of the intervening sequence has no influence upon the splicing-ligation of yeast tRNA, for it is removed before this reaction occurs (see Figure 3). Neither, however, does it appear to provide recognition elements for the splicing endonuclease: in cases, for example, where different intervening sequences are present in different copies of one kind of tRNA gene, for example yeast genes specifying tRNA^{phe} and tRNA^{tyr}, the in vitro splicing system makes no discrimination among the different intervening sequences. More dramatically, it will also remove normally from pre-tRNA$^{leu}_{3}$ an IVS into which 21 bp of DNA of the lac operator of *Escherichia coli* have been inserted.[29]

It therefore seems clear from all that is known about the nature of pre-tRNA molecules carrying intervening sequences and the properties of the reactions whereby they are spliced in yeast extracts, that it is the secondary, and therefore presumably tertiary, structure of these molecules rather than their primary sequence that is recognized by splicing enzymes. In this context, it is worth noting that in a tertiary structure of the kind found in those tRNA species that have been examined by X-ray diffraction,[30,31] the two splice points of a pre-tRNA containing intervening sequences would lie fairly close together on the same side of the molecule. Examination of the structure and splicing of abnormal tRNA precursors that accumulate in certain yeast mutants has emphasized the importance of structure in the splicing reaction: in one such mutant, a122, a pre-tRNA^{tyr} species bearing mature 5′ and 3′ termini but retaining the 14-nucleotide-long IVS accumulates in sevenfold greater concentration than in the parental strain.[32] The precursor is mutated by an A→G transition at the 5′-splice junction. Nevertheless the mature t-tRNA^{tyr} made in a122 cells, albeit in decreased amounts, is spliced correctly. Thus, this base change affects the rate rather than the fidelity of this splicing reaction, presumably as a result of interference with the recognition of the substrate by the enzyme. Other mutations that similarly influence the rate but not the fidelity of the splicing reaction also lie in regions of the pre-tRNA molecule distant from the IVS,[32] a result that suggests that features retained in the mature tRNA, presumably structural, as well as those of the splice junctions, are scanned by the splicing enzyme(s).

D. Enzymes that Splice Yeast tRNA Precursors

The two step excision-ligation mechanism of pre-tRNA splicing in yeast immediately raises the question of whether both steps are performed by a single enzyme, or enzyme complex, or whether the endonuclease specific for pre-tRNA and the unusual RNA ligase exist as separate physical entities in the cell. It now seems clear that in fact the endonuclease and ligase are distinct enzymes: the splicing ligase can be extracted quite efficiently from yeast cells in the presence of high salt, whereas the endonuclease activity remains insoluble under such conditions.[33] It can, however, be released by exposure to nonionic detergents. Further support for the notion that the splicing endonuclease may be membrane-bound comes from the observation that this enzyme bands at the density of light yeast membrane when crude particulate preparations from yeast are subject to equilibrium sucrose-density gradient centrifugation.[33] Although the solubilized endonuclease and ligase have both been partially purified, the ligase to quite a considerable degree, neither has yet been obtained in homogenous form. Consequently, nothing is yet known of their composition or interaction, although it has been established that separated and the partially-purified endonuclease and ligase are both necessary and sufficient for splicing of pre-tRNA species in vitro.[33]

An enzyme (or enzymes) that will correctly splice yeast pre-tRNA^{tyr} and -tRNA$^{leu}_{3}$ has also been partially purified, some 50-fold, from extracts of *Xenopus* oocytes nuclei.[34] The nature and structure of the intermediates of the splicing reaction catalyzed by such preparations are identical to those observed with the yeast enzymes. Similarly, too, the *Xenopus* enzyme(s) requires Mg^{2+} and is inhibited by mature tRNA. It is not inactivated by micrococcal nuclease treatment,[34] suggesting it does not contain an essential RNA component, analogous to RNAase P of *E. coli*.[35,36] Correct splicing of a yeast tRNA$^{leu}_{3}$ precursor that retains both the IVS and 5′ and 3′ flanking sequences (see Figure 2) has also been observed in a HeLa cell transcriptional extract[37] in a reaction that proceeds via half-molecule intermediates to yield correctly spliced tRNA$^{leu}_{3}$ and the IVS.[38] It therefore seems clear that the enzymatic machinery that mediates splicing of tRNA precursors is highly conserved among eukaryotes. Whether or not the *Xenopus* and HeLa endonuclease and ligase activities reside in separable proteins is, however, not yet established.

In summary, it is fair to say that a great deal about the splicing reaction, for example the nature of the intermediates and the mode of substrate recognition, has been learned from the study of splicing pre-tRNA in vitro. The availability of such soluble systems also provides the opportunity to purify and characterize the splicing enzymes: some steps towards this goal have been taken, with interesting results, but there remains a long way to go to attain a complete molecular understanding of this reaction.

III. SPLICING OF INTERVENING SEQUENCES FROM rRNA PRECURSORS

A. Occurrence and Structure of Eukaryotic rRNA Genes with Intervening Sequences

The organization and expression of genes encoding the two large rRNA species (rDNA) has been amenable to study for some time: rRNA genes are moderately reiterated in every eukaryote that has been examined (see Long and Dawid[39]), are frequently clustered in tandem arrays and in some organisms, such as *X. laevis,* can be separated from the bulk of the chromosomal DNA by virtue of their high bouyant density.[40,41] The rRNA species themselves can be obtained in high quantity and easily purified. Despite these advantages, that provided an early opportunity to elucidate the organization and expression of rDNA sequence, the presence of intervening sequences in rRNA genes was not reported until 1977, when they were observed in rDNA units of *D. melanogaster* by several groups of investigators.[42-45] Some 66 and 16% of the rDNA units present in the nucleolar organizers of the X and Y chromosomes, respectively, carry an insertion, which always lies in a similar position, toward the 3′ end of the region encoding 28S rRNA.[42-46] The rDNA insertion sequences are themselves heterogeneous, both in length and sequence.[43,44] The most predominant type of insertion found in rDNA units of the X-chromosome of *D. melanogaster* is, as illustrated in Figure 4, 5.0 kb in length. All rDNA units that include either this 5.0 kb insert or much shorter insertion sequences, 0.5 to 1.0 kb in length, that are homologous to sequences present in the 5.0 kb form have been designated class 1.[47,48] The remaining rDNA insertion sequences, class 2, are homologous neither to class 1 sequences nor necessarily to one another, but characteristically carry an Eco R1 site. Such class 2 intervening sequences predominate in rDNA units present in the Y-chromosome.[46]

Intervening sequences have also been detected in some but not all copies of 28SrRNA coding regions of *D. virilis*[49] and the higher dipteran fly *Calliphora erythrocephela,*[50] but DNA isolated from lower diptera does not hybridize to *Drosophila* ribosomal insertion sequences. The nucleolar rRNA genes of species as diverse as yeast, *Xenopus* and man contain no intervening sequences[49-51] nor apparently do the mitochondrial rDNA sequences of higher eukaryotes.[52-54] However, rDNA intervening sequences are not restricted to certain dipteran flies, but have also been observed in the slime mold *Physarum polycephalum* and certain strains of the ciliated protozoan *Tetrahymena.* In both these organisms, genomic rDNA units are organized in about 150 palindromic extrachromosomal DNA molecules, each of which carries two copies of the rRNA coding sequences.[55,56] As illustrated in Figure 4, the *Physarum* DNA sequences that encode the larger (26S) rRNA species contain two intervening sequences, of 0.7 and 1.2 kb, that divide the coding sequence into segments of 2.4, 0.7, and 0.6 kb.[58-60] At least 88% and most probably all the rDNA units carry such intervening sequences in the two strains of *Physarum* whose rDNA has been examined. At least four strains of *T. pigmentosa* and one of *T. thermophila* possess an 0.4 kb IVS in that region of their macronuclear rDNA units encoding the larger rRNA species (23S)[61,62] (see Figure 4). Two strains of *T. pigmentosa* that have no such rDNA IVS have also been described, but within a strain all rDNA units are homogenous.

The only other reported examples of rRNA genes that contain an IVS have, perhaps surprisingly, been found in the DNA molecules present in cytoplasmic organelles of certain lower eukaryotes. The two copies of the sequences encoding the large rRNA species (23S) of *Chlamydomonas reinhardii* chloroplasts, for example, each contain an 0.87 kb IVS located 0.27 kb from the 3′ end of the coding sequence (see Figure

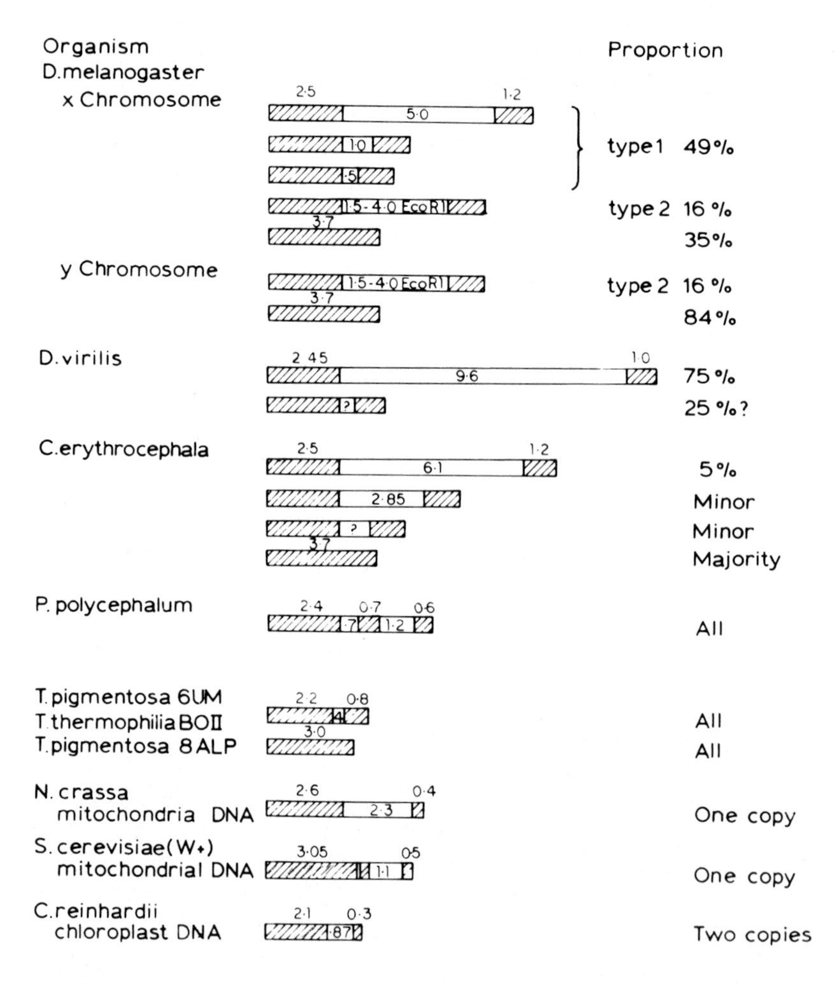

FIGURE 4. Structure of genes encoding large rRNA that contain intervening sequences. For each organisms listed at the left, the structure of the region encoding the large rRNA species is shown by the hatched and open boxes, representing coding and intervening sequences, respectively. In all cases, the structure is drawn such that the 5′ end of the rRNA is to the left. The sizes of the coding and intervening sequences are given above and within the hatched and open regions, respectively. The sources of these data are given in the text.

4).[63] In both *Chlamydomonas* and *Zea mays* chloroplast DNA, the two copies of the rDNA unit are organized in an inverted repeat; the 23S rRNA coding sequences are not, however, interrupted by an IVS in chloroplasts of the latter species,[64] indicating that intervening sequences are not ubiquitous among plant chloroplast rRNA genes. The same can be said of intervening sequences in mitochondrial rRNA genes; to date these have been observed in mitochondrial DNA of only *Neurospora crassa* and certain strains of yeast. The one copy of the sequence encoding the larger (24S) rRNA species in *N. crassa* mitochondrial DNA contains an IVS of 2.3 kb dividing the coding region into segments of 2.6 and 0.4 kb (see Figure 4).[65-67] In those strains of *Saccharomyces cerevisiae* that exhibit polarity for recombination between markers in the region of mitochondrial DNA encoding the larger rRNA species (21S), both an 1.1 kb and a 66 bp IVS are present within the 21S rRNA coding region.[68-72] The presence of the very small IVS was deduced from sequencing data and an analogous IVS may in fact be present in the large rRNA coding regions of the DNA of the other cytoplasmic organelles discussed previously, whose organization has been examined only by restriction enzyme digestion and R-loop electron microscopy. Although the number of fully characterized rDNA units is not enormous, it seems fair to conclude from the currently available data that intervening sequences are present in neither genomic nor organelle DNA of chordates and higher plants. They may be present in either genomic or organelle DNA of lower eukaryotes, but we currently possess too small a sample to discern a coherent pattern in their distribution.

Inspection of the structures of the currently recognized discontinuous rDNA units summarized in Figure 4 reveals no obvious uniformity in the number, location or size of their intervening sequences, beyond the tendency for an IVS to lie near the 3′ end of the region encoding the larger rRNA species. It is, however, obvious that the presence of intervening sequences is not mandatory for the expression of rRNA genes and in some cases, discussed subsequently, may lead to their inactivation. The occurrence of intervening sequences in rRNA genes present both in the genome and the DNA of cytoplasmic organelles may imply the existence of at least two distinct systems for processing rRNA transcripts that contain intervening sequences. However, although rRNA genes were the first recognized to contain intervening sequences, progress towards understanding the mechanisms whereby their transcripts are processed has been relatively slow. Such slow progress seems in part to reflect the variability in rRNA gene organization illustrated in Figure 4, but may also be historical: the first discontinuous rDNA units to be discovered, those of *D. melanogaster,* do not appear to be expressed as functional products.

B. Expression of Discontinuous rDNA Units

It is obvious upon inspection of Figure 4 that a significant fraction of the several hundred copies of *D. melanogaster* rDNA units contain no insertion. Moreover, were all the insertion sequences that have been observed in these rDNA units to be transcribed, then a heterologous population of pre-rRNA molecules must result, a prediction in marked contrast to the homogeneous pre-rRNA species actually observed.[73,74] The inference that *D. melanogaster* rDNA units that contain an insertion are transcriptionally inert has received strong support from the results of hybridization experiments with cloned DNA segments that contain only such insertion sequence DNA: in such experiments, nuclear RNA species as large as 10 kb that contain type 1 sequence transcripts can be detected in embryos[75] or a *Drosophila* cell line.[76] Such transcripts are, however, present at a concentration of only 1 to 3 copies per nucleus, by contrast to the concentrations of some 5000 copies per cell of each of the mature rRNA species. The concentration of RNA sequences complementary to type II insertions is similarly very low.[77] Moreover, although transcripts of the insertion and 28S rRNA coding sequences are linked within individual RNA molecules of up to 10 kb in length, no molecules large enough to comprise transcripts of an rDNA unit that contains the most common, type 1, IVS insertion, 5.0 kb, have been observed.[75] Finally, such transcripts hybridize only to regions comprising the left part of this intervening sequence.[75] It has not proved especially difficult to detect complete transcripts of rDNA units that contain an IVS in other organisms, so it must be concluded that transcription of discontinuous rDNA units does not make a significant contribution to rRNA synthesis in *D. melanogaster,* but probably reflects rare transcriptional aberrations. This conclusion is reinforced by the observation that three cloned rDNA units each containing the major type 1 insertion sequences all have small deletions of the 28S coding sequence at the 5′ junction between coding and insertion sequences.[78] They may also possess other aberrations, compared to the sequence of an uninterrupted rDNA unit.[78] Indeed, several properties of these *Drosophila* insertion sequences, including their presence at numerous chromosomal locations in addition to the nucleolar organizer[79,80] and the presence of 11 to 14 bp directly-repeated sequences flanking their ends,[81] suggest they are derived from transposable elements.

Transcriptional inactivation cannot, however, pertain in those organisms in which all copies of the rDNA unit include an IVS. Indeed, transcripts of such units containing copies of the intervening sequences have been observed. Interestingly, though, there appears to be quite considerable variation in the time during processing at which the IVS is removed by splicing or, in other words, splicing does not exhibit a constant temporal relationship to the other steps that mediate the production of mature rRNA species from the primary product of transcription, such as the endonucleolytic cleavages necessary to liberate the individual rRNA species. In yeast mitochondria, for example, splicing appears to be a late processing event, for a pre-21S rRNA species that comprises the sequences of mature 21S rRNA, additional 3′ trailing sequences and sequences of the 1.2 kb IVS has been described.[82] Other forms of the precursor in which either trimming of the 3′ end or splicing had occurred could also be found, suggesting that the order of the final processing steps in the maturation of 21S rRNA in yeast mitochondria is not absolutely fixed. Splicing is also a late step in the processing of *N. crassa* mitochondrial pre-rRNA[69,83] and of tRNA and mRNA precursors in eukaryotic cells (see Sections II.B and C). Nevertheless, late removal of the IVS from pre-rRNA species is not ubiquitous: splicing of pre-rRNA species that contain transcripts of an IVS has been examined in one other organism and in this case splicing takes place early in processing.

Three species of rRNA precursors can be isolated from *Tetrahymena* macronuclei and exhibit molecular weights of 2.2 to 2.3 $\times$ 10^6, 1.4 $\times$ 10^6, and 0.66 $\times$ 10^6 daltons. These represent the complete transcript of the rDNA unit, pre-26S rRNA and -17S rRNA, respectively.[84] The only transcript to contain sequences complementary to the 0.4 kb IVS is the largest.[61,84] Thus, removal of the IVS by a splicing reaction must take place before the cleavages that release pre-26S rRNA and -17S rRNA from the initial transcript. The half-life of this primary transcript is very short, considerably less than a minute,[85,86] implying that the transcribed intervening sequence must be removed very rapidly. It will be of considerable interest to learn whether this difference in the pathways of processing of nuclear and mitochondrial pre-rRNA transcripts actually reflects different mechanisms of splicing.

C. Splicing of Pre-rRNA In Vitro

To date, splicing of an rRNA precursor has been described in only one in vitro system. When *Tetrahymena* macronuclei, or nucleoli prepared from them, are incubated under appropriate conditions in vitro, transcription and splicing occur, as evidenced by the appearance of a discrete 0.4 kb RNA species.[87,88] This RNA species can comprise as much as 6% of the total RNA made in vitro and hybridizes only to restriction endonuclease fragments derived from the rDNA IVS. Synthesis of the 0.4 kb RNA complementary to the IVS occurs when synthesis of all nonribosomal RNA is inhibited by α-amanitin and only in nuclei isolated from those strains of *Tetrahymena* whose rDNA include an IVS. The IVS is excised as a linear molecule but becomes converted to a circular form during incubation of isolated nuclei in vitro.[89] The same reaction seems to occur in the cell because both linear and circular forms of the transcript of the 0.4 kb IVS are found in *Tetrahymena* macronuclear RNA preparations.[27] Further characterization of the excised intervening sequence itself as well as its mode of excision from rRNA precursors in vitro has led to the formulation of a coherent model for the mechanism of splicing of *Tetrahymena* pre-rRNA: this mechanism is quite different from that discussed in section II.C for pre-tRNA splicing.

When the sequence of the 0.4 kb IVS RNA excised in isolated *Tetrahymena* was compared to that of the corresponding region of rDNA,[27] it became apparent the excised IVS carries a 5′-terminal pG residue that is not represented in the genome, but is otherwise colinear with the DNA sequence. Elucidation of the source of this additional

5′-terminal nucleotide has been possible through the ability to uncouple splicing from transcription in isolated *Tetrahymena* macronuclei: in the presence of a low concentration of monovalent cation 5 m*M* $(NH_4)_2SO_4$, transcription of rRNA genes proceeds at close to optimal rates, but excision of the IVS from the transcripts thus made is substantially inhibited.[90] The pre-rRNA species synthesized under such conditions will, when purified and incubated in vitro in the presence of 120 m*M* $(NH_4)_2SO_4$, excise the intervening sequence in an apparently spontaneous reaction, i.e., in the absence of nuclear extracts, an unexpected observation.[90] The only requirements for this excision reaction are monovalent and divalent cations and a guanosine compound, GTP, GDP, GMP, or G.[90] Because the reaction will accept GMP or GDP, the nucleotide dependence clearly does not reflect an energy requirement. Only guanosine compounds with unblocked 2′ and 3′ OH groups are active in this reaction and Cech and colleagues[90] have demonstrated that the G compound becomes covalently attached via a phosphodiester bond to the 5′-end of the IVS during excision. In in vitro reactions, the precise 5′-terminal sequence of the excised IVS is determined by the guanosine cofactor supplied, (that is, GTP and GMP, for example, yield the 5′ terminal sequences 5′ pppGpAp. . . and 5′pGpAp. . ., respectively, in the excised IVS), but the IVS released following coupled transcription-splicing in isolated nuclei carries the sequence 5′pGpAp. . ., suggesting that GMP is normally the preferred cofactor.

Direct evidence that the in vitro reaction that leads to excision of the IVS from the intermediate accumulated under conditions of low monovalent cation concentration also leads to ligation of the 26S rRNA coding segments is presently lacking. No discrete RNA fragment that corresponds in size to the 3′-terminal portion of the 26S rRNA coding sequence that would be generated by excision of the IVS in the absence of ligation (1070 nucleotides, see Figure 4) is observed when the products of the in vitro reaction are analyzed by electrophoresis[90] but it might not be readily distinguished in the gel system employed. Nevertheless, ligation can take place under these conditions for the linear IVS excised from the purified intermediate in the presence of 120 m*M* $(NH_4)_2SO_4$ and GTP can be converted to the circular form.[90] It therefore seems quite probable that the two segments of the 26S rRNA coding sequence are also ligated in vitro under these conditions, although a more direct demonstration of such a ligation reaction is required.

These features of the in vitro splicing of *Tetrahymena* pre-rRNA suggest a model in which the relevant enzyme acts as a phosphotransferase.[90] Thus, as illustrated in Figure 5, the guanosine compound cofactor would provide a 3′-hydroxyl terminus to which the 5′ phosphate group of the IVS, generated by a specific endonucleolytic cleavage, could be transferred. Such a cleavage might precede or be concomitant with (and dependent upon) transfer. Insufficient information about the structure of the intermediate that accumulates when excision is inhibited is available to distinguish between these possibilities. Thus, this intermediate could be intact or cleaved at the splice junction between coding sequences and the 5′ end of the IVS. The latter possibility seems more likely, for reasons discussed below. A second phosphoester transfer (line C in Figure 5) could then liberate the IVS and join the ends of the coding segments, while a third (D) would circularize the excised intervening sequence. By means of such a mechanism, one enzyme could accomplish the endonucleolytic cleavages and ligation that constitute splicing. Moreover, because the two steps are coupled, any requirement for ATP or GTP is circumvented. Such a series of phosphotransferase reactions is clearly very different from the two-step cleavage and ligation mechanism by which pre-tRNA species are spliced, but is not too dissimilar from reactions performed by DNA nicking-closing enzymes, such as *E. coli* DNA topoisomerase I[91,92] or DNA gyrase[93] neither of which exhibit any energy requirement.

Although the mechanism of *Tetrahymena* pre-rRNA splicing outlined in Figure 5 accounts for the generation of all the products actually observed and is aesthetically

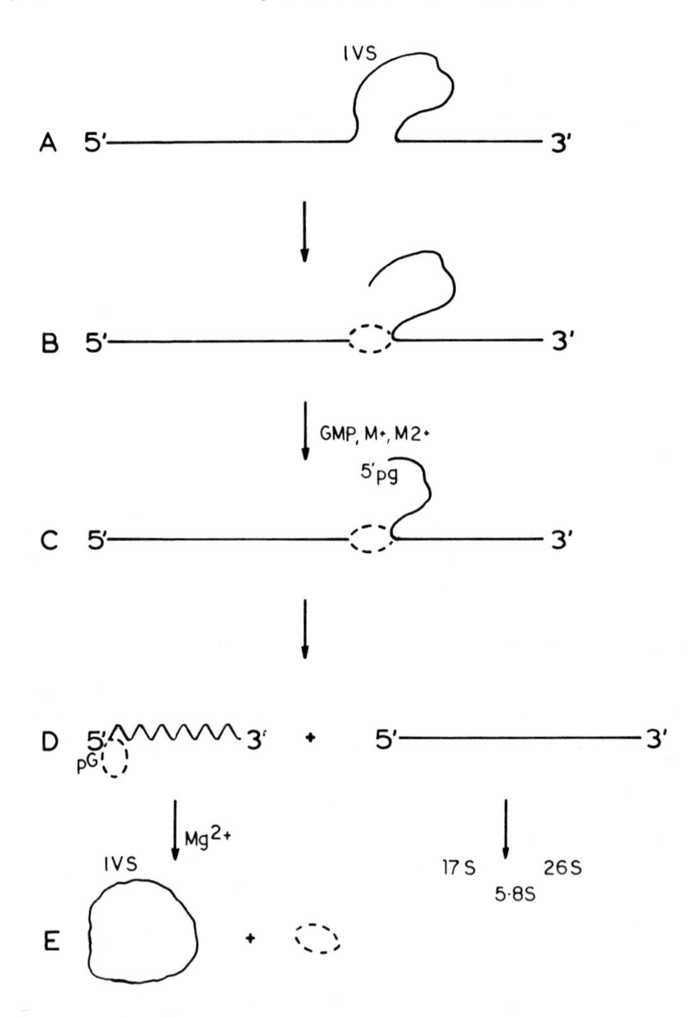

FIGURE 5. Mechanism of splicing of *Tetrahymena* rRNA precursors. The *Tetrahymena* precursor rRNA, in which the IVS, exaggerated in size is represented as ∿∿ is shown in line A. Binding of the splicing enzyme ⬭ and cleavage at the 5′ splice junction creates the intermediate shown in line B and probably isolated by Cech et al.[92] There then follow three phosphotransferase reactions, the first requiring a G compound to which the 5′ end of the IVS is transferred (product C), the second to splice the 26S rRNA coding sequences and excise the IVS (product-D) and finally to circularize the excised IVS (E). This summary is modified from that given by Cech et al.[92]

satisfying, it does not address the question of the nature of the enzyme responsible for performing these reactions. This enzyme has not been identified and no direct physical evidence for its existence has yet been collected. Indeed, the observations that have been made suggest that such an enzyme must be quite peculiar.

The intermediate from which the IVS can be excised in the presence of a guanosine compound and monovalent cations was purified from isolated nuclei by standard deproteinization methods followed by centrifugation in denaturing gradients.[90] The specificity of the reactions performed when the purified intermediate is incubated in vitro and their dependence upon a guanosine compound have been advanced to support the notion that the enzyme that performs these reactions is present in the intermediate. Such an enzyme must be so tightly bound as to survive not only the purification procedures mentioned, but also boiling in the presence of SDS or β-mercaptoethanol.[90] Retention of *activity* after such treatments is certainly unusual, but a covalent linkage between the splicing protein and its substrate would certainly survive. More surprising still, is the resistance of the ability of the purified intermediate to excise the IVS to proteases.[90] Clearly much further work is required to establish that a splicing enzyme is indeed associated with the purified intermediate and to establish the nature of this association. The most plausible scenario at the present time must be that the intermediate in fact represents a transcript that has undergone the first endonucleolytic cleavage to form a molecule in which the 5′ end of the IVS is covalently attached to the splicing enzyme, that is, the intermediate isolated is trapped in the act of the first phosphoester transfer from the 5′ end of the IVS to the guanosine compound. Clear precedent for such intermediates in which nucleic acids become covalently attached to enzymes are provided by the covalent complexes between DNA and DNA topoisomerases formed when a nicking-closing reaction is interrupted[93] and by the A protein of ΦX174, an enzyme that initiates replication of the RF form by introduction of a single-stranded break at a specific site and remains covalently bound to the 5′-terminus thus created.[94,95] The structure of the complex could well contribute to some of the enzymes unusual properties, e.g., its resistance to proteases. It is to be hoped that direct physical evidence for the presence of such a protein in the intermediate will be forthcoming in the near future.

A second important question that is not addressed by the model depicted in Figure 5 concerns the mechanisms whereby the splice junctions are actually recognized and aligned to facilitate ligation. The structure of pre-tRNA molecules than contain an IVS is, as discussed in Section II, such that this problem does not arise; the residues that must ultimately be covalently joined by splicing are held in close proximity (see Figure 3). It seems, unlikely, however, that pre-rRNA species or mRNA precursors for that matter, possess sufficient secondary structure to align splice junction sequences. Small RNA species have been postulated to fill this role in pre-mRNA splicing (see Section IV.C) and some adapter molecule may also be necessary to build an appropriate pre-rRNA substrate structure.

The sequences surrounding the junctions between coding segments and intervening sequences are currently available for rRNA genes of *T. pigmentosa,*[96] *T. thermophilia,*[90] *P. polycephalum,*[60] *S. cerevisiae* mitochondria,[72,97] and *C. reinhardii* chloroplasts and are summarized in Figure 6. The most striking feature of these sequences is a negative one, the lack of an obvious, strong homology among all the sequences listed. Some short regions of homology are apparent, for example a similar sequence in the 5′ rRNA coding segment of the two *Tetrahymena* splice junctions and those of the *Physarcum* and yeast mitochondrial splice junctions, underlined in Figure 6. Nevertheless, there is no sequence common to all the rDNA splice junctions, beyond the occurrence of T and G residues at the 3′ end of the 5′ coding segment and at the 3′ end of the IVS, respectively. This situation is in complete contrast to that observed when a very large number of mRNA splice junction sequences are compared, for all approximate closely to one pair of consensus sequences (see Section IV.B). Such heterology among rDNA splice junction sequences indicates that it is unlikely that one adapter molecule could serve to align all recognized rDNA splice junctions. Even when the sequences of the mitochondrial and chloroplast splice junctions are omitted from the comparison (on the basis that their splicing might be mediated by distinct activities to those present in the nucleus), the picture does not change radically.

The conserved sequences bordering the *Tetrahymena* and other rDNA splice sequences indicated in Figure 6 are complementary to the amino-acid acceptor stem of the methionine initiation tRNA and to sequences of the nucleolar U3 RNA species of the rat[99] and its *Dictyostelium* analogue.[100] It therefore remains possible that small

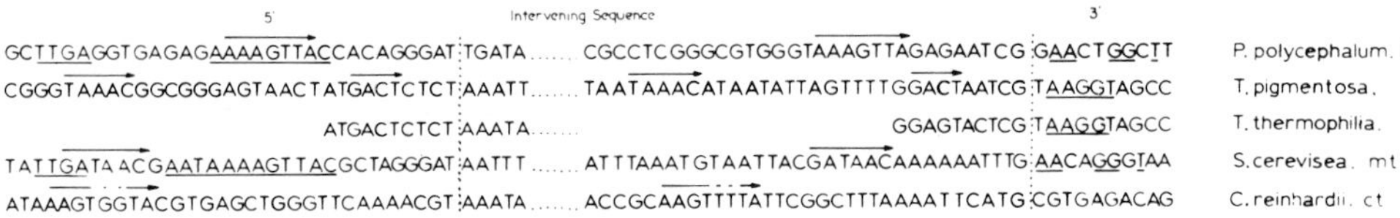

```
                       5'                          Intervening Sequence                        3'
GCTTGAGGTGAGAGAAAAGTTACCACAGGGAT:TGATA...... CGCCTCGGGCGTGGGTAAAGTTAGAGAATCG:GAACTGGCTT   P. polycephalum. I
CGGGTAAACGGCGGGAGTAACTATGACTCTCT:AAATT...... TAATAAACATAATATTAGTTTTGGACTAATCG:TAAGGTAGCC  T. pigmentosa.
                     ATGACTCTCT:AAATA......                        GGAGTACTCG:TAAGGTAGCC  T. thermophilia.
TATTGATAACGAATAAAAGTTACGCTAGGGAT:AATTT...... ATTTAAATGTAATTACGATAACAAAAAATTTG:AACAGGGTAA  S.cerevisea. mt
ATAAAGTGGTACGTGAGCTGGGTTCAAAACGT:AAATA...... ACCGCAAGTTTTATTCGGCTTTAAAATTCATG:CGTGAGACAG  C.reinhardii. ct
```

FIGURE 6. Sequences at splice junctions in rDNA. All sequences are drawn in the 5′ to 3′ direction with the boundaries between coding and intervening sequences marked by the dashed vertical line. Regions of homology between the sequences are underlined, whereas directly repeated sequences within a gene are indicated by the horizontal arrows drawn above the sequence. The sources of these data are given in the text.

RNA species may be responsible for juxtaposition of residues to be ligated during pre-rRNA splicing. There is at present no experimental evidence to support this view. It is also as well to keep in mind that pre-rRNA species become associated with proteins before they are processed[102-104] so it is presently equally plausible that such proteins serve to impart a suitable substrate structure to the rRNA molecule.

Interestingly, all rRNA intervening sequences whose nucleotide sequence has been determined are flanked by one, or more, sets of short, direct repetitions (see Figure 6). These are of the order of 4 to 7 bp in length and are located 5 to 30 nucleotides away from the 5′ and 3′ splice junctions. Whether such repetitive elements are diagnostic of a transposition of the IVS,[98] contribute to the formation of a secondary structure that would juxtapose the nucleotides to be ligated[97,98] or comprise a recognition element for the splicing enzyme(s)[60] remains to be established. Of course these possibilities are by no means mutually exclusive.

At present, then, our picture of pre-rRNA splicing is far from complete: the quite dramatic progress with in vitro splicing of pre-rRNA made recently has provided considerable insight into at least one rRNA splicing mechanism, certainly sufficient to indicate that it is quite different from that discussed previously for pre-tRNA splicing (see section II.C). It is not yet established that splicing of all pre-rRNA species, particularly those of mitochondrial or chloroplast origin, is mediated by phosphotransferase reactions. Indeed, the existence of this enzyme remains to be physically demonstrated. Finally, we currently possess no understanding of how splice junctions in pre-rRNA molecules are recognized or aligned.

IV. SPLICING OF mRNA PRECURSORS

A. The Occurrence and Structure of Genes Specifying Polypeptides that Contain Intervening Sequences

The great majority of genes of higher eukaryotic organisms that specify polypeptide products are constructed from a mosaic of coding and intervening sequences, as are those of many of their viruses (see Breathnach and Chambon[104] and Flint,[105] for reviews). The list of such discontinuous genes is now so long that it seems reasonable to conclude that the absence of an IVS is the exception rather than the rule, at least among chordate species. Example of those rare genes that contain no IVS include those encoding the histones,[106,107] the human interferons,[108-114] and coding sequences of peptide hormones such as ACTH, β-endorphin, α- and β-menotropin, corticotropin, and β-lipotropin.[115,116] Whether or not such a lack of an IVS has any significance, for example permitting unusually rapid expression of these genes, is not established at the present time.

Intervening sequences are also present in the genes of nonchordate eukaryotes, but appear to be less common. The alcohol dehydrogenase[117,118] and actin genes of *D. melanogaster* and the sea urchin,[119,120] for example, contain intervening sequences, but several other *Drosophila* genes, including those specifying most heat-shock proteins,[121,127] do not. Similarly, the presence of intervening sequences in the yeast genes specifying actin[123,124] and a certain ribosomal proteins[125] seems to be exceptional. Despite such apparent differences between higher and lower eukaryotes, the rather general distribution of intervening sequences in protein coding sequences implies that all eukaryotes possess the enzymatic machinery to mediate their removal and such machinery must permit absolutely accurate removal of an IVS, otherwise the coding sequences of the gene beyond the splice point would be destroyed by the splicing process.

B. Splice Junction Consensus Sequences

Many protein-coding genes that contain intervening sequences have now been characterized in considerable detail. It is apparent from the results of such studies that the number of intervening sequences with a gene varies from the obvious of minimum of one, for example, in a rat insulin gene,[126,127] in actin genes of yeast and *D. melanogaster*,[119,120,123,124] some ribosomal protein genes of yeast,[125] and silkmoth chorion genes[128] to a very large number indeed, 33 in the *X. vitellogenin* gene[129] or more than 50 in the chicken α2 type 1 collagen gene.[130,131] The size of any one intervening system that must be removed is also very variable: the smallest intervening sequences described are less than 100 nucleotides, for example that spliced from SV40 early pre-mRNA to fashion the mRNA for small T-antigen,[132-134] whereas the largest, for example in the adenoviral late transcript, are greater than 20 kb (see Flint,[105] for a review). Such an enormous range in the size permitted to intervening sequences implies that the internal segments of an IVS have little role in the splicing process itself, an inference that is confirmed by the number of observations.

Little homology, for example, has been observed in the size or nucleotide sequence of seven intervening sequences of the chicken α2 type 1 collagen gene;[135] the only conserved sequences in the intervening sequences lie at its termini, bordering the splice junctions. Similarly, the members of the α and β globin gene families possess two intervening sequences whose locations within the coding sequence have been maintained yet little conservation of the nucleotide sequences of the intervening sequences has been observed either among members of the α- or β-globin gene families within one species[136-140] or among different organisms.[137] More dramatically, the deletion of an IVS from SV40 early or late genes to within 10 to 12 nucleotide pairs of the splice junctions does not abolish synthesis of the mature, correctly-spliced mRNA species by splicing of the aberrant precursor.[141-146]

These observations pinpoint the splice junction sequences themselves as those essential to splicing. It is therefore not too surprising to find that naturally-occurring mutants of human globin genes that are not expressed in protein, either pseudogenes or genes carried by patients with certain types of thalassemia, differ in nucleotide sequence from their normal, expressed counterparts at splice junctions.[147-151] Similar splice junction aberrations are found in the genome of adenovirus mutants that fail to synthesize one of the three spliced mRNA species complementary to region E1A.[152,153] This is not, however, to deny any influence over splicing reactions to the internal portions of an IVS: the deletions in SV40 intervening sequences mentioned previously can alter the efficiency with which that intervening sequence is removed from an mRNA precursor,[144,146] an observation interpreted in favor of the notion that the conformation surrounding splice junction sequences is an important parameter.

When the nucleotide sequences of the large number of splice junctions now sequenced are aligned at the borders between coding and IVS segments, consensus 5′ and 3′ splice junction sequences can be derived, such that all intervening sequences in an mRNA precursor can be described by the structure shown in Figure 7.[154-157] Each

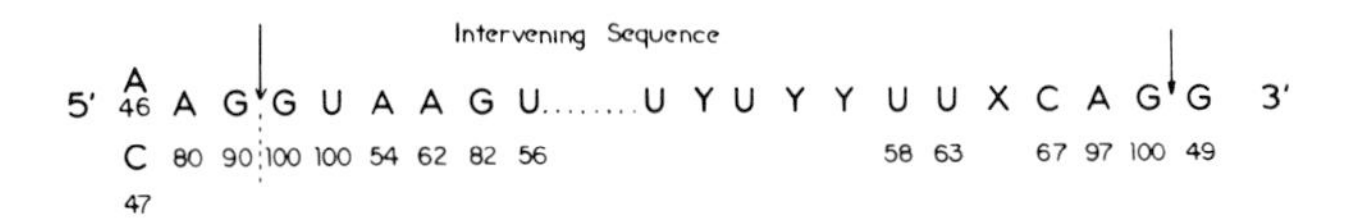

FIGURE 7. The eukaryotic mRNA precursor splice junction consensus sequences. The splice junction consensus sequences are shown in the 5′ to 3′ direction with the boundaries between coding and intervening sequences marked by the vertical arrow. Y and X represent any pyrimidine and any nucleotide, respectively. Below the sequences are shown the frequencies with which the residue shown occurs. The sources of these data are discussed in the text.

pair of splice junction sequences surrounding an IVS relate to these consensus sequences with one or two alterations, the most highly conserved feature being the occurrence of the dinucleotides GU and AG and the 5′ and 3′ ends, respectively, of the intervening sequence itself[158] (see Figure 7). It should be noted that the boundaries of the great majority of intervening sequences and thus the locations of the splice junctions have not been placed with final precision because of the presence of sequence duplications. That the GU-AG boundary rule can be more than a formality is, however, confirmed by the observation that the three intervening sequences of the chicken ovomucoid gene that possess no such duplications conform to the structure depicted in Figure 7.[159] These splices junction consensus sequences are universal down to yeast among animal species[125,126] and apparently are also present in plant genes, for example, the leghemaglobin gene of the soybean[160] and the gene encoding the major storage protein, phaseolin, of the French bean.[161]

The fact that consensus splice junction sequences can be deduced from the sequences of genes of a very diverse set of organisms implies that the splicing apparatus, whatever its nature, must be highly conserved. It is not too surprising, therefore, that numerous examples of trans-species splicing, that is, a pre-mRNA species of one organism spliced correctly when introduced into cells of a second organism, have been reported.[162-166] Although yeast cells failed to make a correctly spliced mRNA from a rabbit β-globin gene introduced in a plasmid vector the β-globin transcripts synthesized in these circumstances were defective in a number of respects.[167] Thus it does not necessarily follow that the yeast splicing machinery is incompatible with rabbit splice junction sequences.

The existence of such consensus sequences might also suggest that all 5′ and 3′ splice junction sequences should be interchangeable, i.e., that a chimeric RNA molecule containing an IVS whose 5′ segment comes from one mRNA species and 3′ portion from a second should be spliced normally, a prediction that has been confirmed with such a chimera constructed from SV40 and mouse β-globin sequences.[168] Nevertheless, as pointed out by Sharp[169] all intervening sequences are not necessarily equivalent, for they interrupt the coding sequence at different points in the amino-acid codon immediately before the splice junction. Assuming the GU-AG rule is followed absolutely, then examples of intervening sequences that interrupt the reading frame at all three possible positions, between the first and second nucleotides within a codon, between the second and third nucleotides of a codon and between codons, have been found (see Sharp,[169]). It might therefore be that each type of 5′ splice junction must be ligated to a compatible 3′ splice site, one that would recreate a complete codon, and thus maintain the integrity of the coding sequence. Incompatability of 5′ and 3′ splice junctions has not been observed experimentally, but as noted by Sharp[169] splicing of few chimeric genes has yet been examined. On the other hand, it may be that this failure is a consequence of the duplicated nucleotides that border splice sites, or in other words, that the GU-AG rule is not absolute. As illustrated in Figure 8, for an ovalbu-

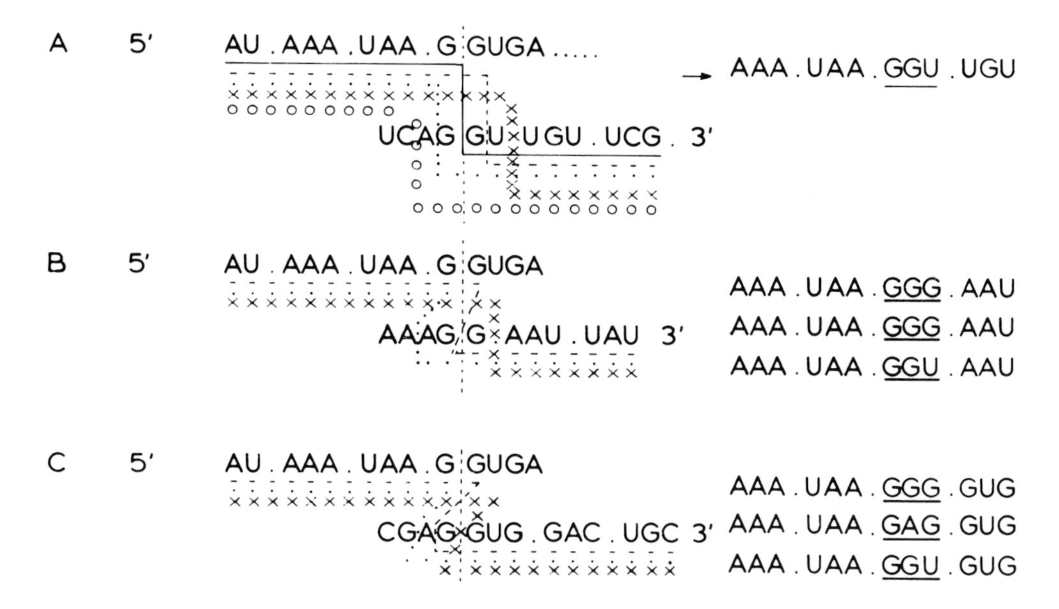

FIGURE 8. Flexibility in the site of ligation during splicing. Part A shows the 5′ and 3′ splice junction sequences surrounding one IVS of the chicken ovalbumin gene, in which the boundaries between coding and intervening sequences are indicated by dashed vertical line and the ends of codons by periods. The site of ligation assuming that the GU..AG rule is followed is shown by the solid line, —. Alternative sites of ligation, all of which generate the same codon, GGU in this case as shown on the right of the figure, are also depicted: these are shifted with respect to the solid line by 1 (——), or 2 (xxxx) nucleotides into the IVS at its 5′ end or 1 (. . .) or 2 (oooo) nucleotides back into the IVS at its 3′ end. Parts B and C show the same ovalbumin 5′ splice junction paired with heterologous 3′ splice junctions, which, according to the GU..AG rule are incompatible, i.e., if the GU..AG rule were followed, the reading frame beyond the splice site would not be conserved. The 3′ splice junctions in parts B and C are from a second ovalbumin IVS and an ovomucoid IVS, respectively, and break the coding sequence between the first and second residues of a codon (B) or between codons (C). The sites of ligation, shifted 1 or 2 nucleotides relative to that used were the GU..AG rule followed that would recreate a complete codon are illustrated as in part A. Shown on the right hand side are the coding sequences thus created, in which the codon regenerated by splicing is underlined.

min IVS such duplications would permit a complete codon to be created even when the 5′ and 3′ junctions appear, by the above criterion, to be incompatible, by use of splice sites one or two nucleotide removed from those defined on the basis of the GU-AG rule, illustrated in the top part Figure 8. Thus, the presence of such duplications suggests that there may in most cases be some flexibility in the joining of the ends of coding sequences bordering an IVS, provided that the splice junction sequences conform to the consensus sequences: as illustrated in Figure 7 nucleotides immediately adjacent to the junctions are those most highly conserved. This kind of play in the joining reactions is somewhat reminiscent of the joining of J and C segments and J and D segments during the DNA rearrangements that form immunoglobulin light and heavy chain genes, respectively.[170-173]

C. Small Nuclear RNAs and Splicing

It is clear that the splice junction consensus sequences shown in Figure 7 do not permit base-pairing between the termini of an IVS to align the coding sequences to be ligated during a splicing reaction. It has in fact been suggested that the small, nuclear RNA (snRNA) species U1 mediates construction of an appropriate substrate structure for splicing of pre-mRNA species: the 5′-terminal sequence of U1 RNA is, as illustrated in Figure 9A, complementary to the splice junction consensus sequences

A

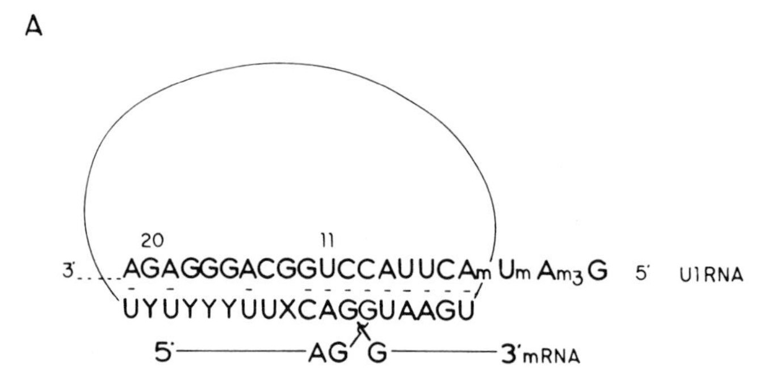

B

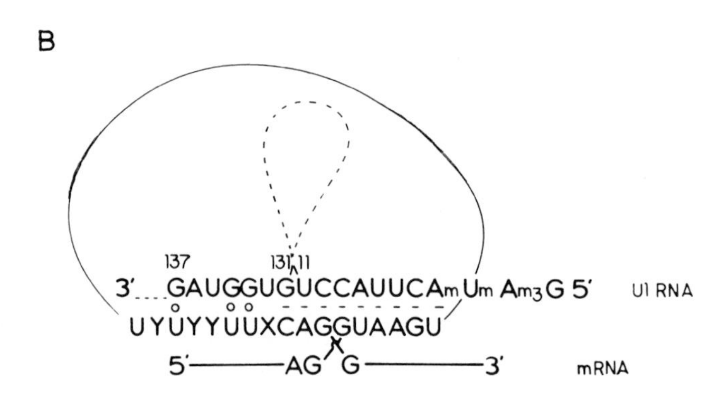

FIGURE 9. Potential base-pairing between U1 snRNA and mRNA splice junction consensus sequences. The mRNA is shown as the solid line with IVS looped out and coding sequences aligned. Part A depicts potential base pairing between this structure and 5′-terminal sequences of U1 RNA as proposed by Lerner et al.[156] Part B illustrates an alternative interaction discussed by Ohshima et al.[175] In both parts of the figure, Watson-Cerick and non-Watson-Crick base pairs are indicated by - and O, respectively.

such that the 5′ splice site could base-pair with nucleotides 2 to 10 and the 3′ site with nucleotides 13 to 22 of U1 RNA.[156,157] An alternative mode of interaction between U1 RNA and splice junctions, based on construction of secondary structure models of U1 RNA from the primary sequence (see Reddy and Busch[174]) has been suggested by Ohshima et al.[175] and is illustrated in Figure 9B; in this model, it is nucleotides 131 to 138 that are postulated to base-pair with sequences at the 3′ end of an IVS for Ohshima and colleagues[175] believe that these residues are less likely to be sequestered by intramolecular base-pairing than nucleotides 13 to 22. When the free energies of the predicted base-paired structures between U1 RNA and each of some 30 pairs of splice junctions is calculated, the original model (Figure 9A) would result in the most stable structure in half the cases examined, whereas interactions involving nucleotides 131 to 137 would appear to be more favorable in the other 50%.[175] While it should be noted that such calculations take no account of the secondary structure of the pre-mRNA molecule or, and possibly more importantly, the influence of the proteins in which both the pre-mRNA and U1 RNA are packaged, it remains quite possible that U1 RNA can interact with the ends of an IVS in more than one fashion.

The sequences of human, rat and chicken U1 RNA are very similar[176] and sucn snRNA species appear to be ubiquitous, abundant, and conserved in all eukaryotic cells examined.[100,156,157,174] In all cases U1 RNA is present in the form of small, nuclear ribonucleoprotein particles, which are more abundant in actively growing cells[156,178] and found associated with structures that contain hnRNA,[156,179-184] properties consistent with a role in RNA processing. Further support for the notion that U1 snRNP participates in splicing comes from the observation that one IVS of the chicken $\alpha 2$ type 1 collagen gene contains three sets of sequences that exhibit potential complementarity to the 5′ terminal sequence of U1 RNA and these are utilized during splicing to create intermediates in which the 3′ end of the IVS is ligated sequentially to the three internal 5′ splice sites.[185] Each splicing event fashions a new 3′ splice junction sequence so that removal of this particular IVS proceeds processively in the 3′ → 5′ direction, each step mediated by interaction of U1 RNA with the appropriate consensus sequence.

Although models implicating U1 RNA in splicing of mRNA precursors have attractive features, potential sequence complementarities between sequences surrounding splice junctions and other small, nuclear RNA species have been noted. Transcripts of the human, highly-repeated Alu sequence and its mouse or hamster cell counterparts[186-188] are, for example, associated with poly(A)-containing nuclear and cytoplasmic RNA.[189-192] Those of hamster and mouse 4.5S RNA include regions that could form short base-paired structures (up to 11 nucleotides in length) with sequences found at eukaryotic pre-mRNA splice junctions, particularly those surrounding 5′ splice sites.[193,194] In many examples of specific splice junction sequences that might base-pair in such fashion, at least 50% of the potentially complementary nucleotides lie in the coding sequence, in marked contrast to the putative interactions between a pre-mRNA molecule and U1 RNA depicted in Figure 9.

Base-pairing interactions between mRNA precursors and a small RNA species that rely heavily on the coding portions of splice junction sequences have also been suggested for both U2 RNA[175] and adenoviral VA-RNA_I.[195] These models are illustrated in Figure 10. Such base-pairing between coding sequences and a small RNA species might complement those between the ends of an IVS and U1 RNA (compare Figures 9 and 10) to provide an additional degree of specificity to the alignment of the correct pair of coding sequences, that is, the involvement of a second small RNA species that would primarily base-pair with coding sequence could provide an explanation for the correct pairing of 5′ and 3′ splice junctions within an mRNA precursor that contains many intervening sequences. Thus, because U1 could primarily base-pair with intervening sequences and the other small RNA species with neighboring coding sequences, Ohshima et al.[175] suggest that the two kinds of interaction are not mutually exclusive, but rather cooperative, imparting a high degree of precision to splicing reactions. In addition, the kind of interaction illustrated in Figure 10 provides, at least in principle, a mechanism to keep the ends of the coding segments together once the endonucleolytic cleavages have occurred; U1 RNA would tend to remain associated with the IVS. Whether there actually exists any requirement to hold liberated but unligated termini of coding sequences in position of course depends on the mechanism of splicing; in a two-step cleavage-ligation, like that seen when pre-tRNA species are spliced, it would be necessary, but in a phosphotransferase reaction like that described previously for pre-rRNA.

It is obvious that not all coding sequences adjacent to the ends of an IVS will possess the appropriate nucleotide sequence to permit base-pairing with U2-RNA. Nevertheless, quite a few mammalian gene sequences could interact in this fashion.[175] As mentioned previously, other small RNA species possess the potential to interact with some splice junction sequences in a similar way. It is therefore possible that different classes of snRNA are responsible for mediating splicing of transcripts of different genes, or classes of genes, in a specific manner, thus providing an element of regulation.[175]

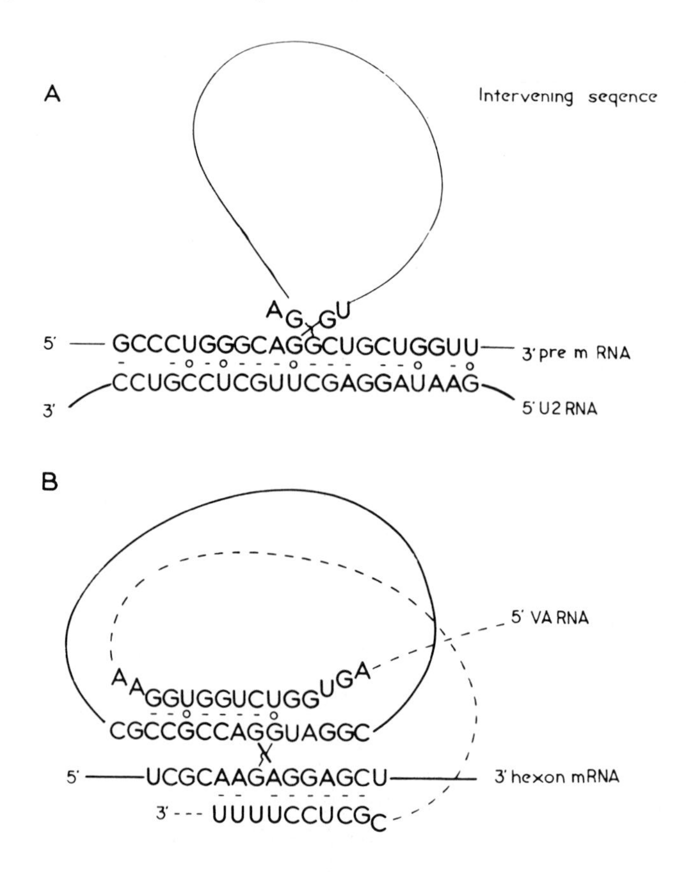

FIGURE 10. Potential base pairing between U2 RNA and adenovirus VA RNA, and mRNA splice junction sequences. Potential base-pairing interactions between U2 RNA and sequences surrounding the first IVS of mouse β^{major}-globin pre-mRNA and adenovirus VA-RNA, and hexon precursor mRNA are illustrated in parts A and B. These summaries are adapted from those given by Ohshima et al.[175] and Murray and Holliday,[195] respectively. Base-pairs are depicted as in Figure 9.

D. Splicing of mRNA Precursors In Vitro

In order to test ideas such as these about the role of snRNA species in splicing and to purify the components of the enzymatic machinery that mediates splicing of mRNA precursors, it is essential to develop in vitro splicing systems. As we have seen, such systems have provided valuable insights into the reactions whereby pre-tRNA and pre-rRNA species are spliced. Progress towards the development of systems that splice pre-mRNA in vitro has been slow, becoming substantial only in the last year.

The first in vitro systems in which correct splicing of pre-mRNA species was observed comprised isolated nuclei in which adenoviral or SV40 RNA species labeled in the cell prior to nuclear isolation or transcribed in vitro were processed faithfully.[196-200] Such nuclear systems vary somewhat in the efficiency with which splicing occurs and their requirement for "cytoplasmic factors", the latter presumably a reflection of leakage of nuclear components during isolation,[197,200] reminiscent of the leakage of the enzymes that splice pre-tRNA species discussed in Section II.C. The precise requirements of the splicing reactions performed in isolated nuclei have not been established, at least in part because transcription and processing are coupled in these systems: indeed, in all cases, conditions similar to those necessary for transcription in vitro, that is provision of an ATP generating system, divalent metal ions and moderate salt concentrations, have been employed. The only reported separation of transcription from splicing observed was in those nuclei that appeared to have leaked components essential to splicing reactions. While this observation suggests a method of preparation of a nuclear fraction enriched in splicing enzyme(s), it is not one that has been pursued.

One such nuclear system in which transcripts of adenovirus type 2 early and late genes are spliced efficiently[199,201] has been exploited to collect experimental evidence for a role of U1 snRNP in splicing. When antibodies from patients with systemic lupus erythymatosus (SLE) that react with U1 snRNP, anti-RNP, and anti-Sm sera[178] are preincubated with such isolated nuclei, splicing in vitro of the products of transcription of at least three adenoviral early genes is substantially inhibited.[201] No such inhibition is observed with SLE antibodies that recognize either small, cytoplasmic RNP species or a distinct class of snRNP that does not include U1 RNP, anti-Ro, and anti-La, respectively.[202] The deduction that binding of anti-U1 snRNP antibodies to the particles prevents their participation in splicing assumes both that the antibodies are indeed specific and that U1 snRNP plays a direct role in splicing. The failure of anti-La antibodies to inhibit splicing and the inhibition observed with a monoclonal antibody with anti-Sm specify[203] argue that the first assumption is justified. Proof for the second, i.e., the exclusion of models in which U1 snRNP mediates some other event upon which splicing is dependent, requires a soluble splicing system, as does determination of the nature of the role U1 snRNP might play. The two obvious, and not necessarily exclusive, possibilities are that U1 snRNP acts as the scaffold upon which the correct substrate structure is constructed or that it in fact contains the enzyme activities that actually perform the splicing reactions. Several enzymes that contain RNA components have been described, most compelling of which in this context is the *E. coli* tRNA processing enzyme RNAase P.[35,36] Three reports of soluble systems that splice pre-mRNA correctly have appeared within the last few months; in all three sets of experiments, the splicing of adenoviral pre-mRNA species has been investigated.

In the first successful demonstration of mRNA splicing in a soluble system[204] a cloned adenoviral DNA fragment containing a complete early gene (E3) was provided to a concentrated extract of uninfected HeLa cells prepared by methods originally devised to permit isolation of RNA polymerase form II[205] and recently modified to obtain accurate initiation of transcription on a naked DNA template in vitro.[206] When the RNA species synthesized under these conditions were compared to authentic adenoviral E3 mRNA species by a nuclease S1 assay, it could be demonstrated that RNA species whose synthesis was initiated at the E3 promoter site in vitro were correctly spliced in the soluble system.[204] Splicing of the products of transcription of the adenoviral ElA gene has also been observed in these extracts.[207] The precise requirements of these splicing reactions have not been reported and may be impossible to determine because transcription and processing are coupled in the extracts. The efficiency of splicing in this extract has not been assessed quantitatively, but appears to be quite high. Others have had little success in demonstrating splicing in similar extracts[206,208] and it is not known whether these differences reflect differences in the methods employed in extract preparation, in the in vitro conditions employed or in the structure of the splicing substrate. It is, for example, clear that in the experiments of Weingartner and Keller[204] at least some fraction, 5 to 20%, of the RNA made in vitro is polyadenylated; although RNA molecules that do not carry poly(A) can be spliced in unusual circumstances[209,210] polyadenylation normally precedes splicing.[210-215]

A second but rather different system in which transcription and processing are coupled has been developed by S. Chien-Kang and J. E. Darnell.[216] When nucleoprotein complexes extracted from adenovirus 2-infected cells during the late phase of infection[217] are incubated in vitro, at least 85% of the RNA made is virus-specific. The

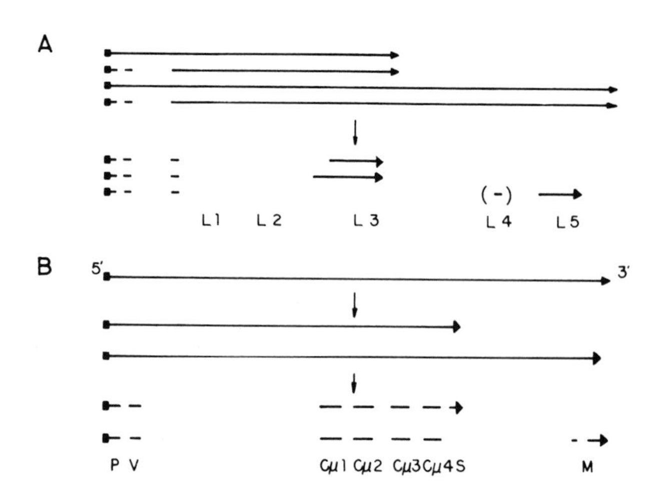

FIGURE 11. Maturation of adenoviral late and immunoglobulin heavy chain μ mRNA species. In both parts of the figure, capped (■) initial transcripts are shown to be processed by the addition of poly(A) (▲) at alternative sites (step a). In the case of the adenoviral late transcript, part A, some splicing events, indicated by gaps in the horizontal line drawn in the direction of transcription that represents the mRNA, can occur in the absence of polyadenylation. The polyadenylated intermediates are then spliced to generate overlapping mRNA species. In part A, the adenoviral late transcript, the members of two of the five late mRNA families are shown and the positions of the other indicated. The various coding segments of the heavy μ chain gene are indicated at the bottom of part B.

length of the RNA chains made increases with time of incubation in vitro, as does the concentration of sequences complementary to distal regions of the major, late transcriptional unit. Indeed, very large RNA species, up to 25 to 30 kb, the distance from the major late promoter site near 16.45 units to the end of the transcriptional unit are made from isolated nucleoprotein complexes. Some 10% of this RNA contains poly(A). Determination of the sizes of the poly(A)-containing RNA molecules labeled in vitro and protected against nuclease S1-digestion by hybridization to specific, cloned, restriction endonuclease fragments of adenovirus DNA has demonstrated that the cleavages of the primary Ad2 late transcript to generate sites of poly(A) addition occur in vitro at the sites utilized in vivo.[218] Moreover, RNA species whose sizes are characteristic of the splicing cleavages that must occur when mature late mRNA species are created are also observed.[216] That transcriptional complexes retain the ability to perform splicing ligations has been demonstrated by showing that RNA species with correctly cleaved 5′ ends of the L1 and L2 mRNA families (see Figure 11) also contain sequences of the first segment of the 5′ tripartite leader located far upstream near 16.4 units[216] (see Figure 12). Thus some splicing reactions seem to be performed in transcriptional complexes, although those that join the three segments of the initial transcript located near 16.4, 19.6, and 26.8 units in the r-strand to form the tripartite leader common to Ad2 late mRNA species[219-221] apparently do not.[216] This is somewhat surprising observation for synthesis of the tripartite leader appears to precede ligation of the mature leader to an mRNA coding sequence in the cell and in isolated nuclei.[206,210,222] The transcriptional complex extract would be expected to be considerably simplified in terms of its protein composition compared to the whole cell extracts mentioned previously and could thus prove to be a useful starting point in the purification of splicing enzymes. In this context, it is encouraging that processing activities remain associated with transcriptional complexes during fractionation in sucrose gradients.[218]

An extract of whole cells, MOPC-315 cells rather than HeLa, has also been employed in the third demonstration of pre-mRNA splicing in vitro. The most striking property of this extract is that when supplied with the precursor to mRNA complementary to the E2A gene of Ad2 made in vivo (28S), it will convert it to a 20S form from which both intervening sequences have been removed.[223] The incubation conditions employed in these conditions essentially mimicked those used for transcription in vitro, that is, 100 mM KC1; 0.5 mM nucleoside triphosphates and an ATP-generating system, and no variations have been described. No success in splicing naked RNA in vitro has been reported previously and, if reproducible, this observation will represent a significant advance. Such separation of synthesis of an appropriate precursor RNA from its splicing should make purification of processing activities a much more manageable task by avoiding the formidable problem of fractionating and then reconstituting all the factors necessary for correct initiation of transcription in vitro with the processing enzymes of interest.

Although Goldenberg and Raskas[223] also observed splicing of the E2A mRNA precursor in HeLa cell extracts, the activit(ies) proved to be unstable during storage. Such instability may well reflect the fact that the protein concentrations attained in HeLa cell extracts were only 20 to 25% of those obtained in MOPC-315 extracts.[223] The ability of the MOPC-315 extracts to splice was optimal at a protein concentration of 0.6 mg/mℓ. under the conditions employed, declining at higher extract concentrations. This property suggests that the extract contains a factor, or factors, that are inhibitory to splicing. These apparently do not bind to DEAE-cellulose columns, whereas the splicing component(s) does.[223]

Little about the mechanisms by which pre-mRNA species are spliced has yet been learned from the three soluble systems discussed in previous paragraphs; the requirements of the splicing reaction have not been established, nor has much attention been paid to determination of the structure of excised intervening sequences. As we have seen, (Section III.C), the observation that the 0.4 kb IVS of *Tetrahymena* pre-rRNA excised as a linear molecule could be circularized in vitro provided an important clue to the enzymatic mechanism of pre-rRNA splicing. Blanchard and colleagues[197] observed a molecule that contained sequences of the larger IVS present in the E2A mRNA precursor and was of the size predicted for excision in linear form in isolated nuclei, but this has yet to be confirmed in soluble systems. Despite the fact that we have not yet learned a great deal from them, the soluble systems developed represent a significant accomplishment, one awaited for 5 years, and provide the opportunities to define the components of the splicing machinery.

Moreover, the observation that concentrated cell extracts will splice naked RNA raises some intriguing questions. The substrate for splicing within a cell is obviously not deproteinized RNA but rather hnRNP structures. Whether equivalent ribonucleoprotein complexes can be formed by association of proteins in the extracts with either exogenously added RNA or RNA transcribed from DNA templates remains to be established, but is not an impossibility given the fate of deproteinized DNA introduced in *Xenopus* oocyte nuclei or extracts made from them.[224,225] This is an important question, for it is generally assumed that an important relationship between RNP structure and pre-mRNA processing must exist (see, for example, discussions in references[226-228]).

E. Regulation of Pre-mRNA Splicing

It is anticipated that soluble splicing systems like those discussed in previous paragraphs will permit the identification and purification of the enzymatic machinery and any necessary cofactors, snRNP particles, e.g., that mediate the removal of one IVS. However, it is likely that knowledge of such components will not prove sufficient to gain a complete understanding of the mechanism of pre-mRNA splicing, a business that is usually more complex than the simple removal of one IVS; as discussed in Section IV.A, the majority of eukaryotic pre-mRNA species that have been characterized contain several intervening sequences. Thus, a central question must concern the means whereby the 3′ end of one coding segment becomes ligated only to the 5′ end of the next coding segment rather than to that of any other coding segment present in a particular transcript. One possible solution, the involvement of a class of snRNAs that interact primarily with coding sequences has been discussed previously (Section IV.B). The most obvious problem with this idea is that it would seem to require a large set of such RNA species to accommodate possible sequences surrounding splice junctions. This model also predicts that each set of splice junction sequences within a transcript that possess several intervening sequences should be complementary to a different small RNA species.

A second relatively simple solution to this problem might be to fix either the splicing machinery and/or the transcript to be spliced such that splicing can proceed in only one direction, with respect to the RNA substrate. By definition, such a mechanism would result in the orderly, processive removal of the intervening sequences encountered in the pre-mRNA substrate. Examples of processive splicing have been reported: one particular IVS of the transcript of the chicken $\alpha 2$ type 1 collagen gene and intervening sequences of the Ad2 E1A gene transcript appeared to be removed in steps, processive in the 3′ → 5′ direction,[185,229] whereas the splicing reactions that synthesize the tripartite leader common to Ad2 late mRNA species appear to be processive with the opposite polarity.[222] However, there is little evidence in support of long-range order, the serial removal of the several intervening sequences that may be present in a pre-mRNA in a 3′ → 5′ or 5′ → 3′ direction. Indeed, the information available indicates not only that splicing is not necessarily processive from one or other ends of the molecule, but also that there frequently exist several splicing pathways. Even, for example, in the case of the relatively simple transcript of the mouse β^{major}-globin gene that contains but two intervening sequences, intermediates whose structures are consistent with at least two pathways of IVs removal have been described.[230] Similarly, removal of intervening sequences from transcripts that include several, those of the ovalbumin and ovomucoid genes, for example, can follow any one of several preferred, but not obligate, pathways in which the first IVS removed may lie near the 3′ or the 5′ end of the pre-mRNA species.[231-233]

A reasonable model might therefore be that the splicing machinery becomes attached to one end of an IVS, works processively until removal of that IVS is completed, and then dissociates to start the process anew at a second site. The order in which individual intervening sequences would be removed would presumably be influenced by the secondary and therefore the tertiary structure of the substrate RNA, as well as the mode of association of proteins in the formation of RNP. It is, for example, not too difficult to imagine circumstances in which the conformation of a pre-mRNA species in its RNP form is such that only one or two of many intervening sequences that must be excised are accessible to the splicing machinery. These would therefore be removed first, an act that might induce quite profound alterations in the conformation of the partially-spliced substrate to render additional intervening sequences accessible. Thus, quite orderly removal of intervening sequences, such that the opportunities for scrambling of coding segments during ligation are reduced, could be achieved in a relatively simple

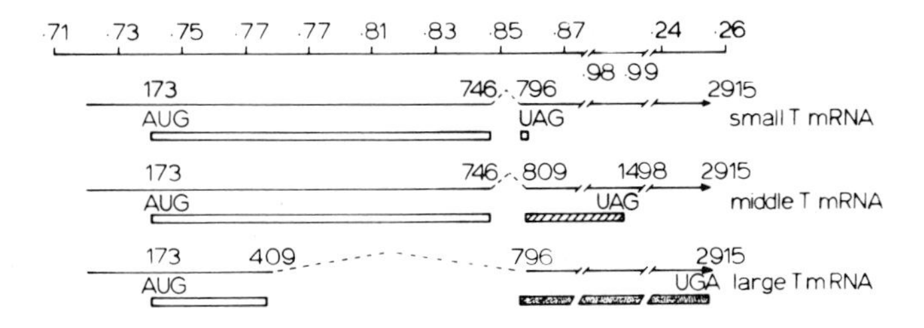

FIGURE 12. Expression of the polyomavirus early region. The polyomaviral genome is represented by the topmost horizontal line, divided into map units. The mRNA species are drawn as horizontal lines with the arrowheads in the direction of transcription. Sequences removed during splicing are indicated by the caret symbols. The regions of each mRNA species actually translated are represented by the boxes drawn below each mRNA species. To illustrate the frames used for translation, that read in the 5′ segment is open, whereas frames shifted by 1 or 2 nucleotides relative to the first are shown as stippled and cross-hatched, respectively. The sources of these data are given in the text.

way. While it seems clear that each RNP has a recognizable structure,[226,227] the consequences of such structure for RNA processing, including splicing, have barely begun to be considered.

A final complexity that must be included when considering splicing of pre-mRNA species is its potential for regulation of gene expression. Among spliced, viral mRNA species, it is common to find that a transcript of a given gene can be processed to produce several mRNA species, encoding partially related polypeptides. A striking example of this phenomenon provided by the polyomavirus early region is illustrated in Figure 12. The early region transcript is spliced in three different ways, joining the sequences in three different reading frames to generate three early mRNA species whose products share N-terminal but not C-terminal sequences.[234-239] This kind of splicing pattern is common among viral genes[105] and is of obvious advantage to viruses whose genomes are of limited size. In such cases, it is the "choice" of one of several potential 5′ or 3′ splice sites that determines which mRNA species is actually fashioned from a given precursor molecule. Interestingly, it appears that in certain cases at least this "choice" is not determined simply by the structure/conformation of the RNA substrate, but may be actively regulated such that different spliced mRNA species predominate during different periods of an infectious cycle.[240-242] No cellular gene products appear to be processed in quite this fashion, although the production of two discrete forms of mRNA differing in their 5′ untranslated region, from one α-amylase gene in mouse liver and salivary gland, possibly mediated by differential processing, has been reported.[243]

A second type of posttranscriptional selection of viral mRNA species from a primary transcript is depicted in Figure 11A; the major late transcript codied from the Ad2 genome is an enormous molecule, some 30 kb, that can be processed to yield one of 17 or 18 mRNA species, each of which is polyadenylated and carries the common 5′ tripartite leader segment (see Tooze,[244] and Flint[105] for reviews). This example, selection of one mRNA species from a precursor that includes the sequences of 17 to 18, is extreme, but similar modes of processing have been observed in other viral systems.[105] There are two critical steps that determine which mRNA species is actually fashioned from the Ad2 late precursor, the "choice" of one of five potential poly(A)-addition sites and the "choice" of 3′ splice site to which the mature tripartite leader is ligated within a polyadenylated intermediate (see Figure 11A). An analogous posttranscriptional selection process regulates the synthesis of membrane-bound or secreted immunoglobulin heavy chains, the IgM μ chain in the example shown in Figure 11B, but also the IgD δ chain, as well as the selection of μ or δ mRNA species from a common precursor in certain exceptional myeloma cell lines that synthesize both IgM and IgD.[245-250] Such examples emphasize the importance of posttranscriptional mechanisms including differential splicing, in the regulation of gene expression in eukaryotic cells. The molecular mechanisms governing these regulatory processes are completely mysterious at the present time.

REFERENCES

1. **Goodman, H. M., Olson, M. V., and Hall, B. D.**, Nucleotide sequence of a mutant eukaryotic gene: the yeast tyrosine-inserting ochne supressor SUP4-0, *Proc. Natl. Acad. Sci. U.S.A.*, 74, 5453, 1977.
2. **Valenzuela, P., Venegas, F., Weinberg, F., Bishop, R., and Rutter, W. J.**, Structure of yeast phenylalanine tRNA genes: an intervening segment in the DNA coding for the tRNA, *Proc. Natl. Acad. Sci. U.S.A.*, 75, 190, 1978.
3. **Ogden, R. C., Beckmann, J. S., Abelson, J., Kang, H. S., Söll, D., and Schmidt, O.**, *In vitro* transcription and processing of a yeast tRNA gene containing an intervening sequence, *Cell*, 17, 399, 1979.
4. **Abelson, J.**, RNA processing and the intervening sequence problem, *Annu. Rev. Biochem.*, 48, 1035, 1979.
5. **Venegas, A., Quiroga, M., Zaldivar, J., Rutter, W. J., and Valenzuela, P.**, Isolation of yeast $tRNA^{leu}$ genes: DNA sequence of a cloned $tRNA^{leu}_3$, *J. Biol. Chem.*, 254, 12306, 1979.
6. **Etcheverry, T., Colby, D., and Guthrie, C.**, A precursor to a minor species of yeast $tRNA^{ser}$ contains an intervening sequence, *Cell*, 18, 11, 1979.
7. **Feldmann, H., Olah, J., and Friedenreich, H.**, Sequence of a yeast DNA fragment containing a chromosomal replicator and a $tRNA^{gly}_3$ gene, *Nucl. Acids Res.*, 9, 2949, 1981.
8. **Page, G. S. and Hall, B. D.**, Characterization of the yeast $tRNA^{ser}_2$ gene family: genomic organization and DNA sequence, *Nucl. Acids Res.*, 9, 921, 1981.
9. **Hopper, A. K., Banks, F., and Evangelidis, V.**, A yeast mutant which accumulates precursor tRNAs, *Cell*, 14, 211, 1978.
10. **Hopper, A. K., Schultz, L. D., and Shapiro, R. A.**, Processing of intervening sequences: a new yeast mutant which fails to excise intervening sequences from precursor tRNAs, *Cell*, 19, 741, 1980.
11. **Muller, F. and Clarkson, S. G.**, Nucleotide sequence of genes coding for $tRNA^{phe}$ and $tRNA^{tyr}$ from a repeating unit of *X. laevis* DNA, *Cell*, 19, 345, 1980.
12. **Robinson, R. R. and Davidson, N.**, Analysis of a *Drosophila* tRNA gene cluster: two $tRNA^{leu}$ genes contain intervening sequences, *Cell*, 23, 251, 1981.
13. **Knapp, G., Beckmann, J. S., Johnson, P. F., Fuhrman, S. A., and Abelson, J.**, Transcription and processing of intervening sequences in yeast tRNA genes, *Cell*, 14, 221, 1978.
14. **O'Farrell, P. Z., Cordell B., Valenzuela, P., Rutter, W. J., and Goodman, H. M.**, Structure and processing of yeast precursor tRNAs containing intervening sequences, *Nature (London)*, 274, 438, 1978.
15. **Schmidt, O., Mao, J., Silverman, S., Hovemann, B., and Söll, D.**, Specific transcription of eukaryotic tRNA genes in *Xenopus* germinal vesicle extracts, *Proc. Natl. Acad. Sci. U.S.A.*, 75, 4819, 1978.
16. **Cortese, R., Melton, D., Tranquilla, T., and Smith, J. D.**, Cloning of nematode tRNA genes and their expression in the frog oocyte, *Nucl. Acids Res.*, 16, 4593, 1978.
17. **Hägenbuchle, O., Larson, D., Hall, G., and Sprague, K.**, The primary transcription product of a silkworm alanine tRNA gene: identification of *in vitro* sites of initiation, termination and processing, *Cell*, 18, 1217, 1979.
18. **Melton, D. A. and Cortese, R.**, Transcription of cloned tRNA genes and the nuclear partitioning of a tRNA precursor, *Cell*, 18, 1165, 1979.
19. **Mattoccia, E., Baldi, M. I., Carrara, G., Fruscoloni, P., Bendetti, P., and Tocchini-Valentini, G. P.**, Separation of RNA transcription and processing activities from *X. laevis* germinal vesicle, *Cell*, 18, 643, 1979.
20. **Garber, G. L. and Gage, L. P.**, Transcription of a cloned *Bombyx mori* $tRNA^{ala}_2$ gene: nucleotide sequence of the tRNA precursor and its processing *in vitro*, *Cell*, 18, 817, 1979.

21. **DeRobertis, E. M. and Olson, M. V.**, Transcription and processing of cloned yeast tyrosine tRNA genes microinjected into frog oocytes, *Nature (London)*, 78, 137, 1979.
22. **Melton, D. A., DeRobertis, E. M., and Cortese, R.**, Order and intracellular location of the events involved in the maturation of a spliced tRNA, *Nature (London)*, 284, 143, 1980.
23. **Garber, R. L. and Altman, S.**, *In vitro* processing of *B. mori* transfer RNA precursor molecules, *Cell*, 17, 389, 1979.
24. **DeRobertis, E. M., Black, P., and Nishikura, K.**, Intranuclear location of the tRNA splicing enzymes, *Cell*, 23, 89, 1981.
25. **Peebles, C. L., Ogden, R. C., Knapp, G., and Abelson, J.**, Splicing of yeast tRNA precursors: a two-stage reaction, *Cell*, 18, 27, 1979.
26. **Knapp, G., Ogden, R. C., Peebles, C. L., and Abelson, J.**, Splicing of yeast tRNA precursors: structure of the reaction intermediates, *Cell*, 18, 37, 1979.
27. **Zaug, A. J. and Cech, T. R.**, unpublished observations, cited in 92, 1982.
28. **Konarska, M., Filipowicz, W., Domdey, H., and Gross, H. J.**, Formation of a 2′ phosphomonoester linkage by a novel RNA ligase in wheat germ, *Nature (London)*, 293, 112, 1981.
29. **Johnson, J. D., Ogden, R. C., Johnson, P., Abelson, J., Dembeck, P., and Itakura, K.**, Transcription and processing of a yeast tRNA gene containing a modified intervening sequence, *Proc. Natl. Acad. Sci. U.S.A.*, 77, 2564, 1980.
30. **Kim, S-H.**, Crystal structures of yeast tRNA^{phe} and general structural features of other tRNAs, in *Transfer RNA Structure Properties and Recognition*, Schimmel, P., Söll, D., and Abelson, J., Eds., Cold Spring Harbor, New York, 1979, 83.
31. **Moras, D., Comarmond, M. B., Fischer, J., Weiss, R., Thierry, J. C., Ebel, J. P., and Giége, R.**, Crystal structure of yeast tRNA^{asp}, *Nature (London)*, 288, 669, 1980.
32. **Colby, D., Leboy, P. S., and Guthrie, C.**, Yeast tRNA precursor mutated at a splice junction is correctly processed *in vitro*, *Proc. Natl. Acad. Sci. U.S.A.*, 78, 415, 1981.
33. **Abelson, J.**, unpublished observations, 1981.
34. **Otsuka, A., dePaolis, A., and Tocchini-Valentini, G. T.**, Ribonuclease 'Xlal', an activity from *Xenopus laevis* oocytes that excises intervening sequences from yeast transfer RNA precursors, *Mol. Cell Biol.*, 1, 269, 1981.
35. **Stark, B. C., Kole, R., Bowman, E. J., and Altman, S.**, Ribonuclease P: an enzyme with an essential RNA component, *Proc. Natl. Acad. Sci. U.S.A.*, 75, 3717, 1978.
36. **Kole, R., Baer, M. F., Stark, B. C., and Altman, S.**, *E. coli* RNAase P has a required RNA component, *Cell*, 19, 881, 1980.
37. **Weil, P. A., Luse, D. S., Segal, J., and Roeder, R. G.**, Selective and accurate initiation of transcription at the adenovirus 2 major late promoter site in a soluble system dependent on RNA polymerase II and DNA, *Cell*, 18, 469, 1979.
38. **Standring, D. W., Venegas, A., and Rutter, W. J.**, Yeast tRNA$^{leu}_{3}$ gene transcribed and spliced in a HeLa extract, *Proc. Natl. Acad. Sci. U.S.A.*, 73, 5963, 1981.
39. **Long, E. O. and Dawid, I. B.**, Repeated genes in eukaryotes, *Annu. Rev. Biochem.*, 49, 727, 1980.
40. **Brown, D. D. and Weber, C. S.**, Gene-linkage by RNA-DNA hybridization 1. Unique DNA sequences homologous to 4S RNA, 5S RNA and ribosomal RNA, *J. Mol. Biol.*, 34, 661, 1968.
41. **Brown, D. D. and Weber, C. S.**, Gene-linkage by RNA-DNA hybridization. II. Arrangement of the redundant gene sequences for 28S and 18S rRNA, *J. Mol. Biol.*, 34, 681, 1968.
42. **Glover, D. M. and Hogness, D. S.**, A novel arrangement of the 18S and 28S sequences in a repeating unit of *D. melanogaster* rDNA, *Cell*, 10, 167, 1977.
43. **Pelligrini, M., Manning, J., and Davidson, N.**, Sequence arrangement of the rDNA of *D. melanogaster*, *Cell*, 10, 213, 1977.
44. **Wellauer, P. K. and Dawid, I. B.**, The structural organization of rDNA in *D. melanogaster*, *Cell*, 10, 193, 1977.
45. **White, R. L. and Hogness, D. S.**, R-Loop mapping of the 18S and 28S sequences in the long and short repeating units of *Drosophila melanogaster* rDNA, *Cell*, 10, 177, 1977.
46. **Wellauer, P. K., Dawid, I. B., and Tartof, K. D.**, X and Y chromosomal rDNA of *Drosophila:* comparison of spacers and insertions, *Cell*, 14, 269, 1978.
47. **Glover, D. M.**, Cloned segments of *D. melanogaster* rDNA containing new types of insertion sequence, *Proc. Natl. Acad. Sci. U.S.A.*, 74, 4932, 1977.
48. **Wellauer, P. K. and Dawid, I. B.**, Ribosomal DNA of *D. melanogaster* II. Heteroduplex mapping of cloned and uncloned rDNA, *J. Mol. Biol.*, 126, 769, 1978.
49. **Meyernick, J. H., Retel, J., Rane, H. A., Planta, R. J., van der Ende, A., and van Bruggen, E. F. J.**, Genetic organization of the ribosomal transcription units of the yeast *S. carlbsbergensis*, *Nucl. Acids Res.*, 5, 2801, 1978.
50. **Phillipsen, P., Thomas, M., Kramer, R. A., and Davis, R. W.**, Unique arrangement of coding sequences for 5S, 18S and 25S rRNA in *S. cerevisiae* as determined by R-loop and hybridization analysis, *J. Mol. Biol.*, 123, 387, 1978.

51. **Wellauer, P. K. and Dawid, I. B.**, Isolation and sequence organization of human ribosomal DNA, *J. Mol. Biol.*, 128, 289, 1979.
52. **Anderson, S., Bankier, A. J., Barrell, B. G., de Bruijn, M. H. L., Coulson, A. R., Drouin, J., Eperon, I. C., Nierlich, D. P., Roe, B. A., Sanger, F., Schreier, P. H., Smith, A. J. H., Staden, R., and Young, I. G.**, Sequence and organization of the human mitochondrial genome, *Nature (London)*, 290, 57, 1981.
53. **Ojala, D., Montoya, J., and Attardi, G.**, tRNA punctuation model of RNA processing in human mitochondria, *Nature (London)*, 290, 470, 1981.
54. **van Etten, R. A., Walberg, M. W., and Clayton, D. A.**, Precise localization and nucleotide sequence of the two mouse mitochondrial rRNA genes and three immediately adjacent novel tRNA genes, *Cell*, 22, 157, 1980.
55. **Karrer, K. M. and Gall, J. G.**, The macronuclear ribosomal DNA of *Tetrahymena pyriformis* is a palindrome, *J. Mol. Biol.*, 104, 421, 1976.
56. **Engberg, J., Andersson, P., Leick, V., and Collins, J.**, The free rDNA molecules from *Tetrahymena pyriformis* GL are giant palindromes, *J. Mol. Biol.*, 104, 455, 1976.
57. **Molgaard, H. V., Mathews, H. R., and Bradbury, E. M.**, Organization of genes for rRNA in *Physarum polycephalum*, *Eur. J. Biochem.*, 68, 541, 1976.
58. **Gübler, U., Wyler, T., and Braun, R.**, The gene for 26S rRNA in *Physarum* contains two insertions, *FEBS Lett.*, 100, 347, 1979.
59. **Campbell, G. R., Littau, V. C., Melera, P. W., Allfrey, V. G., and Johnson, E. M.**, Unique sequence arrangement of ribosomal genes in the palindromic rDNA molecules of *Physarum polycephalum*, *Nucl. Acids Res.*, 6, 1433, 1979.
60. **Nomiyami, H., Sakaki, Y., and Takagi, T.**, Nucleotide sequence of a ribosomal RNA gene intron from slime mold *Physarum polycephalum*, *Proc. Natl. Acad. Sci. U.S.A.*, 78, 1376, 1981.
61. **Cech, T. R. and Rio, D.**, Localization of transcribed regions on the extrachromosomal rRNA genes of *Tetrahymena* thermophilia by R-loop mapping, *Proc. Natl. Acad. Sci. U.S.A.*, 76, 5051, 1979.
62. **Wild, M. A. and Gall, J. G.**, An intervening sequence in the gene coding for 25S rRNA of *Tetrahymena pigmentosa*, *Cell*, 16, 565, 1979.
63. **Rochaix, J. D. and Malnoe, P.**, Anatomy of the chloroplast ribosomal DNA of *C. reinhardii*, *Cell*, 15, 661, 1978.
64. **Bedbrook, J. R., Kolodner, R., and Borograd, L.**, *Zea mays* chloroplast rRNA genes are part of a 22,000 base pair inverted repeat, *Cell*, 11, 739, 1977.
65. **Hahn, V., Lazarus, C. M., Lünsdorf, H., and Küntzel, H.**, Split gene for mitochondrial 24S ribosomal RNA of *Neurospora crassa*, *Cell*, 17, 191, 1979.
66. **Heckman, J. E. and Rajbhandary, V. L.**, Organization of tRNA and rRNA genes in *N. crassa* mitochondria: intervening sequence in the large rRNA gene and strand distribution of the RNA genes, *Cell*, 17, 583, 1979.
67. **Mannella, C. A., Collins, R. A., Green, M. R., and Lambowitz, A. L.**, Defective splicing of mitochondrial rRNA in cytochrome-deficient nuclear mutants of *N. crassa*, *Proc. Natl. Acad. Sci. U.S.A.*, 76, 2635, 1979.
68. **Bos, J. L., Heyting, C., Borst, P., Arnberg, A. C., and van Bruggen, E. F. J.**, An insert in the single gene for the large rRNA in yeast mitochondrial DNA, *Nature (London)*, 275, 336, 1978.
69. **Jacq, C., Kujawa, C., Grandchamp, C., and Netter, P.**, Physical characterization of the difference between yeast mitochondrial DNA alleles w^+ and $w-$, in *Mitochondria: Genetics and Biogenesis of Mitochondria*, Bandlow, W., Schweyen, R. J., Wolf, W., and Kandewitz, F., Eds., DeGruyter, Berlin, 1977, 255.
70. **Heyting, C. and Menke, H. H.**, Fine structure of the 21S rRNA region of yeast mitochondrial DNA III. Physical location of mitochondrial genetic markers and the molecular nature of ω, *Mol. Gen. Genet.*, 168, 279, 1979.
71. **Faye, G., Dennebouy, N., Kujawa, C., and Jacq, C.**, Inserted sequence in the mitochondrial 23S ribosomal RNA gene of the yeast *Saccharomyces cerevisiae*, *Mol. Gen. Genet.*, 168, 101, 1979.
72. **Dujon, B.**, Sequence of the intron and flanking exons of the mitochondrial 21S ribosomal RNA of yeast strains having different alleles at the w and rib-1 loci, *Cell*, 20, 185, 1980.
73. **Levis, R. and Penman, S.**, Processing steps and methylation in the formation of rRNA in cultured *Drosophila* cells, *J. Mol. Biol.* 121, 219, 1978.
74. **Greenberg, J. R.**, Synthesis and properties of ribosomal RNA in *Drosophila*, *J. Mol. Biol.*, 46, 85, 1969.
75. **Long, E. O. and Dawid, I. B.**, Expression of ribosomal DNA insertions in *Drosophila melanogaster*, *Cell*, 18, 1185, 1979.
76. **Jolly, D. J. and Thomas, C. A.**, Nuclear RNA transcripts from *Drosophila melanogaster* rRNA genes containing introns, *Nucl. Acids Res.*, 8, 67, 1980.
77. **Kidd, S. J. and Glover, D. M.**, *Drosophila melanogaster* rDNA containing type II insertions is variably transcribed in different strains and tissues, *J. Mol. Biol.*, 151, 645, 1981.

78. Rae, P. M., Coding region deletions associated with the major form of rDNA interruption in *Drosophila melanogaster, Nucl. Acids Res.*, 9, 4997, 1981.
79. Dawid, I. B. and Botchan, P., Sequences homologous to ribosomal insertions occur in the *Drosophila* genome outside the nucleolar organizer, *Proc. Natl. Acad. Sci. U.S.A.*, 74, 4233, 1977.
80. Dawid, I. B. and Ribbert, M. L., Nucleotide sequences at the boundaries between gene and insertion regions in the rDNA of *Drosophila melanogaster, Nucl. Acids Res.*, 9, 5011, 1981.
81. Dawid, I. B., Long, E. O., DiNocera, P. P., and Pardue, M. L., Ribosomal insertion-like elements in *Drosophila melanogaster* are interspersed with mobile sequences, *Cell*, 25, 399, 1981.
82. Merten, S., Synenki, R. M., Locker, J., Christianson, T., and Rabinowitz, M., Processing of the precursor of 21S ribosomal RNA from yeast mitochondria, *Proc. Natl. Acad. Sci. U.S.A.*, 77, 1417, 1980.
83. Green, M. R., Grimms, M. F., Goewert, R. R., Collins, R. A., Cole, M. D., Lambowitz, A. M., Heckman, J. E., Yin, S., and Rajbhandary, U. L., Transcripts and processing patterns for the ribosomal RNA and transfer RNA region of *N. crassa* mitochondrial DNA, *J. Biol. Chem.*, 256, 2027, 1981.
84. Eckert, W. A., Kaffenberger, W., Krohne, G., and Franke, W. W., Introduction of hidden breaks during rRNA maturation and aging in *Tetrahymena pyriformics, Eur. J. Biochem.*, 87, 607, 1978.
85. Din, N., Engberg, J., Kaffenberg, W., and Eckert, W. A., The intervening sequence in the 26S rRNA coding region of *T. thermaophila* is transcribed within the largest stable precursor for rRNA, *Cell*, 18, 525, 1979.
86. Cech, T., personal communication, 1981.
87. Zaug, A. J. and Cech, T. R., *In vitro* splicing of the rRNA precursor in nuclei of *Tetrahymena, Cell*, 19, 331, 1980.
88. Carin, M., Jensen, B. F., Jentsch, K. D., Leer, J. C., Nielson, O. F., and Westergaard, O., *In vitro* splicing of the rRNA precursor in isolated nucleoli from *Tetrahymena, Nucl. Acids Res.*, 8, 5551, 1980.
89. Grabowski, P. J., Zaug, A. J., and Cech, T. R., The intervening sequence of the rRNA precursor is converted to a circular RNA in isolated nuclei of *Tetrahymena, Cell*, 23, 467, 1981.
90. Cech, T. R., Zaug, A. J., and Grabowski, P. J., Splicing of the rRNA precursor of *Tetrahymena:* involvement of a guanosine nucleotide in the excision or the intervening sequence, *Cell*, 27, 487, 1981.
91. Champoux, J. J., Proteins that affect DNA conformation, *Annu. Rev. Biochem.*, 47, 449, 1978.
92. Kornberg, A., *DNA Replication*, W. H. Freeman, San Francisco, 1980.
93. DePew, R. E., Liu, L. F., and Wang, J. C., Interaction between DNA and *E. coli* protein W., *J. Biol. Chem.*, 253, 511, 1978.
94. Eisenberg, S., Griffith, J., and Kornberg, A., ØX174 cistron A protein is a multifunctional enzyme in DNA replication, *Proc. Natl. Acad. Sci. U.S.A.*, 74, 3198, 1977.
95. van der Ende, A., Langveld, S. A., Teertstra, R., van Arkel, G. A., and Weisbeck, P. J., Enzymatic properties of the bacteriophage ØX174 A*\protein on DNA: a model for the termination of rolling circle replication, *Nucl. Acids Res.*, 9, 2037, 1981.
96. Wild, M. A. and Sommer, R., Sequence of a ribosomal RNA gene intron from *Tetrahymena, Nature (London)*, 283, 693, 1980.
97. Allet, B. and Rochaix, J. D., Structure analysis of the ends of the intervening DNA sequences in the chloroplast 23S ribosomal genes of *C. Reinhardii, Cell*, 18, 55, 1979.
98. Bos, J. L., Osinga, K. A., van der Horst, G., Hecht, N. B., Tabak, H. F., van Omnen, G. J. B., and Borst, P., Splice point sequences and transcripts of the intervening sequence in the mitochondrial 21S rRNA of yeast, *Cell*, 20, 207, 1980.
99. Reddy, R., Henning, D., and Busch, H., Nucleotide sequence of nucleolar U3B RNA, *J. Biol. Chem.*, 254, 11, 097, 1979.
100. Wise, J. A. and Weiner, A. M. Dictyostelium small nuclear RNA D2 is homologous to rat nucleolar RNA U3 and is encoded by a dispersed, multigene family, *Cell*, 22, 109, 1980.
101. Kumar, A. and Warner, J. R., Characterization of ribosomal precursor particles from HeLa cell nucleoli, *J. Mol. Biol.*, 63, 233, 1972.
102. Shepherd, J. and Maden, B. E. H., Ribosome assembly in HeLa cells, *Nature (London)*, 236, 211, 1972.
103. Prestayko, A. W., Klomp, G. R., Schmoll, D. J., and Busch, H., Comparison of proteins of ribosomal subunits and nucleolar preribosomal particles from Novikoff hepatoma as cites cells by 2D gel electrophoresis, *Biochemistry*, 13, 1945, 1974.
104. Breathnach, R. and Chambon, P., Organization and expression of eukaryotic split genes coding for proteins, *Annu. Rev. Biochem.*, 50, 349, 1981.
105. Flint, S. J., Splicing and the regulation of viral gene expression, *Current Topics Microbiol. Immunol.*, 93, 47, 1981.

106. **Kedes, L. H.**, Histone messengers and histone genes, *Cell*, 8, 321, 1976.
107. **Hentschel, C. H. and Birnstiel, M. L.**, The organization and expression of histone gene families, *Cell*, 25, 301, 1981.
108. **Derynck, R., Content, J., DeClerq, E., Volckaert, G., Tavernier, J., Devos, R., and Fliers, W.**, Isolation and structure of a human fibroblast interferon gene, *Nature (London)*, 285, 542, 1980.
109. **Goeddel, D. V., Leung, D. W., Dull, T. T., Gross, M., Lawn, R. M., McCandliss, R., Seeburgh, P. H., Ullrich, A., Yelverton, E., and Gray, P. W.**, The structure of eight distinct cloned human leukocyte interferon cDNAs, *Nature (London)*, 290, 20, 1980.
110. **Streuli, M., Nagata, S., and Weissmann, C.**, At least three human type α interferons: structure of $\alpha 2$, *Science*, 209, 1343, 1980.
111. **Lavin, R. M., Ádelman, J., Dull, T. J.**, Gross, M., Oredell, D., and Ullrich, A., DNA sequence of two closely linked human leukocyte interferon genes, *Science*, 212, 1159, 1981.
112. **Lawn, R. M., Adelman, J., Franke, A. E., Houck, C. M., Gross, M., Najarian, R., and Goeddel, D. V.**, Human fibroblast interferon gene lacks introns, *Nucl. Acids Res.*, 9, 1045, 1981.
113. **Houghton, M., Jackson, I. J., Porter, A. G., Doel, S. M., Catlin, G. H., Barber, C., and Carey, N. H.** The absence of introns within a human fibroblast interferon gene, *Nucl. Acids Res.*, 9, 247, 1981.
114. **Tavernier, J., Derynck, R., and Fiers, W.**, Evidence for a unique human fibroblast interferon (IFN-$\beta 1$) chromosomal gene, devoid of intervening sequences, *Nucl. Acids Res.* 9, 461, 1981.
115. **Droain, J. and Goodman, H. M.**, Most of the coding region of the rat ACTHβ-LPH precursor gene lacks intervening sequences, *Nature (London)*, 288, 610, 1980.
116. **Chang, A. C. Y., Cochet, M., and Cohen, S. N.** Structural organization of human genomic DNA encoding the pro-opiomelanocortin peptide, *Proc. Natl. Acad. Sci. U.S.A.*, 77, 4890, 1980.
117. **Goldberg, D. A.**, Isolation and partial characterization of the *Drosophila* alcohol dehydrogenase gene, *Proc. Natl. Acad. Sci. U.S.A.*, 77, 5794, 1980.
118. **Benyajati, C., Place, A. R., Powers, D. A., and Sofer, W.**, Alcohol dehydrogenase gene of *Drosophila melanogaster:* relationship of intervening sequences to functional domains in the protein, *Proc. Natl. Acad. Sci. U.S.A.*, 78, 2717, 1981.
119. **Durica, D. S., Schloss, J. A., and Crain, R.**, Organization of actin gene sequences in the sea urchin: molecular cloning of an intron-containing DNA sequence coding for a cytoplasmic actin, *Proc. Natl. Acad. Sci. U.S.A.*, 77, 5683, 1980.
120. **Fryberg, E. A., Kindle, K. L., Davidson, N., and Sodja, A.**, The actin gene of *Drosophila:* a dispersed multigene family, *Cell*, 19, 365, 1980.
121. **Mirault, M. E., Goldschmidt-Chermont, M.**, Artavanis-Tsakonis, S., and Schedl, P., Organization of the multiple genes for the 70,000-dalton heat-shock protein in *Drosophila melanogaster, Proc. Natl. Acad. Sci. U.S.A.*, 76, 5254, 1979.
122. **Holmgren, R., Livak, K., Morimoto, R., Freund, R., and Meselson, M.**, Studies of cloned sequences from four *Drosophila* heat shock loci, *Cell*, 18, 1359, 1979.
123. **Gallwitz, D. and Sures, I.**, Structure of a split yeast gene: complete nucleotide sequence of the actin gene in *Saccharomyces cerevisiae Proc. Natl. Acad. Sci. U.S.A.*, 77, 2546, 1980.
124. **Ng, R. and Abelson, J.**, Isolation and sequence of the gene for actin in *Saccharomyces cerevisiae, Proc. Natl. Acad. Sci. U.S.A.*, 77, 3912, 1980.
125. **Rosbash, M., Harris, P. K. W., Woolford, J. L., and Taem, J. L.**, The effect of temperature-sensitive RNA mutants on the transcription products from cloned ribosomal protein genes of yeast, *Cell*, 24, 679, 1981.
126. **Cordell, B., Bell, G., Tischer, E., deNoto, F. M., Ullrich, A., Pictet, R., Rutter, W. J., Goodman, H. M.**, Isolation and characterization of a cloned rat insulin gene, *Cell*, 18, 533, 1979.
127. **Lomedico, P., Rosenthal, N., Efstratiadis, A., Gilbert, W., Kolodner, R., and Tizard, R.**, The structure and evolution of two non-allelic preproinsulin genes, *Cell*, 18, 545, 1979.
128. **Jones, C. W. and Kafatos, F. C.**, Structure, organization and evolution of developmentally regulated chorion genes in a silkmoth, *Cell*, 22, 855, 1980.
129. **Wahli, W., Dawid, I. B., Wyler, T., Weber, R., and Ryffel, U.**, Comparative analysis of the structural organization of two closely-related vitellogenin genes, *Cell*, 20, 107, 1980.
130. **Ohkubo, H., Vogeli, G., Mudry, M., Avvedimento, V. E., Sullivan, M., Pastan, I., and de-Crombrugghe, B.**, Isolation and characterization of overlapping genomic clones containing the chicken $\alpha 2$ (type 1) collagen gene, *Proc. Natl. Acad. Sci. U.S.A.*, 77, 7059, 1980.
131. **Wozney, J., Hanahan, D., Morimoto, R., Boedter, H., and Doty, P.**, Fine structural analysis of the chicken pro $\alpha 2$ collagen gene, *Proc. Natl. Acad. Sci. U.S.A.*, 78, 712, 1981.
132. **Berk, A. J. and Sharp, P. A.**, Spliced early mRNAs of Simian virus 40, *Proc. Natl. Acad. Sci. U.S.A.*, 75, 1274, 1978.
133. **Reddy, V. B., Ghosh, P. K., Leibowitz, P., Piatak, M., and Weissman, S. M.**, SV40 early mRNAs. 1. Genomic localization of 3′ and 5′ termini and two major splices in mRNA from transformed and lytically-infected cells, *J. Virol.*, 30, 279, 1979.

134. Thimmappaya, B. and Weissman, S. M., The early region of SV40 may have more than one gene, *Cell*, 11, 837, 1977.
135. Yamada, Y., Avvidimento, E., Mudryi, M., Ohkubo, H., Vogeli, G., Irani, M., Pastan, I., and deCrombrugghe, B., The collagen gene: evidence for its evolutionary assembly by amplification of a DNA segment contaning an exon of 54 bp, *Cell*, 22, 887, 1980.
136. Hardison, R. C., Butler, E. T., Lacy, E., Maniatis, T., Rosenthal, N., and Efstratiadis, A., The structure and transcription of four linked rabbit-β-like globin genes, *Cell*, 18, 1285, 1979.
137. Efstratiadis, A., Posakony, J. W., Maniatis, T., Lawn, R. M., O'Connell, C., Spritz, R. A., deRiel, J., Forget, B. G., Weissman, S. M., Slightom, J. L., Blechl, A. E., Smithies, O., Baralle, F. E., Shoulders, C. C., and Proudfoot, N. J. The structure and evolution of the human β-globin gene family, *Cell*, 21, 653, 1980.
138. Konkel, D. A., Maizel, J. V., and Leder, P., The evolution and sequence comparison of two recently diverged mouse chromosomal β-globin genes, *Cell*, 18, 856, 1979.
139. Leder, P., Hansen, J. N., Konkel, D., Leder, A., Nishioka, Y., and Talkington, C., Mouse globin system: a functional and evolutionary analysis, *Science*, 209, 1336, 1980.
140. Lawn, R. M., Efstratiadis, A., O'Connell, C., and Maniatis, T., The nucleotide sequence of a human β-gene, *Cell*, 21, 647, 1980.
141. Volckaert, G., Feuntuen, J., Crawford, L. V., Berg, P., and Fiers, W., Nucleotide sequence deletions within the coding region for small t-antigen of SV40, *J. Virol.*, 30, 674, 1979.
142. Thimmappaya, B. and Shenk, T., Nucleotide sequence analysis of viable deletion mutants lacking segments of the SV40 genome coding for small t-antigen, *J. Virol.*, 30, 668, 1979.
143. Contreras, R., Cole, C., Berg, P., and Fiers, W., Nucleotide sequence analysis of two SV40 mutants with deletions in the late region of the genome, *J. Virol.*, 29, 789, 1979.
144. Villareal, L. P., White, R. T., and Berg, P., Mutational alterations within the SV40 leader segment generate altered 16S and 19S mRNAs, *J. Virol.*, 29, 209, 1979.
145. Subramanian, K. N., Segments of SV40 DNA spanning most of the leader sequence of the major late mRNA are dispensible, *Proc. Natl. Acad. Sci. U.S.A.*, 76, 2556, 1979.
146. Khoury, G.,Gruss, P., Dhar, R., and Lai, J. C., Processing and expression of early SV40 mRNA: a role for RNA conformation in splicing, *Cell*, 18, 85, 1979.
147. Proudfoot, N. and Maniatis, T., The structure of a human α-globin pseudogene and its relationship to α-globin gene duplication, *Cell*, 21, 537, 1980.
148. Lacy, E. and Maniatis, T., The nucleotide sequence of a rabbit β-globin pseudogene, *Cell*, 21, 545, 1980.
149. Spritz, R. A., Jagadeeswaran, P., Choudray, P. V., Biro, P. A., Elder, J. T., deRiel, J. K., Manley, J. L., Gefter, M. L., Forget, B. G., and Weissman, S. M., Base substitution in an intervening sequence of a β^+ thalassemia human globin gene, *Proc. Natl. Acad. Sci. U.S.A.*, 78, 2455, 1981.
150. Baird, M., Driscoll, C., Schreiner, H., Sciarratta, V., Sensone, G., Viazi, G., Ramirez, F., and Bank, H., A nucleotide change at a splice junction in the human β-globin gene is associated with β-thalassemia, *Proc. Natl. Acad. Sci. U.S.A.*, 78, 4218, 1981.
151. Orkin, S. H., Goff, S. C., and Hechtman, L. L., Mutation in intervening sequence splice junction in man, *Proc. Natl. Acad. Sci. U.S.A.*, 78, 5041, 1981.
152. N. Jones, personal communication, 1981.
153. Solnick, D., An adenovirus mutant defective in splicing, *Nature (London)*, 291, 508, 1981.
154. Seif, I., Khoury, G., and Dhar, R., BKV splice sequences based on analysis of preferred donor and acceptor sites, *Nucl. Acids Res.*, 6, 3387, 1979.
155. Seif, I., Khoury, G., and Dhar, R., The genome of human papovavirus BKV, *Cell*, 18, 963, 1979.
156. Lerner, M. R., Boyle, J. A., Mount, S. M., Wolin, S. L., and Steitz, J. A., Are snRNPs involved in splicing?, *Nature (London)*, 283, 220, 1980.
157. Rogers, J. and Wall, R., A mechanism of RNA splicing, *Proc. Natl. Acad. Sci. U.S.A.*, 77, 1877, 1980.
158. Breathnach, R., Benoist, C., O'Hare, K., Gannon, F., and Chambon, P., Ovalbumin gene: evidence for a leader segment in mRNA and DNA sequences at exon-intron boundaries, *Proc. Natl. Acad. Sci. U.S.A.*, 75, 4853, 1978.
159. Stein, J. P., Carterall, J. F., Kristo, P., Means, A. R., and O'Malley, B.W., Ovomucoid intervening sequences specify functional domains and generate protein polymorphism, *Cell*, 21, 681, 1980.
160. Jensen, E. O., Paladan, K., Hyldig-Nielsen, J. J., Jorgensen, P., and Marcker, K. A., The structure of a chromosomal leghemoglobin gene from soybean, *Nature (London)*, 291, 677, 1981.
161. Sun, S. M., Slightom, J. L., and Hall, T. C., Intervening sequences in a plant gene: comparison of the partial sequence of cDNA and genomic DNA of French bean phaseolin, *Nature (London)*, 289, 37, 1981.
162. Hamer, D. H., Smith, K. D., Boyer, S. H., and Leder, P., SV40 recombinants carrying rabbit β-globin gene, *Cell*, 17, 725, 1979.

163. **Hamer, D. H. and Leder, P.**, Splicing and the formation of stable RNA, *Cell*, 18, 1299, 1979.
164. **Wold, B., Wigler, M., Lacy, E., Maniatis, T., Silverstein, S., and Axel, R.**, Introduction and expression of a rabbit β-globin gene in mouse fibroblasts, *Proc. Natl. Acad. Sci. U.S.A.*, 76, 5684, 1979.
165. **Breathnach, R., Mantei, N., and Chambon, P.**, Correct splicing of a chicken ovalbumin gene transcript in mouse L cells, *Proc. Natl. Acad. Sci. U.S.A.*, 77, 740, 1980.
166. **Lai, E. C., Woo, S. L. C., Bordelon-Riser, M. E., Fraser, T. H., and O'Malley, B. W.**, Ovalbumin synthesized in mouse cells transformed with the natural chicken ovalbumin gene, *Proc. Natl. Acad. Sci., U.S.A.*, 77, 244, 1980.
167. **Beggs, J. D., van den Berg, J., van Ooyen, A., and Weissmann, C.**, Abnormal expression of chromosomal rabbit β-globin gene in *Saccharomyces cerevisiae*, *Nature (London)*, 283, 835, 1980.
168. **Chu, G. and Sharp, P. A.**, A gene chimera of SV40 and mouse β-globin is transcribed and properly spliced, *Nature (London)*, 289, 379, 1981.
169. **Sharp, P. A.**, Speculations on RNA splicing, *Cell*, 23, 643, 1981.
170. **Max, E. E., Seidman, J. G., and Leder, P.**, Sequences of five potential recombination sites encoded close to an immunoglobulin κ constant region gene, *Proc. Natl. Acad. Sci. U.S.A.*, 76, 3450, 1979.
171. **Sakaro, H., Hüppi, K., Heinrich, G., and Tonegawa, S.**, Sequences at somatic recombination sites of immunoglobulin light chain genes, *Nature (London)*, 280, 288, 1979.
172. **Early, P., Huang, H., Davis, M., Calane, K., and Hood, L.**, An immunoglobulin heavy chain variable region is generated from three segments of DNA, V_H, D and J_H, *Cell*, 19, 981, 1980.
173. **Sakano, H., Mauki, R., Kurosawa, Y., Roeder, W., and Tonegawa, S.**, Two types of somatic recombination are necessary for the generation of complete immunoglobulin heavy-chain genes, *Nature (London)*, 286, 676, 1980.
174. **Reddy, R. and Busch, H.**, UsnRNAs of nuclear snRNPs, *Cell Nucl.*, 8, 261, 1981.
175. **Ohshima, Y., Itoh, M., Okada, N., and Miyata, T.**, Novel models for RNA splicing that involve a small nuclear RNA, *Proc. Natl. Acad. Sci. U.S.A.*, 78, 4471, 1981.
176. **Branlant, C., Krol, A., Ebel, J-P., Lazar, E., Gallinaro, H., Jacob, M., Sri-Wadada, J., and Jeanteur, P.**, Nucleotide sequences of nuclear Ula RNAs from chicken, rat and man, *Nucl. Acids Res.*, 8, 4143, 1980.
177. **Chung, S. Y., Cone, R., and Wooley, J.**, personal communication, 1981.
178. **Lerner, M. R. and Steitz, J. A.**, Antibodies to small nuclear RNAs complexed with proteins are produced by patients with systemic lupus erythematosus, *Proc. Natl. Acad. Sci. U.S.A.*, 76, 5495, 1979.
179. **Zieve, G. and Penman, S.**, Small RNA species of the HeLa cell: metabolism and subcellular location, *Cell*, 8, 19, 1976.
180. **Deimel, D., Louis, C., and Sekeris, C. E.**, The presence of small molecular weight RNAs in nuclear RNP carrying hnRNA, *FEBS Lett.*, 73, 80, 1977.
181. **Northemann, W.**, Scheurlen, M., Gross, V., and Heinrich, P. C., Circular dichroism of ribonucleoprotein complexes from rat liver nuclei, *Biochem. Biophys. Res. Commun.*, 76, 1130, 1977.
182. **Howard, E. F.**, Small, nuclear RNA molecules in nuclear ribonucleoprotein complexes from mouse erythroleukemia cells, *Biochemistry*, 17, 3228, 1978.
183. **Flytzanis, R., Alonso, A., Louis, C., Krieg, L., and Sekeris, C. E.**, Association of small nuclear RNA with HnRNA isolated from nuclear RNP complexes, carrying HnRNA, *FEBS Lett.*, 96, 201, 1981.
184. **Zieve, G. W.**, Two groups of small stable RNAs, *Cell*, 25, 296, 1981.
185. **Avvedimento, V. E., Vogeli, G., Yamada, Y., Maizel, J. V., Pastan, I., and deCrombrugghe, B.**, Correlation between splicing sites within an intron and their sequence complementarity with U1 RNA, *Cell*, 21, 689, 1980.
186. **Jelinek, W. R., Toomey, T. P., Leinwand, L., Duncan, C. H., Biro, P. A., Choudray, P. V., Weissman, S. M., Rubin, C. M., Houck, C. M., Duninger, P. L., and Schmid, C. W.**, Ubiquitous, interspersed, prepeated sequences in mammalian genomes, *Proc. Natl. Acad. Sci. U.S.A.*, 77, 1398, 1980.
187. **Weiner, A. M.**, An abundant cytoplasmic 7S RNA is complementary to the dominant interspersed middle repetitive DNA sequence family in the human genome, *Cell*, 22, 209, 1980.
188. **Haynes, S. R. and Jelinek, W.**, Low molecular weight RNAs transcribed *in vitro* by RNA polymerase III from Alu-type dispersed repeats in Chinese hamster DNA are also found *in vivo*, *Proc. Natl. Acad. Sci. U.S.A.*, 78, 6130, 1981.
189. **Peters, G. G., Harada, F., Dahlberg, J. E., Panet, A., Haseltine, W. A., and Baltimore, D.**, Low molecular weight RNAs of Moloney murine leukemia virus: identification of the primer for RNA-directed DNA synthesis, *J. Virol.*, 21, 1031, 1977.
190. **Jelinek, W. and Leinwand, L.**, Low molecular weight RNAs hydrogen-bonded to nuclear and cytoplasmic poly(A)-terminated RNA from cultured Chinese hamster ovary cells, *Cell*, 15, 205, 1978.
191. **Harada, F. and Ikawara, Y.**, A new series of RNAs associated with the genome of spleen focus forming virus (SFFV) and poly(A)-containing RNA from SFFV-infected cells, *Nucl. Acids Res.*, 7, 895, 1979.

192. **Harada, F., Kato, N., and Hoshino, H.**, Series of 4.5S RNAs associated with poly(A)-containing RNAs of rodent cells, *Nucl. Acids Res.*, 7, 909, 1979.
193. **Harada, F. and Kato, N.**, Nucleotide sequences of 4.5S RNAs associated poly(A)-containing RNAs of mouse and hamster cells, *Nucl. Acids Res.*, 8, 1273, 1980.
194. **Krayev, A. S., Kramerov, D. A., Skryabin, K. G., Ryskov, A. P., Bayer, A. A., and Georgiev, G. P.**, The nucleotide sequence of the ubiquituous, repetitive DNA sequence B1 complementary to the most abundant class of mouse fold-back RNA, *Nucl. Acids Res.*, 8, 120, 1980.
195. **Murray, V. and Holliday, R.**, Mechanism for RNA splicing of gene transcripts, *FEBS Lett.*, 106, 5, 1979.
196. **Goldenberg, C. J. and Raskas, H. J.**, *In vitro* processing of intervening sequences in the precursors of messenger RNA for adenovirus 2 DNA binding protein, *Biochemistry*, 19, 2719, 1980.
197. **Blanchard, J. M., Weber, J., Jelinek, W., and Darnell, J. E.**, *In vitro* RNA-DNA splicing in adenovirus mRNA formation, *Proc. Natl. Acad. Sci. U.S.A.*, 75, 5344, 1978.
198. **Manley, J. L., Sharp, P. A., and Gefter, P. L.**, RNA synthesis in isolated nuclei: identification and comparison of adenovirus 2 transcripts synthesized *in vitro* and *in vivo*, *J. Mol. Biol.*, 135, 171, 1979.
199. **Yang, V. W. and Flint, S. J.**, Synthesis and processing of adenoviral RNA in isolated nuclei, *J. Virol.*, 32, 394, 1979.
200. **Hamada, H., Igarashi, T., and Muramatsu, M.**, *In vitro* splicing of SV40 late mRNA in isolated nuclei from CV-1 cells, *Nucl. Acids Res.*, 8, 587, 1980.
201. **Yang, V. W., Lerner, M. R., Steitz, J. A., and Flint, S. J.**, A small, nuclear ribonucleoprotein is required for splicing of adenoviral early RNA sequences, *Proc. Natl. Acad. Sci. U.S.A.*, 78, 1371, 1981.
202. **Lerner, M. R., Boyle, J. A., Hardin, J. A., and Steitz, J. A.**, Two novel classes of small ribonucleoproteins detected by antibodies associated with lupus erythematosus, *Science*, 211, 400, 1981.
203. **Yang, V. W., Lerner, M. R., Steitz, J. A., and Flint, S. J.**, unpublished observations, 1981.
204. **Weingartner, B. and Keller, W.**, Transcription and processing of adenoviral RNA by extracts from HeLa cells, *Proc. Natl. Acad. Sci. U.S.A.*, 78, 4092, 1981.
205. **Sugden, B. and Keller, W.**, Mammalian DNA-dependent RNA polymerases 1. Purification and properties of an α-amanitin sensitive RNA polymerase and stimulatory factors from HeLa and KB cells, *J. Biol. Chem.*, 248, 3777, 1973.
206. **Manley, J. L., Fire, A., Campo, A., Sharp, P. A., and Gefter, M. L.**, DNA-dependent transcription of adenovirus genes in a soluble whole-cell extract, *Proc. Natl. Acad. Sci. U.S.A.*, 77, 3855, 1980.
207. **Keller, W.**, personal communication, 1981.
208. **Cepko, C. L., Hansen, U., Handa, H., and Sharp, P. A.**, Sequential transcription-translation of SV40 by using mammalian cell extracts, *Mol. Cell. Biol.*, 1, 919, 1981.
209. **Zeevi, M., Nevins, J. R., and Darnell, J. E.**, Nuclear RNA is spliced in the absence of poly(A) addition, *Cell*, 26, 39, 1981.
210. **Yang, V. W. and Flint, S. J.**, unpublished observations, 1980.
211. **Curtis, P. J., Mantei, N., and Weissmann, C.**, Characterization and kinetics of synthesis of 15S β-globin RNA, a putative precursor of β-globin mRNA, *Cold Spring Harbor Symp. Quant. Biol.*, 42, 971, 1977.
212. **Nevins, J. R. and Darnell, J. E.**, Steps in the processing of Ad2 mRNA: poly(A)$^+$ nuclear sequences are conserved and poly(A) addition precedes splicing, *Cell*, 15, 1477, 1978.
213. **Schibler, U., Marcu, K. B., and Perry, R. B.**, The synthesis and processing of mRNAs specifying heavy and eight chain immunoglobins in MPC-11 cell, *Cell*, 15, 1495, 1978.
214. **Gilmore-Herbert, M. and Wall, R,.** Nuclear RNA precursors in the processing pathway to MOPC 21K light chain messenger RNA, *J. Mol. Biol.*, 135, 879, 1979.
215. **Weber, J., Blanchard, J. M., Ginsberg, H. S., and Darnell, J. E.**, Order of polyadenylic acid addition and splicing events in early adenovirus mRNA formation, *J. Virol.*, 33, 286, 1980.
216. **Chien-Kang, S., and Darnell, J. E.**, personal communication, 1981.
217. **Brison, O., Kédinger, C., and Chambon, P.**, Adenovirus DNA template for late transcription is not a replicative intermediate, *J. Virol.*, 32, 91, 1979.
218. **Chien-Kang, S., Wolgemuth, D. J., Hsu, M-T., and Darnell, J. E.**, Transcription and accurate polyadenylation *in vitro* of RNA from the major late adenovirus-2 transcription unit, in press.
219. **Berget, S. M., Moore, C. and Sharp, P. A.** Spliced segments at the 5′ terminus of adenovirus 2 late mRNA, *Proc. Natl. Acad. Sci. U.S.A.*, 74, 317, 1977.
220. **Chow, L. T., Gelinas, K. E., Broker, T. R., and Roberts, R. J.**, An amazing sequence arrangement at the 5′ ends of adenovirus 2 messenger RNA, *Cell*, 12, 1, 1977.
221. **Klessig, D. F.**, Two adenovirus mRNAs have a common 5′ terminal leader sequence encoded at least 10kb upstream from their main coding regions, *Cell*, 12, 9, 1977.
222. **Berget, S. M. and Sharp, P. A.**, Structure of the late adenovirus 2 heterogeneous nuclear RNA, *J. Mol. Biol.*, 129, 547, 1979.
223. **Goldenberg, C. J. and Raskas, H. J.**, *In vitro* splicing of purified precursor RNAs specified by early region 2 of the adenovirus 2 genome, *Proc. Natl. Acad. Sci. U.S.A.*, 78, 5430, 1981.

224. **Laskey, R. A., Mills, A. D., and Norris, N. R.,** Assembly of SV40 chromatin in a cell-free system from *Xenopus* eggs, *Cell,* 10, 237, 1977.
225. **Laskey, R. A. and Earnshaw, W. C.,** Nucleosome assembly, *Nature (London),* 286, 763, 1980.
226. **Beyer, A. L., Miller, O. L., and McKnight, S.,** Ribonucleoprotein structure in nascent RNA is non-random and sequence dependent, *Cell,* 20, 75, 1980.
227. **Beyer, A. L., Bouton, A. H., and Miller, O. L.,** Correlation of hnRNP structure and nascent transcript cleavage, *Cell,* 26, 155, 1981.
228. **Steitz, J. A. and Kamen, R.,** Arrangement of 30S heterogeneous nuclear ribonucleoprotein on polyoma virus late nuclear transcripts, *Mol. Cell Biol.,* 1, 21, 1981.
229. **Carlock, L. and Jones, N.,** unpublished observations, 1981.
230. **Kinniburgh, A. J. and Ross, J.,** Processing of the mouse β-globin mRNA precursor: at least two cleavage-ligation reactions are necessary to excise the larger intervening sequence, *Cell,* 17, 915, 1979.
231. **Nordstrom, J. L., Roop, D. R., Tsai, M.-J., and O'Malley, B. W.,** Identification of potential ovomucoid mRNA precursors in chick oviduct nuclei, *Nature (London),* 278, 328, 1979.
232. **Ryffel, G. U., Wyler, T., Muellner, D. B., and Weber, R.,** Identification, organization and processing intermediates of the pututive precursors of *Xenopus vitellogenin* messenger RNA, *Cell,* 19, 53, 1980.
233. **Tsai, M.-J., Ting, A. C., Nordstrom, J. L., Zimmer, W., and O'Malley, B. W.,** Processing of high molecular weight ovalbumin and ovomucoid precursor RNAs to messenger RNA, *Cell,* 22, 219, 1980.
234. **Soeda, E., Arrand, J. R., Smolar, N., Walsh, J. E., and Griffin, B.,** Coding potential and regulatory signals of the polyoma virus genome, *Nature (London),* 283, 445, 1980.
235. **Friedman, T., Esty, A., LaPorte, P., and Deininger, P.,** The nucleotide sequence and genome organization of the polyoma early region: extensive nucleotide and amino acid homology with SV40, *Cell,* 17, 715, 1979.
236. **Ito, Y., Spurr, N., and Dulbecco, R.,** Characterization of polyoma virus T-antigen, *Proc. Natl. Acad. Sci. U.S.A.,* 74, 1259, 1977.
237. **Schaffhausen, B. S., Silver, B. E., and Benjamin, T. J.,** Tumor antigens in cells productively-infected by wild-type polyoma virus and mutant Ng-18, *Proc. Natl. Acad. Sci. U.S.A.,* 75, 79, 1978.
238. **Hunter, T., Hutchinson, M. A., and Eckhart, W.,** Translation of polyoma virus T-antigens, *Proc. Natl. Acad. Sci. U.S.A.,* 75, 5917, 1978.
239. **Kamen, R., Favaloro, J., Parker, J.,** Treissman, R., Lanig, L., Fried, M., and Mellor, A., Comparison of polyoma virus transcription in productively-infected mouse cells and transformed rodent cell lines, *Cold Spring Harbor Symp. Quant. Biol.,* 44, 63, 1980.
240. **Spector, D. J., McGrogan, M., and Raskas, H. J.,** Regulation of the appearance of cytoplasmic RNAs from region 1 of the adenovirus genome, *J. Mol. Biol.,* 126, 395, 1978.
241. **Chow, L. T., Broker, T. R., and Lewis, J. B.,** Complex splicing patterns of RNAs from the early regions of adenovirus 2, *J. Mol. Biol.,* 134, 265, 1979.
242. **Esche, H., Matthews, M. B., and Lewis, J. B.,** Proteins and messenger RNAs of the transforming region of wild-type and mutant adenoviruses, *J. Mol. Biol.,* 142, 399, 1980.
243. **Young, R. A., Hagenbüchle, O., and Schibler, U.,** A single mouse α-amylase gene specifies two different, tissue-specific mRNAs, *Cell,* 23, 451, 1981.
244. **Tooze, J., Ed.,** *Molecular Biology of Tumor Viruses: Part 2, DNA Tumor Viruses,* 2nd ed., Cold Spring Harbor, New York, 1980.
245. **Early, P., Rogers, J., Davis, M., Calame, K., Bond, M., Wall, R., and Hood, L.,** Two mRNAs can be produced from a single immunoglobulin μ gene by alternative RNA processing pathways, *Cell,* 20, 313, 1980.
246. **Roger, J., Early, P., Carter, C., Calame, K., Bond, M., Hood, L., and Wall, R.,** Two mRNAs with different 3′ ends encode membrane-bound and secreted forms of immunoglobulin μ chain, *Cell,* 20, 303, 1980.
247. **Singer, P. A., Singer, H. H., and Williamson, A. R.,** Different species of messenger RNA encode receptor and sensory IgM μ chains differing at their carboxy termini, *Nature (London),* 285, 294, 1980.
248. **Liu, C. P., Tucker, P. W., Mushinski, J. F., and Blattner, F. R.,** Mapping of the heavy chain genes for mouse immunoglobulins M and D, *Science,* 209, 1348, 1980.
249. **Moore, K. W., Rogers, J., Hunkapiller, T., Early, P., Nottenburg, C., Weissman, I., Bazin, I., Wall, R., and Hood, L.,** Expression of IgD may use both DNA rearrangement and RNA splicing mechanisms, *Proc. Natl. Acad. Sci. U.S.A.,* 78, 1800, 1981.
250. **Maki, R., Roeder, W., Traunecker, A., Sidman, C., Waki, M., Raschke, W., and Tonegawa, S.,** The role of DNA rearrangement and alternative RNA processing in the expression of immunoglobulin dleta genes, *Cell,* 24, 353, 1981.

Chapter 4

5′-TERMINAL MODIFICATION OF mRNAs BY VIRAL AND CELLULAR ENZYMES

Jerry M. Keith

TABLE OF CONTENTS

I. INTRODUCTION

The formation of eukaryotic messenger ribonucleic acid (mRNA) involves a complex series of enzyme-catalyzed biochemical reactions. In contrast to prokaryotic mRNAs, which are characteristically polycistronic transcriptional units containing a 5′-terminal monophosphate or triphosphate, monocistronic eukaryotic mRNA is synthesized in the cell nucleus from a long precursor transcriptional unit known as heterogeneous nuclear RNA (hnRNA). The biosynthesis of an active eukaryotic mRNA from its hnRNA precursor involves several post transcriptional modification and processing events, including cleavage of the precursor, 5′-terminal capping, 5′-terminal and internal methylation, 3′-terminal polyadenylation, and splicing. Considerable information related to the cellular biosynthesis of active mRNA molecules has been obtained from structural analysis and radioisotopic labeling studies of hnRNA and viral and cellular mRNA; however, isolation and characterization studies of the individual enzymes responsible for the molecular modification processes are required for a more complete understanding of the complex mRNA processing events.

One early event in the modification process is formation of the capped 5′-terminus of the mRNA molecule. During the past few years, several laboratories have been particularly interested in the isolation and characterization of the enzymes which catalyze the formation of cap structures. These investigations, utilizing both viral and cellular systems, have greatly added to our understanding of the temporal order of molecular events involved in the biosynthesis and modification of mRNA.

The majority of eukaryotic viral and cellular mRNAs, as well as hnRNAs, contain a 5′-terminal cap structure. In its simplest configuration, the cap structure consists of a 7-methylguanosine residue linked at its 5′-position through a triphosphate bridge to the 5′-position of the penultimate nucleoside in the RNA molecule. This configuration, known as cap *O*, is represented by the general structure $m^7G(5')pppN_pN_2pN_3p$ . . ., where N is a purine or pyrimidine base. In more complex cap structures, the 5′-penultimate nucleoside (N_1) and the adjacent nucleoside (N_2) can also be methylated at the 2′-*O*-position of the ribose sugar to form $m^7G(5')pppN_1^m pN_2$-pN_3p... (cap I) and $m^7G(5')pppN_1^m pN_2^m pN_3p$ . . . (cap II) 5′-terminal structures, respectively. In both viral and cellular mRNAs, the 5′-penultimate 2′-*O*-methyladenosine nucleoside (N_1^m) can be further methylated to N^6, 2′-*O*-dimethyladenosine, resulting in the 5′-terminal structure $m^7G(5')pppm^6A^mp$ . . . The chemical structure of this blocked 5′-terminus is shown in Figure 1.

This review deals primarily with the enzymes involved in the modification of the 5′-terminus of mRNAs resulting in the formation of cap structure. Several excellent reviews by Shatkin,[1] Filipowicz,[2] and Banerjee[3] have dealt with the cap structure and its biological function. In addition, a review by Moss et al.[4] and Banerjee's review[3] include excellent sections on the viral and cellular enzymes involved in the capping reactions. Reviews by Rottman,[5] Adams,[6] Busch et al.,[7] and Revel and Groner[8] on eukaryotic mRNA structure and function also include excellent discussions of the cap structure.

Recently, considerable progress has been made in elucidating the enzymatic mechanism of the key enzyme in cap formation, i.e., the RNA guanylyltransferase or "capping" enzyme. In this review, the author has attempted to make a thorough search of available literature related to the capping enzymes; however, unintentional omissions may possibly have occurred, for which the author is solely responsible. Included also, are both published and unpublished data relating to the various enzymes involved in cap formation with an emphasis on comparing functionally related enzymes isolated or purified from various sources. From the substrate specificity studies of these various enzymes, a temporal sequence of molecular events for the formation of both capped viral and cellular mRNA molecules is presented.

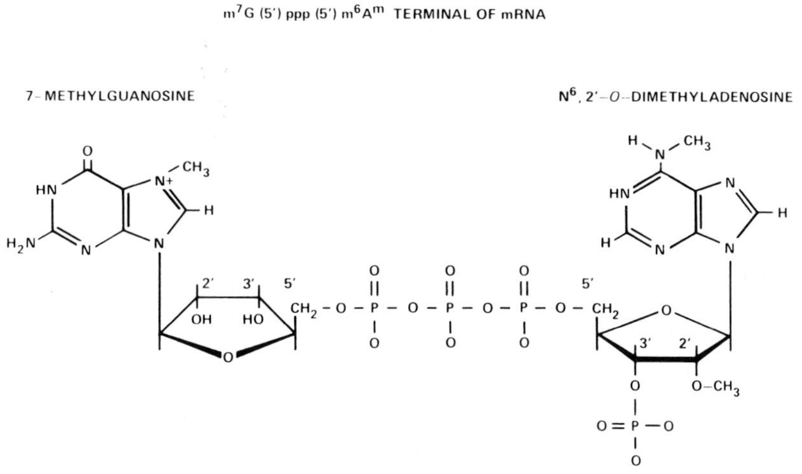

m^7G (5′) ppp (5′) m^6A^m TERMINAL OF mRNA

7-METHYLGUANOSINE

N^6, 2′-*O*-DIMETHYLADENOSINE

FIGURE 1. Chemical structure of the $m^7G(5')pppm^6A^m$ terminus of mRNA.

II. MECHANISM OF VIRAL mRNA CAP FORMATION

A. Vaccinia Virus

The identification and elucidation of the methylated 5′-terminal capped structure of viral[9,10] and eukaryotic[11-13] mRNAs quickly led to a search for the enzymes involved in their biosynthesis. Early enzyme studies in Moss' laboratory by Ensinger et al.[14] demonstrated that both RNA guanylyltransferase and methyltransferase activities could be solubilized from vaccina virus cores. Using labeled GTP and a soluble extract which had been passed through DEAE-cellulose to remove nucleic acids, they showed that a GMP residue was specifically transferred from GTP to the 5′-terminus of an unmethylated vaccinia mRNA substrate to form the structure G(5′)pppGp- and G(5′)pppAp-. Furthermore, in the presence of *S*-adenosylmethionine (AdoMet), the blocked 5′-termini were converted to the cap I structures $m^7G(5')pppG^mp$- and $m^7G(5')pppA^mp$-. Similarly, the enzyme extract modified the 5′-end of synthetic poly(A) to form the structure $m^7G(5')pppA^mp$-.

The solubilized GTP:mRNA guanylyltransferase or "capping" activity demonstrated by Ensinger et al.[14] was subsequently purified by Moss and co-workers.[15] In this same study, it was also shown that the viral capping enzyme copurifies with an AdoMet:mRNA(guanine-7-)-methyltransferase responsible for adding the methyl group to the 7-position of the guanosine in the cap structure. Venkatesan and co-workers[16] later demonstrated that a third enzyme activity, i.e., a RNA triphosphatase, was also associated with the capping and methylating activities forming a 127,000 mol wt multienzyme complex. The ability to use the vaccinia virus guanylyltransferase to specifically label the 5′-ends of RNA has led to considerable interest in this enzyme as an analytical tool.[17-19] The AdoMet:mRNA(nucleoside-2′-)-methyltransferase activity in the soluble extract, which is responsible for the formation of the 2′-*O*-methylnucleoside in the penultimate position of the cap structure, was later purified from vaccinia virus particles by Barbosa and Moss.[20,21]

In viral and cellular mRNAs synthesized in vivo, the adenosine residue in cap structures exists primarily as the unusual dimethylated nucleoside, N^6,2′-*O*-dimethyladenosine (m^6A^m), as shown in Figure 1.[11,22,23] Messenger RNAs synthesized in vitro by enzymes contained within the vaccinia virions[9] and vesicular stomatitis virions[24] contain $m^7G(5')pppA^m$, suggesting that the enzyme responsible for modification of A^m to m^6A^m may be present in the cytoplasm of animal cells.[23,25] Indeed, an

AdoMet:RNA(2′-*O*-methyladenosine-N^6-)-methyltransferase which specifically catalyzes the transfer of a methyl group from AdoMet to the N^6-position of a 2′-*O*-methyladenosine residue located within the capped 5′-end of mRNA has been isolated from the cytoplasm of HeLa cells.[26]

Based on the properties reported for the cellular and vaccinia modification enzymes mentioned above and other studies on purified vaccinia enzymes reported by Martin and Moss[27,28] and Hurwitz' laboratory,[29-32] as well as transcriptional studies by Moss et al.[33] using vaccinia virus cores and specifically labeled ribonucleoside triphosphates, the following temporal sequences of reactions was proposed for the formation of the capped 5′-terminus of vaccinia virus mRNA.[4] In this series of reactions, pppN- is the 5′-terminus of a RNA molecule where N is either a guanosine or adenosine residue and AdoHcy is the abbreviation for *S*-adenosylhomocysteine.

$$\overset{\gamma\beta\alpha}{\text{pppN-}} \rightarrow \overset{\beta\alpha}{\text{ppN}} + \overset{\gamma}{\text{Pi}} \qquad (1)$$

$$\overset{*}{\text{pppG}} + \overset{\beta\alpha}{\text{ppN-}} \rightleftarrows \overset{*\beta\alpha}{\text{G(5')pppN-}} + \text{PPi} \qquad (2)$$

$$\text{AdoMet} + \text{G(5')pppN-} \rightarrow \text{m}^7\text{G(5')pppN-} + \text{AdoHcy} \qquad (3)$$

$$\text{AdoMet} + \text{m}^7\text{G(5')pppN-} \rightarrow \text{m}^7\text{G(5')pppN}^{\text{m}}\text{-} + \text{AdoHcy} \qquad (4)$$

$$\text{AdoMet} + \text{m}^7\text{G(5')pppA}^{\text{m}}\text{-} \rightarrow \text{m}^7\text{G(5')pppm}^6\text{A}^{\text{m}}\text{-} + \text{AdoHcy} \qquad (5)$$

In this modification scheme for vaccinia virus mRNAs, reactions (1), (2), and (3) are catalyzed by the vaccinia capping enzyme complex consisting of the following activities: a RNA triphosphatase,[16] a GTP:mRNA guanylyltransferase, and AdoMet: mRNA(guanine-7-)-methyltransferase.[15] A vaccinia virus AdoMet: mRNA(nucleoside-2′-)-methyltransferase[20] catalyzes reaction (4), whereas a cellular AdoMet:mRNA(2′-*O*-methyladenosine-N^6-)-methyltransferase[26] catalyzes reaction (5).

B. Reovirus

Studies by Furuichi et al.[34] show that a similar sequence of reactions exists for the formation of the capped 5′-ends of reovirus mRNAs. In these studies, it was demonstrated that the diphosphate-terminated dinucleotide ppGpC functions as a substrate for a core-associated guanylyltransferase and is converted to G(5′)pppGpC by the addition of GMP from GTP. The monophosphate-terminated dinucleotide pGpC was not a substrate for the reaction; however, pppGpC was utilized in the reaction after removal of the γ-phosphate by a core-associated nucleotide phosphohydrolase. It was also shown that methyltransferases associated with the reovirus cores transferred methyl groups from AdoMet to form the cap I structure m^{7}G(5′)pppG^{m}pC in the same sequence of reactions as in vaccinia virus. Furthermore, using transcriptional studies with virus cores, it was shown that 5′-terminal caps are formed during initiation of RNA synthesis.

C. Cytoplasmic Polyhedrosis Virus

Studies utilizing cytoplasmic polyhedrosis virus (CPV)[35] and vesicular stomatitis virus (VSV)[24,36,37] indicate that the mechanism of capping or methylation or both appears to be different than in vaccinia and reovirus. Transcriptional and structural analysis studies[38,39] indicate that CPV is a double-stranded RNA virus which, like reovirus,[40] transcribes ten genome RNA segments in vitro, resulting in CPV mRNAs containing the 5′-terminal cap structure m^{7}G(5′)pppA^{m}pGp-.[41] However, in contrast to vaccinia virus[42] and reovirus,[43] CPV mRNA synthesized in vitro is greatly stimulated by the addition of AdoMet.[44] Further CPV studies indicate that the methyltransferase competitive inhibitor, AdoHcy, also effectively stimulates in vitro mRNA synthesis, resulting in mRNAs with G(5')pppA- and ppA- 5'-ends.[45] These results suggest that AdoMet and AdoHcy may have some influence on a regulatory mechanism for the RNA polymerase and capping enzymes.

In another study using the β, γ-imido analog of ATP (AMP-pNHp), Furuichi[35] demonstrated a relationship between cap formation and CPV mRNA synthesis. In this investigation, ribonucleoside triphosphates were replaced by the corresponding β, γ-imido ribonucleoside triphosphate analogs in a RNA-synthesizing reaction mixture containing AdoMet. No RNA synthesis was observed when AMP-pNHp was substituted for ATP, even though mRNA synthesis occurred in the presence of either UMP-pNHp or GMP-pNHp. Since β, γ-imido analogs are resistant to nucleotide phosphohydrolase (an enzyme involved in cap formation) and because the ATP molecule that becomes the 5'-terminal nucleotide of CPV mRNA must be cleaved by this enzyme at the β,γ-position during the capping reaction, it was concluded from these results that in the CPV transcriptional system cap formation is a pretranscriptional event which is a prerequisite for mRNA synthesis.

D. Vesicular Stomatitis Virus

Transcriptional and structural analysis studies of VSV mRNA indicate that capping and methylation are tightly coupled to transcription[46] and that the mechanism for the formation of the VSV cap structure is significantly different than for reovirus, CPV, or vaccinia virus mRNAs.[24,36] In these experiments, a ppG residue was transferred from GTP to the 5'-end of the viral mRNA, resulting in the cap structure $G(5')\overset{\alpha}{p},\overset{\beta}{p},\overset{\alpha'}{p}Ap$, in which both the α and β-phosphate in the cap were derived from the donor GTP molecule. This mechanism is suitable for capping the monophosphate end of a RNA molecule formed by nuclease cleavage. Thus far, an eukaryotic cellular capping activity with this specificity and mechanism has not been detected.

A recent study by Moyer[47] in which cycloleucine was used to inhibit methylation in VSV-infected cells suggests that within infected cells the sequential pathway of viral mRNA methylation occurs in the same order as in vaccinia and reovirus mRNA, i.e., G(5')pppA- → $m^7G(5')pppA$- → $m^7G(5')pppA^m$-. Based on in vitro transcriptional studies, the reverse order of methylation was previously proposed.[37] It is not known if the vesicular stomatitis virion-associated methyltransferases are coded for by the virus or the host cell; and, since in the Moyer[47] study the VSV inoculum used to infect the cycloleucine-treated cells was grown in untreated cells, it is not clear if the differences in the methylation pathway observed in vitro result from an unknown in vivo effect of cycloleucine or occurs because a different methyltransferase is utilized within the cell.

III. MECHANISM OF CELLULAR mRNA CAP FORMATION

Cellular mRNAs are capped and methylated in a series of reactions very similar to those proposed for vaccinia and reovirus. However, some important differences exist in the enzymes involved in the reactions. Elucidation of the mechanism of the cellular capping reactions is important for understanding mRNA processing. Various studies imply that the capping reactions occur within the nucleus; i.e., heterogeneous nuclear RNAs are capped[48,49] and subcellular fractions containing nuclei synthesize capped mRNA in vitro.[50,51] Studies utilizing a subcellular system suggest that in mouse L cell nuclei, cap formation is closely linked to transcription since capping was sensitive to α-amanitin and all four ribonucleoside triphosphates were required.[51] Other studies showed that a crude fraction prepared from HeLa cell nuclei could cap synthetic polynucleotides containing diphosphate or triphosphate 5′-ends.[52,53] In a later study, a GTP:mRNA guanylyltransferase responsible for cap formation was purified from HeLa cell nuclei.[54] Characterization studies demonstrate that this enzyme catalyzes the transfer of a GMP residue from GTP to the 5′-diphosphate end of RNA by a mechanism similar to that for the vaccinia virus capping enzyme.[55,56] However, in contrast to the vaccinia enzyme, neither a RNA triphosphatase activity nor a RNA(guanine-7-)-methyltransferase activity is associated with the purified enzyme. An analogous RNA guanylyltransferase with similar specificities has been partially purified from rat liver nuclei[57] and calf thymus.[58] Recent reports on the purification and characterization of a RNA guanylyltransferase from wheat germ have shown that this enzyme has a catalytic mechanism similar to that described for the HeLa cell capping enzyme and, like the HeLa guanylyltransferase, is not associated with either the RNA(guanine-7-)-methyltransferase or the RNA triphosphatase.[59,60] In fact, the RNA(guanine-7-)-methyltransferase has also recently been partially purified from wheat germ.[60,61]

From the results of these various studies, the cellular RNA guanylyltransferase appears to be smaller than the capping enzyme isolated from vaccinia virus; furthermore, it is separable from both RNA triphosphatase and RNA (guanine-7-)-methyltransferase activities. Indeed, a RNA(guanine-7-)-methyltransferase has been purified as a separate enzyme from both HeLa cells[62] and wheat germ.[60] The additional enzymes involved in cap formation, including a RNA(2′-*O*-methyladenosine-N^6-)-methyltransferase[26] and two separate RNA(nucleoside-2′-)-methyltransferases,[63] have also been purified from HeLa cells. In addition, methyltransferase activities which convert the dinucleoside triphosphate G(5′)pppG to m^7G(5′)pppG^m have been detected in an embryonic chick lens extract[64] and a guanine-7-methyltransferase activity responsible for the in vitro methylation of tobacco mosaic virus specific RNA has been reported.[65]

Based on the substrate specificities reported for the capping and methylating enzymes isolated from HeLa cells, Langberg and Moss[63] proposed the sequence of reactions shown in Figure 2 for modification of the 5′-ends of HeLa cell mRNAs. The enzymes which catalyze the indicated reactions are

1. RNA triphosphatase
2. RNA guanylyltransferase[54]
3. RNA(guanine-7-)-methyltransferase[62]
4. cap I RNA(nucleoside-2′-)-methyltransferase[63]
5. RNA(2′-*O*-methyladenosine-N^6-)-methyltransferase[26]
6. cap II RNA(nucleoside-2′-)-methyltransferase[63]

As discussed in the report by Langberg and Moss,[63] this order of capping and methylation steps, shown in Figure 2, is consistent with studies of the isolated enzymes. The first step in this scheme is the removal of the 5′-terminal γ-phosphate after initiation of transcription. The RNA triphosphatase responsible for this reaction has not yet been isolated from HeLa cells; however, this step is implied since the purified HeLa RNA guanylyltransferase shows a clear substrate specificity for RNA with a diphosphate 5′-end.[55] This same diphosphate 5′-end specificity has also been demonstrated with the RNA guanylyltransferases isolated from rat liver nuclei[57] and wheat germ,[59] as well as vaccinia virus.[16] However, in vaccinia virus, the RNA guanylyltransferase exists as a 127,000 dalton multienzyme complex with RNA(guanine-7-)-methyltransferase[15,29] and RNA triphosphatase;[16] whereas in HeLa cells[54] and wheat germ,[59] the RNA triphosphatase appears to be separated from the capping activity at an early stage of enzyme purification. The RNA (guanine-7-)-methyltransferase from both HeLa cells[62] and wheat germ[60] is also clearly separated from the purified RNA guanylyltransferase.

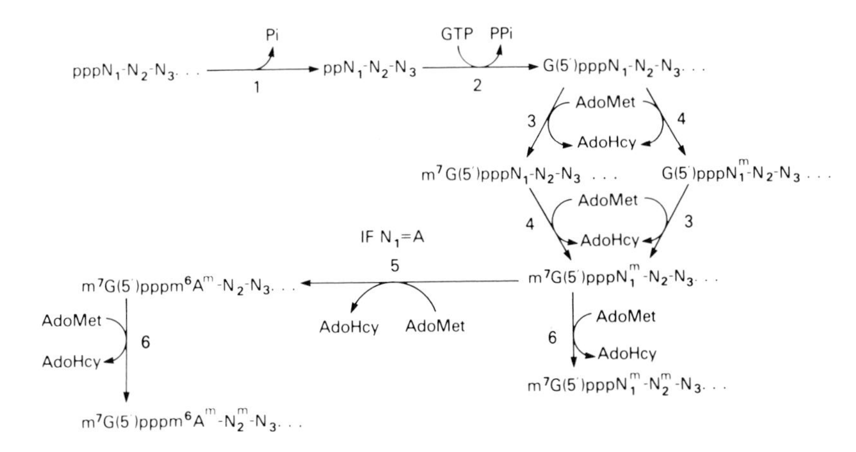

FIGURE 2. Proposed sequence of reactions involved in modification of the 5′-ends of HeLa cell mRNA. The enzymes that catalyze the indicated reactions are (1) RNA triphosphatase, (2) RNA guanylyltransferase, (3) RNA(guanine-7-)-methyltransferase, (4) cap I RNA(nucleoside-2′-)-methyltransferase, (5) RNA(2′-*O*-methyladenosine-N^6-)-methyltransferase, and (6) cap II RNA(nucleoside-2′-)-methyltransferase.

Since oligonucleotides with diphosphate 5′-ends can be capped[55,57,66] and subsequently methylated,[60-63] it has been suggested that these events begin soon after initiation of RNA synthesis.[63] Furthermore, the inability of the HeLa RNA guanylyltransferase to cap a ribonucleoside diphosphate, such as ADP, implies that pretranscriptional capping is an unlikely event in this system.[55] In vitro studies with HeLa enzymes indicate that a cap structure is required for both guanine-7-methylation[62] and 2′-*O*-methylation;[63] however, since the affinities of cap I methyltransferase for polyribonucleotides ending in G(5′)pppN- and m^7G(5′)pppN- were virtually identical, it could not be determined which methylation step occurs first.[63] Substrate specificity studies with HeLa cell RNA (2′-*O*-methyladenosine-N^6-)-methyltransferase suggest that methylation of the N^6-position of adenosine residue in cap I structures follows the 2′-*O*-methylation step.[26]

The final step in the mRNA modification scheme presented in Figure 2 is the 2′-*O*-methylation of the nucleotide adjacent to the 5′-end of the RNA molecule forming the cap II structure. A separate RNA(nucleoside-2′-)-methyltransferase (cap II methyltransferase) responsible for this modification has been purified from a HeLa cell cytoplasmic fraction.[63] The properties and specificities of the enzymes referred to in the viral and cellular mRNA capping schemes are presented in Table 1 and are described in more detail below.

IV. VIRAL ENZYMES INVOLVED IN mRNA CAP FORMATION

A. GTP:mRNA Guanylyltransferase·AdoMet:mRNA(guanine-7-)-Methyltransferase·RNA Triphosphatase Capping Enzyme Complex

The mRNA guanylyltransferase and the mRNA(guanine-7-)-methyltransferase were first isolated from vaccinia virions by Martin et al.[15] They demonstrated that both enzyme activities remained inseparable throughout a 200-fold purification procedure involving sequential chromatography on DEAE-cellulose, DNA-agarose (to which the enzymes did not bind), poly(U)-Sepharose, poly(A)-Sepharose, and Sephadex® G-200 and also during sedimentation through sucrose density gradients. A molecular weight

Table 1
ENZYMES INVOLVED IN mRNA CAP FORMATION

Enzyme	Source	Molecular weight	Preferred acceptor substrate	K_m Value - Substrate	Ref.
RNA triphosphatase	Vaccinia virus	ND[a,b]	$pppN(pN)_n$	ND	16
GTP:mRNA guanylyltransferase	Vaccinia virus	95,000[b,c]	$ppN_1(pN)_n$[d]	5.2 μM $ppA(pA)_n$	15
				360 μM ADP	16, 27, 28, 31, 32
	HeLa cell nuclei	48,500[e]	$ppN(pN)_n$	1.1 μM GTP (donor)	54
		65,000[c]		285 nM ppApGp	55, 56
				19 nM ppA $(pA)_n$	52
	Rat liver nuclei	65,000[f]	$ppN(pN)_n$	ND	57
	Wheat germ	65,000[e]	$ppN(pN)_n$	2.7 μM GTP (donor)	59, 60
		85,000[c]		14.2 nM ppA(pA)n	84
AdoMet:mRNA(guanine-7-) -methyltransferase[g]	Vaccinia virus	ND[b]	G(5′)pppNp-	0.21 μM G(5′)pppA$(pA)_n$	15
				120 μM G(5′)pppG	27, 28
				530 μM GTP	16
	HeLa cell cytoplasm	56,000[e,f]	G(5′)pppNp-	ND	62
	Rat liver nuclei	130,000[f]	G(5′)pppNp-	ND	57
	Wheat germ	ND	G(5′)pppNp-	ND	60, 61
AdoMet:mRNA(nucleoside-2′-) -methyltransferase	Vaccinia virus	38,000[e,f,h]	m^7G(5′)pppAp-	2.0 μM AdoMet (donor)	20, 21
				5 nM BMV RNA[g]	
AdoMet: mRNA(nucleoside-2′-) -methyltransferase (cap I methyltransferase)	HeLa cell nuclei	ND	G(5′)pppNpNp- or m^7G(5′)pppNpNp-	0.6 μM AdoMet (donor)	63
				1.4 nM G(5′)ppA$(pA)_n$	
				1.6 nM m^7G(5′)pppA$(pA)_n$	
AdoMet:mRNA(nucleoside-2′-) -methyltransferase (cap II methyltransferase)	HeLa cell cytoplasm	ND	m^7G(5′)pppN^{m}pNp-	ND	63

AdoMet:mRNA(2′-*O*-methyladenosine-N^6-)-methyltransferase	HeLa cell cytoplasm	65,000[e]	$m^7G(5')pppA^mpNp$-	0.5 μM AdoMet (donor) 2—3 nM $m^7G(5')pppA^mpN$-	26

[a] ND, not determined.

[b] In vaccinia virus, the triphosphatase, guanylyltransferase, and guanine-7-methyltransferase activities exist as a 127,000 dalton complex consisting of two polypeptides with molecular weights of 95,000 and 31,400 (see References 15 and 16). A native molecular weight of 120,000 has also been reported for the complex with molecular weights of 95,800 and 26,400 for the two polypeptides.[31] During the capping reaction, the 95,000 dalton polypeptide forms a covalent intermediate with GMP; however, the other enzyme activities have not yet been assigned to a specific polypeptide.[32] A 120,000 dalton capping enzyme complex consisting of three polypeptides with molecular weights of 95,000; 59,000; and 28,000 has also been reported.[29,30]

[c] Denatured enzyme-GMP complex/gel electrophoresis.

[d] N_1 = Purine.

[e] Native/sedimentation gradient.

[f] Native/gel filtration.

[g] Abbreviation: BMV, brome mosaic virus; AdoMet, *S*-adenosylmethionine.

[h] Denatured/gel electrophoresis.

of 127,000 for the purified enzyme complex was determined by gel filtration on Sephadex® G-200 and sedimentation in sucrose density gradients. Polyacrylamide gel electrophoresis of the purified enzyme under denaturing conditions suggested that the enzyme complex consisted of two polypeptides with molecular weights of 95,000 and 31,400. Later studies by Venkatesan et al.[16] showed that a third enzyme activity, a RNA triphosphatase, cochromatographed with the vaccinia virus capping and methylating enzyme complex on seven different ion exchange or affinity columns and also cosedimented with it on a glycerol gradient. They also showed that the RNA triphosphatase removes the γ-phosphate from triphosphate-ended polyribonucleotides before capping and, on a molar basis, the triphosphatase was about 100 times more active than the associated RNA guanylyltransferase.

These data explain the similar rates of capping diphosphate- and triphosphate-ended polyribonucleotides also reported in this study. Furthermore, the data indicate that cleavage of the γ-phosphate and capping of the RNA are consecutive but not necessarily coupled reactions. A slight stimulation of RNA triphosphatase activity was found upon the addition of GTP to the reaction mixture but not the obligatory coupling of the activities reported by Monroy and co-workers.[30] Furthermore, varying the salt concentration or the pH of the reaction mixture did not cause a dependence of the triphosphatase on GTP. In addition, a much greater capping activity with triphosphate-ended polyribonucleotides, compared with diphosphate-ended polyribonucleotides, which was reported by Monroy et al.,[30] was not found by Venkatesan et al.[16] Although Tutas and Paoletti[67] cited unpublished data regarding a separation of RNA triphosphatase from RNA guanylyltransferase by chromatography on ADP-agarose, Venkatesan et al.,[16] using similar conditions, repeatedly found that the two activities coeluted.

These studies indicate that the vaccinia capping enzyme complex specifically catalyzes the modification of nascent viral RNA and synthetic homopolyribonucleotides in the following sequence:[16,27,28]

$$\overset{\gamma\beta\alpha}{\text{pppN}}(\text{pN})_n \rightarrow \overset{\beta\alpha}{\text{ppN}}(\text{pN})_n + \overset{\gamma}{\text{Pi}} \quad (6)$$

$$\overset{\beta\alpha}{\text{ppN}}(\text{pN})_n + \text{GTP} \rightleftarrows \text{G(5')}\overset{\beta\alpha}{\text{pppN}}(\text{pN})_n + \text{PPi} \quad (7)$$

$$\text{G(5')pppN}(\text{pN})_n + \text{AdoMet} \rightarrow \text{m}^7\text{G(5')pppN}(\text{pN})_n + \text{AdoHcy} \quad (8)$$

The physical association of these three capping and methylating activities in a single complex is an efficient arrangement within the virion to ensure rapid capping and methylation of initiated mRNA molecules. The formation of the 5′-capped end protects the RNA from exonuclease degradation[68,69] while guanine-7-methylation occurring in step (8) greatly inhibits reversal of capping step (7).[27-30]

Further studies by Martin and Moss[27,28] demonstrated that although all the homopolyribonucleotides tested, i.e., $pp(A)_n$, $pp(G)_n$, $pp(I)_n$, $pp(U)_n$, and $pp(C)_n$, are acceptor substrates for the mRNA guanylyltransferase, those containing purines are preferred. Both GTP and dGTP are donor substrates in the capping reaction; however, 7-methylguanosine triphosphate was not utilized, indicating that guanylylation must occur before the guanine-7-methylation step in cap formation. In these studies, Martin and Moss also showed that the guanylyltransferase requires Mg^{++} for activity and is preferred over Mn^{++} and that the pH optimum is 7.8. As expected from the reaction mechanism, PPi is a strong inhibitor. Furthermore, in the presence of PPi, the reverse reaction (that is, the formation of GTP from PPi and RNA containing the capped 5′-

terminal G[5′]pppN-) readily occurs. Similar experiments demonstrated that in the presence of PPi the mRNA guanylyltransferase also catalyzes the pyrophosphorolysis of the dinucleoside triphosphate G(5′)pppA, but does not cleave the methylated configuration $m^7G(5')pppA$.

The characteristics of the AdoMet:mRNA(guanine-7-)-methyltransferase were also reported in these studies. The preferred acceptor of this enzyme was the product of the guanylyltransferase reaction, i.e., a polyribonucleotide with the 5′-terminal sequence G(5′)pppN-. The enzyme catalyzes the transfer of a methyl group from AdoMet to the 7-position of the terminal guanosine residue in the cap structure; internal or conventional 5′- terminal guanosine residues are not methylated. The reaction is inhibited by AdoHcy and is not dependent on divalent cations or GTP. Optimal enzyme activity is observed in a broad pH range around neutrality. Dinucleoside triphosphates with the structure G(5′)pppN were less efficiently methylated by the enzyme, as were GTP, dGTP, ITP, GDP, GMP, and guanosine. The enzyme will not catalyze the transfer of methyl groups to ATP, XTP, CTP, UTP, or guanosine-containing compounds with phosphate groups in either the 2′ or 3′ positions or in 3′-5′ phosphodiester linkages.

Monroy et al.[29,30] reported an approximately 10,000-fold purification of the guanylyltransferase and guanine-7-methyltransferase from purified vaccinia virus. In these studies, different properties were observed than previously reported by Moss and coworkers.[15,27,28] In the Monroy et al.[29,30] studies, both enzyme activities copurified through chromatography on DNA-agarose, phosphocellulose, and sedimentation through glycerol gradients. A molecular weight of 120,000 was determined for the native enzyme activities by sedimentation through glycerol gradients. Analysis of the 120,000 mol wt native protein by gel electrophoresis under denaturing conditions indicated that a major band of 95,000 is released along with minor bands having molecular weights of 59,000 and 28,000. Characterization studies of the purified guanylyltransferase suggested that this enzyme specifically uses a 5′-triphosphate-ended RNA. The apparent K_m reported for triphosphate-ended poly(A) is 0.2 μM.

The enzyme catalyzes the transfer of a GMP residue from GTP to form the triphosphate 5′-5′ bridge of the cap structure. The α- and β-phosphates of the bridge are derived from the RNA acceptor molecule. It was also reported that during the guanylylation reaction in which triphosphate-terminated poly(A) was used as the acceptor molecule Pi and PPi both were released in a one-to-one stoichiometry with cap formation. The discrepancy concerning the results reported by Monroy et al.[29,30] and those reported by Moss and colleagues[15,27,28] is presently unclear. However, as described above, Venkatesan et al.[16] reported that a RNA triphosphatase activity is associated with the vaccinia capping enzyme complex.

Shuman et al.,[31] in Hurwitz' laboratory, reported that a GTP-pyrophosphate exchange activity is associated with the vaccinia capping enzyme complex. This exchange activity catalyzes a nucleotide-pyrophosphate exchange with GTP. The enzyme requires Mg^{++} and has an alkaline pH optimum. This exchange activity cochromatographs with the vaccinia capping enzyme complex through successive steps of DEAE-cellulose, DNA-cellulose, and phosphocellulose and through sedimentation in a glycerol gradient. In this report, they also demonstrated that under high salt conditions the GTP-PPi exchange, capping, and methylating activities cosedimented with a RNA triphosphatase activity and a nucleoside triphosphate phosphohydrolase activity as a 6.5 S multifunctional enzyme complex. This complex consists of two major polypeptides with molecular weights of 96,000 and 26,000. Venkatesan et al.[16] demonstrated the association of a RNA triphosphatase with the capping and methylating enzymes in a 127,000 mol wt complex, which was previously reported to contain only two major polypeptides of 95,000 and 31,400 mol wt.[15]

Shuman and Hurwitz[32] have recently demonstrated that in vaccinia virus the guanylylation reaction is a series of at least two partial reactions:

$$\text{GTP} + \text{E} \rightleftarrows \text{E-pG} + \text{PPi} \tag{9}$$

$$\text{E-pG} + \text{(p)ppN(pN)}_n\text{-} \rightleftarrows \text{G(5')pppN(pN)}_n\text{-} + \text{E} \tag{10}$$

In this reaction scheme, the guanylyltransferase first reacts with GTP in the absence of an RNA acceptor to form a covalent enzyme-guanylate intermediate. Analysis of the capping enzyme-GMP intermediate by polyacrylamide gel electrophoresis under denaturing conditions indicated that the GMP residue was covalently linked to the 95,000 mol wt polypeptide of the 127,000 mol wt complex. It is suggested that this linkage is a phosphoamide bond, as judged by the acid-labile, alkali-stable nature of the covalent bond and by the susceptibility of the linkage to cleavage by hydroxylamine at pH 4.75. A divalent cation (Mg^{++} or Mn^{++}, but not Ca^{++}) and either GTP or dGTP is required for the formation of the enzyme-guanylate complex. The presence of a cap acceptor is not required. As shown in reaction (10) above, Shuman and Hurwitz[32] demonstrated that the purified enzyme-guanylate complex will catalyze the transfer of the GMP moiety to the 5′-end of RNA to form a G(5′)pppN- cap structure. Venkatesan and Moss[56] have recently shown that reaction (10) is reversible with the vaccinia guanylyltransferase but not with the HeLa cell enzyme.

It is suggested that the RNA triphosphatase activity of the vaccinia capping enzyme complex remains associated with the enzyme-guanylate complex since the associated intermediate has the ability to cap triphosphate-terminated poly(A).[16,31] In the presence of PPi, the enzyme-GMP intermediate will form GTP; transfer of the GMP residue to either PPi or the 5′-end of RNA requires Mg^{++}. From the studies presented above, it appears that three enzyme activities (γ-phosphate cleavage, transguanylation, and transmethylation) catalyze the enzymatic reactions carried out by the vaccinia capping enzyme complex and that this complex consists of two major subunits of 95,800 and 26,400 mol wt, as reported by Shuman et al.,[31] or 95,000 and 31,400 mol wt, as reported by Martin et al.[15]

The results reported by Shuman and Hurwitz[32] clearly implicate the 95,000 mol wt subunit in the guanylyltransferase reaction. Although, as the have suggested, the 95,000 subunit is necessary for capping according to the proposed mechanism, it is not clear whether it acts alone to catalyze the transfer of the GMP residue or whether it requires the presence of the 26,000 mol wt protein. It is also not clear from these studies where the functional domains of the RNA triphosphatase or methyltransferase activities are located. These, as well as other structure-function relationship studies, should be facilitated by the ability to specifically label the 95,000 mol wt subunit with [α-^{32}P]GTP.

This mechanism of cap formation involving a covalent enzyme-guanylate intermediate has now been demonstrated with the mRNA guanylyltransferase isolated from HeLa cells[56] and with the guanylyltransferase isolated from wheat germ,[60] indicating a common catalytic mechanism for modifying animal, plant, and vaccinia virus mRNAs. Preliminary results from Shatkin's laboratory[70] with reovirus and insect CPV suggest that a similar mechanism for cap formation involving an enzyme-GMP intermediate also exists in these systems. In these studies a single polypeptide band was labeled when [α-^{32}P]GTP was included in a reaction mixture with reovirus cores or CPV. In similar experiments, [β-^{32}P]GTP was not incorporated into the polypeptide band. The labeled polypeptide is presumably the guanylyltransferase-GMP intermediate since the ^{32}P-labeled GMP was transferred to dinucleoside acceptors to form nuclease P_1 and phosphatase-resistant caps. In reovirus, this activity apparently resides in a λ polypeptide. It is quite likely that the rat liver guanylyltransferase has the same reaction mechanism since, like vaccinia enzyme, partially purified rat liver enzyme catalyzes PPi exchange with GTP.[57] Thus, it appears likely that formation of a covalent guanylyltransferase-GMP intermediate is a general mechanism for the formation of cap.

An enzyme mechanism similar to that described for the mRNA guanylyltransferase has been shown to participate in the DNA ligase reaction.[71] In both *Escherichia coli* DNA ligase and T4 DNA ligase, a covalent enzyme-AMP intermediate is formed in which the AMP residue is linked via a phosphoamide bond to the ε-amino group of a single lysine residue of the enzyme.[72] Shabarova[73] has suggested that a general mechanistic feature of nucleotidyltransfer reactions may be the formation of enzyme-nucleotide intermediates in which the nucleotide is linked to the enzyme through a phosphoamide bond.

B. mRNA(Nucleoside-2′-)-Methyltransferase

Barbosa and Moss[20] reported the isolation and characterization of a mRNA(nucleoside-2′-)-methyltransferase from vaccinia virus cores. A 350-fold enzyme purification was achieved by sequential chromatography on columns of DEAE-cellulose, CM-Sephadex®, and ADP-agarose. Brome mosaic virus RNA, which contains the cap structure $m^7G(5')pppG$- at the 5′ end,[74] was used as a substrate for the vaccinia methyltransferase. The enzyme catalyzed the transfer of a methyl group from AdoMet to the 2′-*O*-position of the penultimate nucleoside in the cap structure, resulting in $m^7G(5')pppG^m$-. Analysis of the purified enzyme by gel electrophoresis under denaturing conditions revealed a single polypeptide with a molecular weight of 38,000. A similar molecular weight was determined by analysis using gel filtration and sedimentation through sucrose gradients. An isoelectric point of pH 8.4 was reported for the purified enzyme.

Substrate specificity studies indicated that RNAs ending in pN-, ppN-, or even G(5′)pppN- are not methyl acceptors.[21] These results demonstrate that the final step in the formation of the $m^7G(5')pppN^m$- cap structure is the addition of the 2′-*O*-methyl to the penultimate nucleoside. In these studies, viral RNAs containing both adenosine and guanosine in the penultimate position were methylated, although adenosine appears to be methylated more rapidly by the purified enzyme.

Methylation studies utilizing homopolyribonucleotides containing $m^7G(5')pppN$- ends indicate that the best methyl acceptors are poly(A) and poly(I). Significant, but much less, activity was obtained using poly(G), poly(U), and poly(C) as the methyl acceptors. Dinucleoside triphosphates with a structure such as $m^7G(5')pppN$ are poor substrates and do not compete with capped RNA in the methylase reaction. Additional studies indicated that the RNA(nucleoside-2′-)-methyltransferase is inhibited by AdoHcy. The pH optimum for the enzyme is 7.5 and the catalyzed reaction does not require divalent cations. A K_m value of 2.0 μM for AdoMet and approximately 5 n*M* for brome mosaic virus RNA was reported in this study.

V. CELLULAR ENZYMES INVOLVED IN mRNA CAP FORMATION

A. GTP:mRNA Guanylyltransferase and AdoMet:mRNA(Guanine-7-)-Methyltransferase: Rat Liver Nuclei

Mizumoto and Lipmann[57] reported the isolation of a guanylyltransferase and guanine-7-methyltransferase from rat liver nuclei. These enzymes were purified from a sonicated extract by column chromatography. The two activities coeluted from a hydroxylapatite column and were separated by Sephadex® G-150 column chromatography. The molecular weight of these enzymes, as estimated by gel filtration, was approximately 65,000 for the guanylyltransferase and 130,000 for the guanine-7-methyltransferase. The methylase activity was assayed using a synthetic dinucleoside triphosphate G(5')pppG as the methyl acceptor and AdoMet as the methyl donor. Further purification of the guanylyltransferase was obtained by column chromatography using CM-Sephadex®. In this study, it was shown that the guanylyltransferase catalyzed the incorporation of the α-phosphate of GTP into cap structures; however, neither the β- nor the γ-phosphate were incorporated.

It was also demonstrated that the purified rat liver guanylyltransferase catalyzes a pyrophosphate exchange with GTP similar to the vaccinia capping enzyme reaction previously discussed. Furthermore, they demonstrated that dithiothreitol was essential for the reaction and that the enzyme was twice as active at the optimal Mn^{++} concentration of 2 m*M* as it was at the optimal Mg^{++} concentration of 8 m*M*.

B. GTP:mRNA Guanylyltransferase: HeLa Cell Nuclei

Venkatesan et al.[54] reported an extensive study on isolation and characterization of the mRNA guanylyltransferase isolated from HeLa cell nuclei. The enzyme catalyzes the transfer of a GMP residue from GTP to the 5'-end of RNA to form the 5'-cap structure. A 1000-fold purification was achieved by sequential column chromatography utilizing DEAE-cellulose, phosphocellulose, Cibacron® blue-agarose, and GTP-agarose. During this purification procedure, RNA triphosphatase and mRNA(guanine-7-)-methyltransferase were separated from the guanylyltransferase. Analysis of the purified native enzyme by sedimentation through sucrose gradients revealed a molecular weight of 48,500. Optimal enzyme activity was demonstrated at pH 7.5. Optimal cap formation was obtained with 2 m*M* Mn^{++}, whereas at the optimal 2 to 5 m*M* concentration range for Mg^{++}, the enzyme was approximately one fourth as active.

Donor and acceptor substrate specificity studies demonstrated that diphosphate-ended polyribonucleotides are the best acceptors for the purified enzyme.[55] Triphosphate-ended poly(A) was a poor acceptor in the capping reaction and monophosphate-ended poly(A) was not capped at all. Utilizing λ-phage C17 RNA containing ppC- at the 5'-end, this study showed that pyrimidine as well as purine-ended RNAs could serve as acceptors in the capping reaction. The diphosphate-ended dinucleoside tri- and tetraphosphates ppGpC and ppApGp were also capped by the enzyme and were 5 to 10 times, respectively, better acceptors than the triphosphate-ended configurations; however, ADP was not utilized. These results suggest that the minimal requirement for capping to occur is one phosphodiester bond, indicating that pretranscriptional capping is an unlikely event in the HeLa cell system. Furthermore, it should be noted that the shorter dinucleoside triphosphate was not as effectively capped by the enzyme as poly(A), suggesting that the natural substrate may be a longer oligoribonucleotide. The K_m values reported for the acceptor substrates were: dinucleoside tetraphosphate ppApGp K_m = 285 n*M* and diphosphate-ended poly(A) K_m = 19 n*M*.

Both GTP and ITP served as donor substrates in the capping reaction, whereas ATP, CTP, UTP, dGTP, m^7GTP, and GDP were not utilized as donors in the formation of the cap structure. The K_m value for GTP was 1.1 μ*M*. These results suggest that methylation in the 7-position of guanine must occur after capping of the RNA.

Venkatesan and Moss[56] recently demonstrated that, in the presence of GTP and without an acceptor substrate, the purified guanylyltransferase isolated from HeLa cells forms a covalent enzyme-GMP complex similar to that described above for the vaccinia guanylyltransferase.[32] In the HeLa cell study,[56] the enzyme-GMP complex was purified from the unreacted GTP in the reaction mixture by repeated chromatography on phosphocellulose. It was demonstrated that this purified enzyme-GMP complex catalyzes the transfer of the covalently-bound GMP moiety to either PPi, regenerating GTP, or to the 5'-diphosphate end of poly(A), forming the cap structure

G(5′)pppA(pA)$_n$. These results suggest that the capping mechanism in HeLa cells consists of at least two partial reactions:

$$GTP + E \rightleftarrows E\text{-}pG + PPi \qquad (11)$$

$$E\text{-}pG + ppA(pA)_n \rightarrow E + G(5')pppA(pA)_n \qquad (12)$$

Reaction (11) is specific for GTP. Although Shuman and Hurwitz[32] reported that dGTP was nearly as effective as GTP in the formation of the vaccinia guanylyltransferase-GMP intermediate, by contrast only traces of the intermediate were formed with dGTP and purified HeLa cell enzyme.[56]

Venkatesan and Moss[56] also found that the enzyme-GMP intermediate complex was readily formed when capped poly(A), i.e., G(5′)pppA(pA)$_n$, was incubated with vaccinia capping enzyme. However, when purified HeLa cell capping enzyme was used in the reaction, only traces of the intermediate were found. These results suggested that in the HeLa cell system, reaction (12) is not reversible. This difference between the HeLa and the vaccinia guanylyltransferaue is consistent with their earlier study in which they demonstrated the different relative abilities of the two enzymes to catalyze the complete reverse capping reaction (12 to 11).[55]

The chemical reactivity of the covalent bond between the HeLa guanylyltransferase and the GMP residue strongly suggests a phosphoamide linkage.[56] The amino acid to which the GMP residue is covalently linked was not identified; however, trypsin digestion of the enzyme-GMP complex and analysis by two-dimensional electrophoresis and chromatography demonstrated that a single GMP-peptide was involved.

Analysis of the enzyme-GMP complex by polyacrylamide gel electrophoresis under denaturing conditions indicated that the complex has a molecular weight of 65,000. However, as discussed above for the 95,000 mol wt subunit in the vaccinia capping enzyme complex, it is possible that the HeLa nuclei 65,000 mol wt enzyme-GMP complex interacts with other proteins involved in the synthesis and modification of mRNA. Bajszar and co-workers[66] demonstrated that 30S ribonucleoprotein particles from rat liver contain RNA guanylyltransferase as well as RNA(guanine-7-)-methyltransferase and RNA(nucleoside-2′-)-methyltransferase, suggesting that all three activities exist as a complex in vivo. Darnell and co-workers[75] have suggested that there may be an association of capping enzymes with initiation complexes for RNA polymerase II transcription. This is suggested by: (1) in vivo data which indicate that capping and methylation occur soon after initiation of RNA synthesis,[75,76] (2) in vitro data which indicate that efficient capping occurs in crude extracts which initiate transcription from defined templates,[77] and (3) analogy with cytoplasmic polyhedrosis virus.[35] However, in preliminary experiments with purified calf thymus RNA polymerase, no capping activity was detected.[56]

Earlier studies with HeLa cells which had been infected with vaccinia virus indicated that the guanylyltransferases and methylases that modified 5′-termini of the viral mRNAs are synthesized after viral infection.[78] The results suggest that these enzyme activities are "early" or prereplicative viral gene products. In their recent report, Venkatesan and Moss[56] also presented evidence that the vaccinia mRNA guanylyltransferase is viral coded. First, the enzyme-GMP complex from vaccinia virus grown in HeLa cells has a molecular weight of 95,000,[32] whereas the HeLa cell enzyme-GMP complex has a molecular weight of 65,000.[56] Secondly, after digesting both enzyme-GMP complexes with trypsin, the two GMP-peptides were clearly resolved from each other by two-dimensional electrophoresis and chromatography.

C. GTP:mRNA Guanylyltransferase: Wheat Germ

Keith and co-workers[59] have recently reported the purification and characterization of the mRNA guanylyltransferase from wheat germ. A 2000-fold enzyme purification was obtained from a crude wheat germ extract by sequential column chromatography on DEAE-cellulose, phosphocellulose, Cibacron® blue-agarose, and ADP-agarose, with an overall recovery of approximately 17%.

The purified enzyme catalyzes the transfer of a GMP residue from GTP to the 5′-end of RNA or synthetic polyribonucleotides. Polymers containing diphosphate ends were capped more efficiently than triphosphate-ended molecules. However, molecules with only monophosphate ends were not capped by the purified enzyme.

Analysis of the final purified fraction by polyacrylamide gel electrophoresis under denaturing conditions indicated that the enzyme was not yet homogeneous. However, the purified wheat germ guanylyltransferase does not appear to be associated with the RNA triphosphatase or the guanine-7-methyltransferase activities, in contrast to the vaccinia capping enzyme system.[15,16] Preliminary results suggest that the wheat germ RNA triphosphatase activity is removed during chromatography on Cibacron® blue-agarose.[59] In a later study, it was demonstrated that the RNA(guanine-7-)-methyltransferase was clearly separated from the guanylyltransferase activity during chromatography on phosphocellulose.[60] Analysis of the purified enzyme by sedimentation through sucrose gradients revealed a broad peak of enzyme activity sedimenting at a position corresponding to a molecular weight of approximately 65,000.[59]

Wheat embryo mRNAs contain cap structures that lack a 2′-*O*-methyl group on the penultimate nucleoside, i.e., $m^7G(5')pppN$-.[79] Using the vaccinia virus RNA(nucleoside-2′-)-methyltransferase to specifically label wheat germ RNA with ^3H-methyl from AdoMet, Muthukrishnan[80] has shown that the penultimate nucleoside may be cytidine as well as guanosine and adenosine. Similar results regarding the 5′-end structure of plant RNAs have been shown in a report on the posttranscriptional modifications of oat coleoptile RNA.[81] Recent evidence indicates that in eukaryotic animal systems, RNA polymerase II initiation of transcription may also begin with a pyrimidine triphosphate, i.e., in adenovirus type 2 early mRNA, data suggest UTP is utilized[82] and in early simian virus 40 RNA, CTP is utilized.[83]

These data suggest that RNAs with pyrimidines in their cap structure are either initiated with pyrimidines and then capped or derived from cleavage at a pyrimidine residue and then capped by a mechanism which has not been demonstrated in cellular systems. Thus, it is important to determine if pyrimidine-ended as well as purine-ended RNAs could be capped by the wheat germ RNA guanylyltransferaue. Using λ-phage C17 RNA containing ppC- at the 5′-end, it was shown that the purified wheat germ enzyme can cap RNA molecules containing pyrimidines at the 5′-terminus.[59] It appears from these results that the wheat germ guanylyltransferase demonstrates little sequence specificity since RNAs with purine ends also serve as acceptors. Further experiments demonstrated that a divalent cation requirement is satisfied by either Mn^{++} at an optimum of 0.5 m*M* or Mg^{++} at an optimum of 5 m*M*. The enzyme demonstrated a pH optimum around neutrality. The reported K_m values for the purified enzyme were 2.7 μM for GTP and 14.2 n*M* for diphosphate-ended poly(A). Inorganic pyrophosphate, which is a putative product of the guanylyltransferase reaction, inhibited the enzyme by approximately 50% at a concentration of 5 μM and almost total inhibition was observed at 50 μM, whereas similar concentrations of inorganic phosphate had much less effect on enzyme activity.

Further investigations from Keith's laboratory[60,84] have demonstrated that the purified RNA guanylyltransferase from wheat germ catalyzes the transfer of a GMP residue from GTP to the 5′-end of synthetic polyribonucleotides by a two-step mechanism involving the formation of a covalent enzyme-GMP intermediate. This complex is similar to that described above for the vaccinia and HeLa cell capping enzyme.

The catalyzed reaction for the formation of the enzyme-GMP intermediate is specific for GTP. However, traces of the intermediate were formed with dGTP and the purified wheat germ capping enzyme. In these studies, the enzyme-GMP complex was purified from unreacted GTP in the reaction mixture by gel filtration. The intermediate can also be purified by repeated chromatography on phosphocellulose.[85]

Optimal conditions for formation of the enzyme-GMP complex are significantly different than those determined for optimal capping enzyme activity, i.e., as judged by formation of the cap structure at the 5′-end of RNA molecules.[84] In formation of the complex, a divalent cation is required with an optimal Mg^{++} concentration of 15 m*M* or Mn^{++} at 6 mM. The optimal pH is 8.5, whereas in the overall reaction for formation of the cap, optimal conditions are 5 m*M* Mg^{++} and a sharp optimum of 0.5 m*M* for Mn^{++}, with higher or lower concentrations of Mn^{++} being much less effective. Optimal pH for capping activity is around neutrality.[59]

It was also demonstrated that the purified enzyme-GMP intermediate catalyzes the transfer of the covalently-bound GMP moiety to the 5′-end of poly(A), forming the cap structure G(5′)pppA$(pA)_n$.[60,84] These results indicate that the capping mechanism in wheat germ consists of at least two partial reactions similar to the vaccinia[32] and HeLa guanylyltransferase.[56] Chemical reactivity of the covalent bond between the enzyme and the GMP residue strongly suggests a phosphoamide linkage similar to that proposed for the vaccinia and HeLa cell guanylyltransferase.[85] The amino acid to which the GMP residue is covalently linked has not yet been identified.

Analysis of the purified enzyme-GMP complex by gel electrophoresis under denaturing conditions indicated the presence of two very closely migrating polypeptide bands.[60] The mobility of these bands is just slightly ahead of the σ-subunit of *E. coli* RNA polymerase, which has a molecular weight of approximately 85,000. Reanalysis of these two bands after elution from the gel indicated that they again migrated as two bands in the same location as in the original gel.[85] Further studies are underway to determine the number of unique GMP-peptides in each of the two polypeptide bands and if both have similar tryptic GMP-peptides.

D. mRNA(Guanine-7-)-Methyltransferase: HeLa Cell

Ensinger and Moss[62] reported the isolation and characterization of a mRNA(guanine-7-)-methyltransferase from the cytoplasm of HeLa cells. This enzyme specifically catalyzes the transfer of a methyl group from AdoMet to the 7-position of the 5′-terminal guanosine residue of RNAs ending in G(5′)pppN-. In these studies, unmethylated vaccinia mRNA was used as the methyl acceptor. It was demonstrated that in HeLa cells which were disrupted by Dounce homogenization approximately two thirds of the methylase activity was located in the cytoplasm, with 30% of this activity associated with ribosomes. The ribosome-associated methylase activity was disassociated from the ribosomes by washing with 0.5 *M* KCl. A 165-fold enzyme purification was achieved by a scheme involving phase partition to remove nucleic acids followed by ammonium sulfate precipitation and sequential column chromatography on DEAE-cellulose, denatured DNA-agarose, and CM-Sephadex®.

Heterologous tRNAs as well as vaccinia mRNAs were methylated by the partially purified enzyme. However, the tRNA methyltransferase was separated from the mRNA enzyme activity by sucrose gradient sedimentation and gel filtration on Sephadex® G-200. Using vaccinia mRNA as a substrate, the partially purified enzyme exclusively methylates the 7-position of guanine located in the terminal cap structure, resulting in the formation of m^7G(5′)pppN-. It was also shown that the guanine-7-methyltransferase could methylate the dinucleoside triphosphate G(5′)pppG. However, GTP, GDP, and G(5′)ppppG were not utilized as methyl acceptors by the enzyme. A molecular weight of 56,000 was determined by sedimentation through sucrose gradients and gel filtration column chromatography using Sephadex® G-200.

E. mRNA(Guanine-7-)-Methyltransferase: Wheat Germ

The mullifunctional capping enzyme complex from vaccinia virus has been shown to contain an associated RNA triphosphatase, a RNA guanylyltransferase, and a RNA(guanine-7-)-methyltransferase.[15,16] Current studies by Keith and coworkers[59,60,86] indicate that similar activities exist in wheat germ but as separate enzymes, although, in vivo, all three activities may exist as a complex. During the purification of the wheat germ guanylyltransferase, the RNA triphosphatase activity is apparently separated from the capping activity during column chromatography on Cibacron® blue-agarose.[59] The guanine-7-methyltransferase does not bind to phosphocellulose and is clearly separated from the capping activity which is tightly bound during this purification step.[86]

When the partially purified guanine-7-methyltransferase was tested for methyltransferase activity using ^{3}H-methyl-labeled AdoMet and tRNA, rRNA, unmethylated or methylated vaccinia mRNA, the only significant incorporation of ^{3}H-labeled methyl groups was with the unmethylated vaccinia mRNA. Analysis of the product of the reaction indicates that the enzyme specifically catalyzes the transfer of a methyl group from AdoMet to the 7-position of a guanine residue within the cap structure, resulting in m^{7}G(5′)pppN-. No internal methylation was detected with these RNAs as methyl acceptors. Further experiments with methyl acceptors, such as the dinucleoside triphosphate G(5′)pppG, suggest the enzyme also catalyzes, but with less efficiency, the methylation of these cap derivatives.[60,86] Similar results have been found with the guanine-7-methyltransferase from vaccinia virus,[28] HeLa cell,[62] and rat liver nuclei.[57]

Locht et al.[61] have also reported on the partial purification of a mRNA guanine-7-methyltransferase from wheat germ. In this study, the dinucleoside triphosphate G(5′)pppG was used as a methyl acceptor. An enzyme purification of approximately 3000-fold was reported by sequential column chromatography on DEAE-cellulose and DNA-agarose. It was reported that the purified methylase activity does not require divalent cations and is, in fact, inhibited by even small amounts of Mg^{++}. KCl stimulated enzyme activity with an optimal concentration of 100 m*M*; however, NaCl had no effect. No increase in enzyme activity was shown when GDP, GTP, and diphosphate-ended poly(A) were added to the reaction mixture in the presence of Mg^{++}, Mn^{++}, and GTP. The RNA(guanine-7-)-methyltransferase has also been partially purified from *Neurospora crassa*.[87]

F. Cap I and Cap II RNA(Nucleoside-2′-)-Methyltransferase: HeLa Cell

Langberg and Moss[63] have recently reported the purification and characterization of two HeLa cell RNA(nucleoside-2′-)-methyltransferases responsible for the conversion of cap 0 [m^{7}G(5′)pppNpN-] to cap I [m^{7}G(5′)pppN^{m}pN-] and cap I to cap II [m^{7}G(5′)pppN^{m}pN^m-]. The cap II methyltransferase activity was found almost exclusively in cytoplasmic fractions, whereas cap I methyltransferase activity was found in the nucleus, its apparent biological site of action, as well as in the cytoplasmic fraction. The two enzyme activities were purified by column chromatography on DEAE-cellulose and separated from each other by chromatography on phosphocellulose.

In these studies, it was demonstrated that the capped terminus with at least two additional nucleotides, e.g., G(5′)pppNpN, were required for enzyme activity with the cap I methyltransferase. These results indicate that 2′-*O*-methylation is preceded by transcription, minimal chain extension, and capping. However, it is important to note that these studies show the affinity of the cap I methyltransferase for polyribonucleotides terminating with either G(5′)pppN- or m^{7}G(5′)pppN- is virtually identical. Thus, no conclusion regarding the order of guanine-7-methylation and 2′-*O*-methylation was determined from these studies. However, Langberg and Moss[63] suggest that the guanine-7-methyltransferase activity acts first, since very high levels of this enzyme are present in HeLa cell extracts[62] compared to the cap I methyltransferase. They also suggest that cap II methyltransferase follows cap I methylation since the cap II methyltransferase was found almost exclusively in the cytoplasmic extract. However, their assay procedure did not permit them to determine whether the unnatural structure $m^7G(5')pppNpN^m$ could be formed in vitro.[63]

Both purine and pyrimidine nucleotides located in the penultimate position of cap structures are methylated by the purified cap I methyltransferase. It was also reported that the K_m values for the purified cap I methyltransferase are 1.6 n*M* for polyribonucleotides ending in $m^7G(5')pppA(pA)_n$, 1.4 n*M* for polyribonucleotides ending in $G(5')pppA(pA)_n$, and 0.6 μ*M* for AdoMet. As expected, the reaction was strongly inhibited by AdoHcy. Cap I methyltransferase activity was stimulated by KCl with an optimum concentration of 140 to 180 m*M*. No pH optimum was demonstrated; however, enzyme activity increased continuously between pH 6 and pH 9.

G. RNA(2'-*O*-methyladenosine-N^6-)-Methyltransferase: HeLa Cell

Keith et al.[26] have purified and characterized an enzyme responsible for the posttranscriptional modification of the 5'-end of mRNA. The methyltransferase was purified approximately 500-fold from a HeLa cell cytoplasmic postribosomal supernatant by ammonium sulfate fractionation and successive chromatography columns utilizing DEAE-cellulose, CM-Sephadex®, and phosphocellulose. This enzyme specifically catalyzes the transfer of a methyl group from AdoMet to the N^6-position of a 2'-*O*-methyladenosine residue located in the penultimate position of the capped 5'-end of mRNA, resulting in the formation of the 5'-end structure shown in Figure 1. The enzyme failed to methylate internal RNA segments, suggesting that a separate methyltransferase catalyzes the formation of internal m^6A^m in U_2 RNA and internal m^6A in mRNA.

RNA acceptor substrates ending in $m^7G(5')pppA^m$- were shown to be the best acceptor molecules, as judged by their relatively high activity in the reaction mixture. Less activity was found with RNA molecules ending in $m^7G(5')pppA$-, and little activity was demonstrated when RNAs ending in $G(5')pppA$- were used as methyl acceptors in the reaction mixture. No enzyme activity was found with RNAs ending in pppA-. Furthermore, no enzyme activity was detected with dinucleoside triphosphates such as $m^7G(5')pppA$ and $m^7G(5')pppA^m$, or oligonucleotides such as $m^7G(5')pppA^mpN$. These results suggest that a longer 5'-end segment of RNA is required for enzyme activity.

A molecular weight of 65,000 was determined by sedimentation through sucrose gradients. Analysis of the partially purified enzyme by polyacrylamide gel electrophoresis under denaturing conditions revealed several protein bands, including one major band at a molecular weight of 65,000.

From the substrate specificity studies, it was concluded that the formation of the N^6,2'-*O*-dimethyladenosine located in the cap structure follows the biosynthesis of RNA molecules containing $m^7G(5')pppA^mpN$-. Further studies demonstrated that a specific sequence beyond the cap structure is not required for enzyme activity. The purified enzyme had a pH optimum of 7.25 and was inhibited by salts and AdoHcy. The apparent K_m values reported for the enzyme are 0.5 μ*M* for AdoMet and 2 to 3 n*M* for mRNAs.

VI. CONCLUSIONS

Several modification and processing events are involved in the biosynthesis of eukaryotic mRNA from an hnRNA precursor. The formation of the capped 5'-terminus of the mRNA molecule is one early event in this process. Isolation and characterization of the enzymes responsible for cap formation have significantly increased our understanding of the sequence of cellular events.

The RNA guanylyltransferase, or "capping" enzyme, is the key enzyme involved in the formation of the 5′-terminal cap structure. The enzymatic mechanism involved in this process has been extensively investigated in viral, animal, and plant systems and considerable progress has been made in elucidating the catalytic mechanisms. The sequence of capping and methylation reactions shown in Figure 2 for the modification of the 5′-terminus of HeLa cell mRNA is consistent with substrate specificity studies utilizing the isolated enzymes.

All of the enzymes involved in these reactions have been isolated or partially purified, with the exception of the RNA triphosphatase. However, the existence of such an activity is implied since the purified HeLa cell mRNA guanylyltransferase is specific for RNA with a diphosphate 5′-terminus. The guanylyltransferase from rat liver and wheat germ exhibits similar diphosphate 5′-ended specificity.

Vaccinia and reovirus mRNAs are capped and methylated in a series of reactions very similar to the initial reactions proposed for cellular mRNA. However, in contrast to the cellular capping enzymes, the vaccinia virus RNA triphosphatase, RNA guanylyltransferase, and RNA guanine-7-methyltransferase are tightly associated in a multifunctional enzyme complex. This physical association of capping and methylating activities in a single enzyme complex ensures efficient 5′-end modification of initiated mRNA molecules. Although the guanylyltransferase isolated from rat liver, HeLa cell, and wheat germ has been easily separated from the RNA triphosphatase and guanine-7-methyltransferase activities during their purification, it is possible that, in vivo, all three activities may be associated as a multifunctional enzyme complex.

It appears likely that a common catalytic mechanism for the formation of capped 5′-ends exists for animal, plant, and some viral mRNAs. In each system studied, the guanylyltransferase catalyzes the transfer of a GMP residue from GTP to the 5′-diphosphate end of RNA by a two-step mechanism involving formation of a covalent enzyme-GMP intermediate. Chemical studies suggest that the covalent bond between the enzyme and the nucleotide involves a phosphoamide linkage similar to that described for DNA ligase.

Studies utilizing cytoplasmic polyhedrosis virus and vesicular stomatitis virus (VSV) indicate that an alternative mechanism may exist for capping, or methylation, or both. Formation of the capped end of VSV mRNA appears to occur through the transfer of a GDP residue to a monophosphate at the 5′-end of the RNA. The formation of capped ends at internal cleavage sites within cellular mRNA precursors could occur by a similar mechanism. Thus far, an eukaryotic cellular capping activity with a 5′-monophosphate specificity has not been demonstrated; however, the cellular and viral guanylyltransferases that have been isolated could cap RNA molecules at internal cleavage sites if one or two additional phosphates (depending on the cleavage mechanism) were added to the cleaved 5′-end of the RNA, generating a diphosphate end. Schibler and Perry[88] have suggested such a mechanism. A polyribonucleotide kinase capable of adding a single phosphate to the 5′-OH end of RNA has been isolated.[89] In addition, a 5′-phosphate-polyribonucleotide kinase which specifically adds phosphates to the 5′-monophosphate end of RNA molecules has been isolated from vaccinia virus.[90] However, such an activity has not yet been reported in eukaryotic cells.

Thus far, there is no direct biological evidence for capping at RNA cleavage sites in eukaryotic cells. However, such a mechanism has been considered since pyrimidine nucleotides have been identified in the capped ends of mRNA. Recent evidence suggests that in eukaryotic animal systems, RNA polymerase II may initiate with pyrimidines as well as purines and it has been clearly demonstrated that viral and cellular guanylyltransferase can cap RNAs containing 5′-terminal pyrimidine diphosphates.

A variation in the methylation of the terminal guanosine residue has been identified in blocked 5′-ends of low-molecular-weight nuclear RNAs from Novikoff hepatoma cells. However, in contrast to mRNAs and hnRNAs, the analogous blocked 5′-terminus is characterized by a 2,2,7-trimethylguanosine residue linked through a triphosphate bridge to the penultimate 2′-*O*-methylnucleoside residue.[91,92]

ACKNOWLEDGMENTS

I am grateful to Bernard Moss for his valuable comments and discussions. I would like to thank Aaron J. Shatkin for supplying preliminary data, and I would also like to thank numerous other colleagues who supplied their reprints and preprints. I am especially grateful and indebted to Marla Keith for her excellent assistance in preparation of this manuscript.

REFERENCES

1. **Shatkin, A. J.**, Capping of eucaryotic mRNAs, *Cell*, 9, 645, 1976.
2. **Filipowicz, W.**, Function of the 5′-terminal ^{7}mG cap in eukaryotic mRNA, *FEBS Lett.*, 96, 1, 1978.
3. **Banerjee, A. K.**, 5′-terminal cap structure in eukaryotic messenger ribonucleic acids, *Microbiol. Rev.*, 44, 175, 1980.
4. **Moss, B., Martin, S. A., Ensinger, M. J., Boone, R. F., and Wei, C-M.**, Modification of the 5′-terminals of mRNAs by viral and cellular enzymes, *Prog. Nucl. Acid Res. Mol. Biol.*, 19, 63, 1976.
5. **Rottman, F. M.**, Methylation and polyadenylation of heterogeneous nuclear and messenger RNA, in *Biochemistry of Nucleic Acids*, Vol. 17, Clark, B. F. C., Ed., University Park Press, Baltimore, 1978, 45.
6. **Adams, J. M.**, Messenger RNA, in *The Ribonucleic Acids*, Letham, D. S., Ed., Springer-Verlag, New York, 1977, 81.
7. **Busch, H., Hirsch, F., Gupta, K. K., Rao, M., Spohn, W., and Wu, B. C.**, Structural and functional studies on the "5′-cap": a survey method for mRNA, *Prog. Nucl. Acid Res. Mol. Biol.*, 19, 39, 1976.
8. **Revel, M. and Groner, Y.**, Post-transcriptional and translational controls of gene expression in eukaryotes, in *Annual Review of Biochemistry*, Vol. 47, Snell, E. E., Boyer, P. D., Meister, A., and Richardson, C. C., Eds., Annual Reviews, Palo Alto, Calif., 1978, 1079.
9. **Wei, C. M. and Moss, B.**, Methylated nucleotides block 5′-terminus of vaccinia virus messenger RNA, *Proc. Natl. Acad. Sci. U.S.A.*, 72, 318, 1975.
10. **Furuichi, Y., Morgan, M., Muthukrishnan, S., and Shatkin, A. J.**, Reovirus messenger RNA contains a methylated, blocked 5′-terminal structure, m^{7}G(5′)ppp(5′)GpCp, *Proc. Natl. Acad. Sci. U.S.A.*, 72, 362, 1975.
11. **Wei, C. M., Gershowitz, A., and Moss, B.**, N^6,2′-O-dimethyladenosine, a novel methylated ribonucleoside next to the 5′-terminal of animal cell and virus mRNAs, *Nature (London)*, 257, 251, 1975.
12. **Perry, R. P., Kelley, D. E., Frederici, K., and Rottman, F.**, The methylated constituents of L cell messenger RNA: evidence for an unusual cluster at the 5′ terminus, *Cell*, 4, 387, 1975.
13. **Adams, J. M. and Cory, S.**, Modified nucleosides and bizarre 5′-termini in mouse myeloma mRNA, *Nature (London)*, 255, 28,1975.
14. **Ensinger, M. J., Martin, S. A., Paoletti, E., and Moss, B.**, Modification of the 5′-terminus of mRNA by soluble guanylyl and methyl transferases from vaccinia virus, *Proc. Natl. Acad. Sci. U.S.A.*, 72, 2525, 1975.
15. **Martin, S. A., Paoletti, E., and Moss, B.**, Purification of mRNA guanylyltransferase and mRNA(guanine-7-)-methyltransferase from vaccinia virions, *J. Biol. Chem.*, 250, 9322, 1975.
16. **Venkatesan, S., Gershowitz, A., and Moss, B.**, Modification of the 5′ end of mRNA: association of RNA triphosphatase with the RNA guanylyltransferase-RNA (guanine-7-)-methyltransferase complex from vaccinia virus, *J. Biol. Chem.*, 255, 903, 1980.
17. **Moss, B.**, Utilization of the guanylyltransferase and methyltransferases of vaccinia virus to modify and identify the 5′-terminals of heterologous RNA species, *Biochem. Biophys. Res. Commun.*, 74, 374, 1977.
18. **Moss, B., Keith, J. M., Gershowitz, A., Ritchey, M. B., and Palese, P.**, Common sequence at the 5′ ends of the segmented RNA genomes of influenza A and B viruses, *J. Virol.*, 25, 312, 1978.

19. **Smith, R. E. and Clark, J. M.**, Effect of capping upon the mRNA properties of satellite tobacco necrosis virus ribonucleic acid, *Biochemistry*, 18, 1366, 1979.
20. **Barbosa, E. and Moss, B.**, mRNA (nucleoside-2′-)-methyltransferase from vaccinia virus: purification and physical properties, *J. Biol. Chem.*, 253, 7692, 1978.
21. **Barbosa, E. and Moss, B.**, mRNA (nucleoside-2′-)-methyltransferase from vaccinia virus: characteristics and substrate specificity, *J. Biol. Chem.*, 253, 7698, 1978.
22. **Dubin, D. T. and Taylor, R. H.**, The methylation state of poly A-containing messenger RNA from cultured hamster cells, *Nucl. Acids Res.*, 2, 1653, 1975.
23. **Moyer, S. A. and Banerjee, A. K.**, *In vivo* methylation of vesicular stomatitis virus and its host-cell messenger RNA species, *Virology*, 70, 339, 1976.
24. **Abraham, G., Rhodes, D. P., and Banerjee, A. K.**, The 5′-terminal structure of the methylated messenger RNA synthesized *in vitro* by vesicular stomatitis virus, *Cell*, 5, 51, 1975.
25. **Boone, R. F. and Moss, B.**, Methylated 5′-terminal sequences of vaccinia virus mRNA species made *in vivo* at early and late times after infection, *Virology*, 79, 67, 1977.
26. **Keith, J. M., Ensinger, M. J., and Moss, B.**, HeLa cell RNA(2′-*O*-methyladenosine-N^6-)-methyltransferase specific for the capped 5′-end of messenger RNA, *J. Biol. Chem.*, 253, 5033, 1978.
27. **Martin, S. A. and Moss, B.**, Modification of RNA by mRNA guanylyltransferase and mRNA(guanine-7-)-methyltransferase from vaccinia virions, *J. Biol. Chem.*, 250, 9330, 1975.
28. **Martin, S. A. and Moss, B.**, mRNA guanylyltransferase and mRNA (guanine-7-)-methyltransferase from vaccinia virions, *J. Biol. Chem.*, 251, 7313, 1976.
29. **Monroy, G., Spencer, E., and Hurwitz, J.**, Purification of mRNA guanylyltransferase from vaccinia virions, *J. Biol. Chem.*, 253, 4481, 1978.
30. **Monroy, G., Spencer, E., and Hurwitz, J.**, Characteristics of reaction catalyzed by purified guanylyltransferase from vaccinia virus, *J. Biol. Chem.*, 253, 4490, 1978.
31. **Shuman, S., Surks, M., Furneaux, H., and Hurwitz, J.**, Purification and characterization of a GTP-pyrophosphate exchange activity from vaccinia virions, *J. Biol. Chem.*, 255, 11588, 1980.
32. **Shuman, S. and Hurwitz, J.**, Mechanism of mRNA capping by vaccinia virus guanylyltransferase: characterization of an enzyme-guanylate intermediate, *Proc. Natl. Acad. Sci. U.S.A.*, 78, 187, 1981.
33. **Moss, B., Gershowitz, A., Wei, C-M., and Boone, R.**, Formation of the guanylylated and methylated 5′-terminus of vaccinia virus mRNA, *Virology*, 72, 341, 1976.
34. **Furuichi, Y., Muthukrishnan, S., Tomasz, J., and Shatkin, A. J.**, Mechanism of formation of reovirus mRNA 5′-terminal blocked and methylated sequence, ^{7}mGpppGmC, *J. Biol. Chem.*, 251, 5043, 1976.
35. **Furuichi, Y.**, "Pretranscriptional capping" in the biosynthesis of cytoplasmic polyhedrosis virus mRNA, *Proc. Natl. Acad. Sci. U.S.A.*, 75, 1086, 1978.
36. **Abraham, G., Rhodes, D. P., and Banerjee, A. K.**, Novel initiation of RNA synthesis *in vitro* by vesicular stomatitis virus, *Nature (London)*, 255, 37, 1975.
37. **Testa, D. and Banerjee, A. K.**, Two methyltransferase activities in the purified virions of vesicular stomatitis virus, *J. Virol.*, 24, 786, 1977.
38. **Miura, K., Watanabe, K., and Sugiura, M.**, 5′-terminal nucleotide sequences of the double-stranded RNA of silkworm cytoplasmic polyhedrosis virus, *J. Mol. Biol.*, 86, 31, 1974.
39. **Shimotohno, K. and Miura, K-I.**, 5′-terminal structure of messenger RNA transcribed by the RNA polymerase of silkworm cytoplasmic polyhedrosis virus containing double-stranded RNA, *J. Mol. Biol.*, 86, 21, 1974.
40. **Miura, K-I., Watanabe, K., Sugiura, M., and Shatkin, A. J.**, The 5′-terminal nucleotide sequences of the double stranded RNA of human reovirus, *Proc. Natl. Acad. Sci. U.S.A.*, 71, 3979, 1974.
41. **Furuichi, Y. and Miura, K-I.**, A blocked structure at the 5′-terminus of messenger RNA from cytoplasmic polyhedrosis virus, *Nature (London)*, 253, 374, 1975.
42. **Wei, C. M. and Moss, B.**, Methylation of newly synthesized viral messenger RNA by an enzyme in vaccinia virus, *Proc. Natl. Acad. Sci. U.S.A.*, 71, 3014, 1974.
43. **Shatkin, A. J.**, Methylated messenger RNA synthesis *in vitro* by purified reovirus, *Proc. Natl. Acad. Sci. U.S.A.*, 71, 3204, 1974.
44. **Furuichi, Y.**, Methylation-coupled transcription by virus-associated transcriptase of cytoplasmic polyhedrosis virus containing double-stranded RNA, *Nucl. Acids Res.*, 1, 802, 1974.
45. **Furuichi, Y. and Shatkin, A. J.**, Stimulation of CPV mRNA synthesis *in vitro* by S-adenosyl methionine and 5′-capping during initiation, in *Transmethylation*, Usdin, E., Borchardt, R. T., and Creveling, C. R., Eds., Elsevier/North-Holland, New York, 1979, 351.
46. **Abraham, G. and Banerjee, A. K.**, The nature of the RNA products synthesized *in vitro* by subviral components of vesicular stomatitis virus, *Virology*, 71, 230, 1976.
47. **Moyer, S. A.**, Alteration of the 5′ terminal caps of the mRNAs of vesicular stomatitis virus by cycloleucine *in vivo*, *Virology*, 112, 157, 1981.
48. **Perry, R. P., Kelley, D. E., Frederici, K. H., and Rottman, F. M.**, Methylated constituents of heterogeneous nuclear RNA: presence in blocked 5′ terminal structures, *Cell*, 6, 13, 1975.

49. **Salditt-Georgieff, M., Jelinek, W., Darnell, J. E., Furuichi, Y., Morgan, M., and Shatkin, A.,** Methyl labeling of HeLa cell HnRNA: a comparison with mRNA, *Cell,* 7, 227, 1976.
50. **Groner, Y. and Hurwitz, J.,** Synthesis of RNA containing a methylated blocked 5′-terminus by HeLa nuclear homogenates, *Proc. Natl. Acad. Sci. U.S.A.,* 72, 2930, 1975.
51. **Winicov, I. and Perry, R. P.,** Synthesis, methylation and capping of nuclear RNA by a subcellular system, *Biochemistry,* 15, 5039, 1976.
52. **Wei, C. M. and Moss, B.,** 5′-terminal capping of RNA by guanylyltransferase from HeLa cell nuclei, *Proc. Natl. Acad. Sci. U.S.A.*, 74, 3758, 1977.
53. **Groner, Y., Gilboa, E., and Aviv, H.,** Methylation and capping of RNA polymerase II primary transcripts by HeLa nuclear homogenates, *Biochemistry,* 17, 977, 1978.
54. **Venkatesan, S., Gershowitz, A., and Moss, B.,** Purification and characterization of mRNA guanylyltransferase from HeLa cell nuclei, *J. Biol. Chem.,* 255, 2829, 1980.
55. **Venkatesan, S. and Moss, B.,** Donor and acceptor specificities of HeLa cell mRNA guanylyltransferase, *J. Biol. Chem.,* 255, 2835, 1980.
56. **Venkatesan, S. and Moss, B.,** Eukaryotic mRNA capping enzyme-guanylate covalent intermediate, *Proc. Natl. Acad. Sci. U.S.A.,* 79, 340, 1982.
57. **Mizumoto, K. and Lipmann, F.,** Transmethylation and transguanylylation in 5′-RNA capping system isolated from rat liver nuclei, *Proc. Natl. Acad. Sci. U.S.A.,* 76, 4961, 1979.
58. **Laycock, D. E.,** Purification of eukaryotic mRNA guanylyltransferase, *Fed. Proc.,* 36, 770, 1977.
59. **Keith, J. M., Venkatesan, S., Gershowitz, A., and Moss, B.,** Purification and characterization of the messenger ribonucleic acid capping enzyme GTP:RNA guanylyltransferase from wheat germ, *Biochemistry,* 21, 327, 1982.
60. **Keith, J. M., Galer, D., and Westreich, L.,** Wheat germ mRNA capping and methylating enzymes, in *Transmethylation,* Usdin, E., Borchardt, R. T., and Creveling, C. R., Eds., in press.
61. **Locht, C., Bouchet, H., and Delcour, J.,** Partial purification of the mRNA(guanine-7-)methyltransferase from wheat germ, in *Transmethylation,* Usdin, E., Borchardt, R. T., and Creveling, C. R., Eds., in press.
62. **Ensinger, M. J. and Moss, B.,** Modification of the 5′ terminus of mRNA by an RNA(guanine-7-)-methyltransferase from HeLa cells, *J. Biol. Chem.,* 251, 5283, 1976.
63. **Langberg, S. R. and Moss, B.,** Post-transcriptional modification of mRNA: purification and characterization of cap I and cap II RNA(nucleoside-2′-)-methyltransferases from HeLa cells, *J. Biol. Chem.,* 256, 10054, 1981.
64. **Lavers, G. C.,** Detection of methyltransferase activities which modify GpppG to m^7GpppGm in embryonic chick lens, *Mol. Biol. Rep.,* 3, 275, 1977.
65. **White, J. L.,** *In vitro* enzymatic methylation of TMV RNA, *FEBS Lett.,* 119, 219, 1980.
66. **Bajszar, G., Szabo, G., Simoncsits, A., and Molnar, J.,** Methylated cap formation by enzymes bound to nuclear informer particles, *Mol. Biol. Rep.,* 4, 93, 1978.
67. **Tutas, D. J. and Paoletti, E.,** Purification and characterization of core-associated polynucleotide 5′-triphosphatase from vaccinia virus, *J. Biol. Chem.,* 252, 3092, 1977.
68. **Furuichi, Y., LaFinandra, A., and Shatkin, A. J.,** 5′-terminal structure and mRNA stability, *Nature (London),* 266, 235, 1977.
69. **Lockard, R. E. and Lane, C.,** Requirement for 7-methylguanosine in translation of globin mRNA, *in vivo, Nucl. Acids Res.,* 5, 3237, 1978.
70. **Smith, R., Furuichi, Y., and Shatkin, A.,** personal communication, 1981.
71. **Lehman, I. R.,** DNA ligase: structure, mechanism, and function, *Science,* 186, 790, 1974.
72. **Gumport, R. I. and Lehman, I. R.,** Structure of the DNA ligase-adenylate intermediate: lysine (ε-amino)-linked adenosine monophosphoramidate, *Proc. Natl. Acad. Sci. U.S.A.,* 68, 2559, 1971.
73. **Shabarova, Z. A.,** Synthetic Nucleotides-peptides, *Prog. Nucl. Acid Rec. Mol. Biol.,* 10, 145, 1970.
74. **DasGupta, R., Harada, F., and Kaesberg, P.,** Blocked 5′ termini in brome mosaic virus RNA, *J. Virol.,* 18, 260, 1976.
75. **Salditt-Georgieff, M., Harpoid, M., Chen-Kiang, S., and Darnell, J. E.,** The addition of 5′ cap structure occurs early in hnRNA synthesis and prematurely terminated molecules are capped, *Cell,* 19, 69, 1980.
76. **Babich, A., Nevins, J. R., and Darnell, J. E., Jr.,** Early capping of transcripts from the adenovirus major late transcription unit, *Nature (London),* 287, 246, 1980.
77. **Weil, P. A., Luse, D. S., Segall, J., and Roeder, R. G.,** Selective and accurate initiation of transcription at the Ad2 major late promotor in a soluble system dependent on purified RNA polymerase II and DNA, *Cell,* 18, 469, 1979.
78. **Boone, R. F., Ensinger, M. J., and Moss, B.,** Synthesis of mRNA guanylyltransferase and mRNA methyltransferases in cells infected with vaccinia virus, *J. Virol.,* 21, 475, 1977.
79. **Haffner, M. H., Chin, M. B., and Lane, B. G.,** Wheat embryo ribonucleates. XII. Formal characterization of terminal and penultimate nucleoside residues at the 5′-ends of "capped" RNA from imbibing wheat embryos, *Can. J. Biochem.,* 56, 729, 1978.

80. Muthukrishnan, S., personal communication, 1981.
81. Haugland, R. A. and Cline, M. G., Post-transcriptional modifications of oat coleoptile ribonucleic acids, *Eur. J. Biochem.*, 104, 271, 1980.
82. Hashimoto, S., Pursley, M. H., Wold, W. S. M., and Green, M., Characterization of distinct 5′-terminal cap structures of adenovirus type 2 early messenger ribonucleic acid and KB cell messenger ribonucleic acid, *Biochemistry*, 19, 294, 1980.
83. Gidoni, D., Kahana, C., Canaani, D., and Groner, Y., Specific *in vitro* initiation of transcription of simian virus 40 early and late genes occurs at the various cap nucleotides including cytidine, *Proc. Natl. Acad. Sci. U.S.A.*, 78, 2174, 1981.
84. Westreich, L. A., The Partial Purification and Characterization of mRNA Guanylyltransferase from Raw Wheat Germ, Masters thesis, New York University, New York, 1981.
85. Westreich, L. A. and Keith, J. M., unpublished data, 1982.
86. Galer, D. and Keith, J. M., unpublished data, 1982.
87. Germershausen, J., Goodman, D., and Somberg, E. W., 5′-cap methylation of homologous poly A(+) RNA by a RNA (guanine-7-) methyltransferase from *Neurospora crassa*, *Biochem. Biophys. Res. Commun.*, 82, 871, 1978.
88. Schibler, U., and Perry, R. P., Characterization of the 5′-termini of HnRNA in mouse L cells: implications for processing and cap formation, *Cell*, 9, 121, 1976.
89. Winicov, I., RNA phosphorylation: a polynucleotide kinase function in mouse L cell nuclei, *Biochemistry*, 16, 4233, 1977.
90. Spencer, E., Loring, D., Hurwitz, J., and Monroy, G., Enzymatic conversion of 5′-phosphate-terminated RNA to 5′-di- and triphosphate-terminated RNA, *Proc. Natl. Acad. Sci. U.S.A.*, 75, 4793, 1978.
91. Ro-Choi, T. S., Yong, C. C., Henning, D., McCloskey, J., and Busch, H., Nucleotide sequence of U-2 ribonucleic acid, *J. Biol. Chem.*, 250, 3921, 1975.
92. Cory, S. and Adams, J. M., Modified 5′-termini in small nuclear RNAs of mouse myeloma cells, *Mol. Biol. Rep.*, 2, 287, 1975.

Chapter 5

POLY(A) POLYMERASE FROM EUKARYOTES

Samson T. Jacob and Kathleen M. Rose

TABLE OF CONTENTS

I. INTRODUCTION

Despite the discovery of an enzymatic activity catalyzing poly(A) synthesis as early as 1960,[1] the functional significance of this enzyme was not known until early 1970s when poly(A) was found associated with mRNA (see review by Brawerman[2]). Poly(A) synthesizing activity has been reported in both prokaryotes and eukaryotes; the eukaryotic poly(A) polymerase (E.C. 2.7.7.19) is present in an array of tissues and cells such as calf thymus, sea urchin embryos, guinea-pig brain, HeLa cells, rat liver, Ehrlich ascites cells, human lymphocytes, hamster embryo fibroblast, quail oviduct, Krebs ascites cells, Landschutz ascites cells, chick chorioallantoic membrane, chick embryonic heart or liver, yeast, tobacco leaf, and corn seedlings (see reviews by Edmonds and Winters[3] and by Jacob and Rose[4]). The enzyme is also present in virions (for a review of viral poly(A) polymerases, see Reference 3). It has been identified in vaccinia virus [5-7] vesicular stomatitis virus,[8,8a,8b] Pichinde virus,[9] vaccinia virus-infected HeLa cells,[10,11] and encephalomyocarditis virus-infected L cells.[12] Extensive purification of the animal or viral poly(A) polymerase has been achieved only in a limited number of cases which include nuclei derived from calf thymus,[13,14] rat liver,[15,16] Morris hepatoma 3924A,[15] and beef liver,[17] Ehrlich ascites cells,[17a] hepatoma mitochondria,[18] cytosol fractions of calf thymus,[19] Artemia dormant embryos,[20] HeLa cells,[21] and vaccinia virus cores.[21]

An earlier review by Edmonds and Winters[3] has given excellent historical documentation of poly(A) polymerase. Another review has dealt with the extraction, purification, and properties of poly(A) polymerases from normal and cancer cells.[4] The present chapter will therefore focus on the more recent developments such as alterations in poly(A) polymerase in response to various physiological and pathological stimuli, the posttranslational modification of poly(A) polymerase, immunological aspects of the enzyme, and the characteristics and functional implications of a nuclease activity closely associated with purified poly(A) polymerase.

II. GENERAL CHARACTERISTICS OF POLY(A) POLYMERASE

The general characteristics of poly(A) polymerase have been discussed in great detail in previous reviews (Edmonds and Winters,[3] Jacob and Rose[4]). Only the salient points concerning intracellular distribution, substrate and primer specificities, metal ion requirements, inhibitors, molecular weight, and subunit composition will be presented in the present chapter.

A. Intracellular Distribution of Poly(A) Polymerase

Poly(A) polymerase has been identified in the nucleus,[13,15] ribonucleoprotein particles (RNP),[22] mitochondria,[18,23-27] microsomes,[28,29] ribosomes,[30-32] and postmicrosomal fractions.[19,21,33,34] Almost all the enzyme activity in the nucleus is present in the extranucleolar fraction[35] which is consistent with the lack of polyadenylation of nucleolar RNA in vivo. The nuclear enzyme exists as a chromatin-bound and free population.[35-39] The two nuclear enzyme entities were first identified in rat liver nuclei.[35,36] The close association of the bound form of the enzyme with chromatin coupled with the selective inhibition of this enzyme fraction by very low levels of cordycepin triphosphate (3′dATP) indicate that the bound population of the enzyme might be involved in the initial polyadenylation reaction.[40,41] In contrast, the free form of the enzyme most probably catalyzes the extension of poly(A) chains known to occur in vivo.[40,41] A large proportion of the postmicrosomal enzyme appears to be of nuclear origin, which is released into the soluble fraction of the cell by homogenization in isotonic or hypotonic buffers.[34] Nevertheless, a significant fraction of poly(A) polymerase is present in the 105,000 × g supernatant fraction of the cell even after homogenization in hypertonic sucrose.[34] However, recent studies using glycerol for homogenization, in order to minimize leakage of nuclear components, have shown that the enzyme in this cellular fraction is largely derived from the nucleus.[31] Due to problems of redistribution of poly(A) polymerase upon cell disruption, the exact cellular distribution of poly(A) polymerase can be evaluated only by using a radioactive enzyme-specific inhibitor in vivo followed by autoradiography or by performing indirect immunofluorescence of specific enzyme-antibody complexes.

B. Primer and Substrate Specificities and Metal Ion Requirements

Eukaryotic poly(A) polymerase generally requires an exogenous primer for its function, utilizes ATP almost exclusively as the substrate, and catalyzes the reaction: Primer + nATP → Primer-$(pA)_n$ + nPPi. The enzyme can use either Mg^{2+} or Mn^{2+}, or both, depending upon several factors such as the enzyme source, substrate and primer concentrations. In general, the Mg^{2+} optima is much higher than the Mn^{2+} optima. The activation of enzyme activity in calf thymus nuclear extracts by Mg^{2+} is also related to the ATP concentration.[3] A Mg^{2+}-activated poly(A)-hydrolytic activity present in the partially purified poly(A) polymerase preparation might be responsible for the lack of the Mg^{2+}-dependent polymerization reactions.[3] However, the nuclease associated with highly purified polymerase can function with both Mg^{2+} and Mn^{2+}.[42] These observations imply that the use of different divalent metals by poly(A) polymerase may not be due to different enzyme forms or species, but rather to other factors. The preference for Mg^{2+} or Mn^{2+} appears to be related to the primer concentrations; at least with calf thymus cytosolic enzyme, the ability to use Mg^{2+} increases as the concentration of primer, $p(A)_4$, is increased.[19] It is not known whether similar data will ensue with natural primers such as poly(A)-deprived mRNA. The requirement of Mg^{2+} might also be related to the chromatin-associated factors. Indeed, poly(A) polymerase, when associated with the chromatin, can utilize Mg^{2+} efficiently and the optimal concentration of Mg^{2+} is even lower than that of Mn^{2+} (1 m*M* vs. 2 m*M*).[36] However, once released from the chromatin, even under mild conditions, the enzyme from most sources is virtually ineffective in the presence of Mg^{2+} and exogenous poly(A) as primer.

The Mn^{2+}-activated nuclear enzyme itself can be resolved into three distinct forms by chromatography on CM-cellulose. Such apparent chromatographic resolution of poly(A) polymerase into different peaks is most likely due to the difference in the phosphorylated state of the enzyme (see Section IV).

Unlike RNA polymerases, the nuclear, mitochondrial, and cytoplasmic poly(A) polymerases are inhibited at relatively low salt concentrations (100 m*M* KCl).[3] Complete inhibition of nuclear poly(A) polymerase can be achieved at KCl concentrations above 250 m*M*.[15] The absence of poly(A) synthesis at high salt concentrations appears to be due to dissociation of the enzyme-primer complexes at high ionic strength.[42a]

Of the synthetic polyribonucleotides, poly(A) itself is the best primer. The Mg^{2+}-activated enzyme from calf thymus uses short poly(A) much less effectively than the poly(A) of about 40 nucleotides[13] whereas the Mn^{2+}-activated enzyme can function equally well with short or long poly(A) on a molar basis.[19] In addition to poly(A) or oligo(A), solubilized poly(A) polymerase preparations can polyadenylate a wide variety of polyribonucleotides with free 3′OH termini as primers. As opposed to the chromatin-associated poly(A) polymerase, enzyme solubilized from the chromatin or from whole nuclei can utilize any RNA or synthetic polyribonucleotide as primer. The primer specificity is also lost following solubilization of poly(A) polymerase from ribosomes.[31a] The reason for the lack of primer specificity of the solubilized enzyme is not clear. Perhaps, the enzyme is deficient in a factor or factors following its release from the organelle.

The mitochondrial enzyme, especially from rat liver, prefers an endogenous primer which is closely associated with the enzyme even after enzyme solubilization.[23,24] The primer can be separated from the enzyme by phosphocellulose chromatography where the primer is eluted in the wash fractions.[25] Based on an A_{260}/A_{280} ratio of 2.0, resistance to RNase A and T_1, and its relatively low sedimentation coefficient, the primer appears to be poly(A) of relatively small size.[25] Subsequent studies[43] have suggested that the major mitochondrial primer might be an oligo adenylate of three nucleotides in length. It has not been established whether the poly(A) or oligo(A) fractions occur naturally as free entities uncomplexed with RNA or are derived from poly(A)$^+$ RNA as a result of the solubilization procedure.

C. Inhibitors

The nuclear, mitochondrial, and cytosolic poly(A) polymerases can be inhibited by a variety of rifamycin derivatives as a result of direct interaction of these compounds with the enzyme.[4,15,44,46] The inhibition by the rifamycin derivative AF/013 appears to be competitive with respect to ATP.[45] The nuclear enzyme is inhibited by 1, 10 phenanthroline[47] which suggests that this enzyme is a zinc-containing protein. It has been proposed that 1, 10 phenanthroline can combine with trace amounts of cupric ion present in reagent grade chemicals and sulfhydryl compounds used in enzyme assays and that the inhibition of enzyme activity is, in fact, due to the removal of the thiol group by the 1,10-phenanthroline-cuprous ion complex. This conclusion is based either on the requirement of a thiol for inhibition by the chelating agent or on the reversal of this inhibition by the noninhibitory cuprous ion-specific chelating agent, 2,9-dimethyl 1,10-phenanthroline.[48] The inhibition of nuclear poly(A) polymerase by 1,10 phenanthroline is not dependent on the presence of thiol,[47] nor does the presence of dimethyl analogue influence inhibition by 1,10 phenathroline.[48a] This observation coupled with actual measurements of zinc by atomic absorption spectrometry and radiolabeling with [^{65}Zn] in vivo[47] have established that Zn is an integral part of poly(A) polymerase molecule. This metal appears to be involved in the interaction of the enzyme with primer and seems to stabilize the enzyme during extensive purification.[47] The mitochondrial poly(A) polymerase is also inhibited by 1,10 phenanthroline and this inhibition is not reversed by the dimethyl analogue, which indicate that this enzyme may also be a zinc-containing polypeptide.[48b]

Another inhibitor of the reaction catalyzed by poly(A) polymerase is the substrate analog, cordycepin triphosphate (3′-dATP). At relatively low concentrations, this compound can selectively inhibit the "chromatin-bound" poly(A) polymerase activity. This reaction is at least 30 times more sensitive than that catalyzed by the "free" enzyme population.[40,41] Interestingly, the former reaction is not competitive with respect to ATP.[41] Following solubilization of the "bound" enzyme, the sensitivity of the released enzyme to 3′dATP becomes identical to that of the free enzyme.[36] This observation suggests that a chromatin-associated factor is involved in conferring the selectivity of this inhibitor and is consistent with the lack of selective inhibition of partially purified poly(A) polymerases to cordycepin 5′ triphosphate.[49,50] The extreme sensitivity of the "bound" poly(A) polymerase to 3′dATP is analogous to the selective inhibition of the initial nuclear polyadenylation reaction in cell culture by cordycepin. These data suggest that the chromatin-associated enzyme is involved in the initial polyadenylation of mRNA precursor.[40,41]

Unlike 3′dATP, the natural compound 2′dATP has no significant effect and at very high concentrations, the latter ATP analog inhibits both "bound" and "free" poly(A) polymerases in the nuclei to nearly the same degree.[41] Interestingly, inhibition of polyadenylation in isolated mitochondria occurs only at relatively high concentrations of 3′dATP 50% inhibition occurring at 300 μM, a value even greater than the K_m^{ATP}.[51] This finding is in accord with the relative insensitivity of mitochondrial polyadenylation to cordycepin in vivo.[52]

Another selective inhibitor of the poly(A) polymerase reaction is Ara-ATP, the ATP analog derived from Ara-A or vidarabine (9-β-D-arabinofuranosyladenine).[53] The calculated $K_i^{Ara\text{-}ATP}$ (4 μM) for the chromatin-bound poly(A) polymerase is greater than the $K_i^{3'dATP}$(1.3 μM). However, unlike 3′dATP, Ara-ATP does not inhibit RNA synthesis even at very high concentrations. The $K_i^{Ara\text{-}ATP}$ for the free enzyme is similar to that observed for $K_i^{3'dATP}$ (40 to 60 μM). Contrary to the noncompetitive inhibition of the chromatin-bound enzyme by 3′dATP, Ara-ATP inhibits this enzyme in a manner competitive with ATP. The remarkable sensitivity of the initial polyadenylation reaction to low levels of Ara-ATP in vitro, coupled with a complete lack of inhibition of RNA synthesis, suggest that the posttranscriptional addition of poly(A) to mRNA may be partially responsible for the action of Ara-A in vivo. Although Ara-A inhibits DNA synthesis, this effect is not alleviated by deoxyadenosine but rather by adenosine.[54,55] It is therefore plausible that cessation of DNA synthesis is secondary to inhibition of processes requiring adenine nucleotides, specifically, mRNA polyadenylation.

Another class of inhibitors of the polyadenylation reaction in vitro is the polyamines. The activity of highly purified nuclear poly(A) polymerase with exogenous primers decreases progressively with increasing concentrations of polyamines.[56] The reaction using nuclear RNA as the primer is most sensitive to spermine. The concentrations of spermine necessary to achieve 50% inhibition with nuclear RNA, poly(A) and tRNA as primers are 0.075, 0.5, 0.7 m*M*, respectively. To obtain the same degree of inhibition with poly(A) as primer, much higher concentrations of spermidine (3 m*M*) and putrescine (12 m*M*) are required. The polyamines block polyadenylation by interaction with the primer. At concentrations inhibitory to polyadenylation in vitro, spermine could stimulate DNA-dependent RNA synthesis catalyzed by RNA polymerases.[57-59] It is not known whether the stimulation of RNA synthesis with concurrent inhibition by polyamines is indeed part of a regulatory process. The extreme sensitivity of the polyadenylation of nuclear RNA, as opposed to other primers, to inhibition by polyamines suggests that the polyamines might indeed affect the polyadenylation of RNA in vivo and consequently alter the availability of functional poly(A)$^+$ RNA in the polysomes.

D. Molecular Weight and Subunit Composition

Poly(A) polymerase purified from almost all eukaryotic cells appears to consist of a single polypeptide with molecular weight ranging from 48,000 for rat liver nuclear enzyme[15] to 120,000 for calf thymus nuclear enzyme.[13] In contrast, the enzyme from vaccinia virus cores contains two subunits of M_r 51,000 to 57,000 and 35,000 to 37,000. It is possible that the highest molecular weight reported for calf thymus nuclear enzyme is due to enzyme aggregation.[13] Alternatively, the enzyme could be synthesized as a high molecular weight precursor and processed to a catalytically active component with varying molecular weights depending upon the source and/or nature of purification procedure. In earlier studies, we have observed that poly(A) polymerase purified from Morris hepatoma 3924A nuclei exhibits a major band on denaturing gels corresponding to a molecular weight of 60,000 with a minor band corresponding to a 48,000 mol wt polypeptide.[15] Interestingly, rat liver nuclear poly(A) polymerase purified under identical conditions has a molecular weight of 48,000. It is possible that proteolytic cleavage may occur during enzyme purification. Although poly(A) polymerase is a phosphoprotein[60] and charge modifications could give rise to slightly different mobilities on polyacrylamide gels (even in the presence of sodium dodecyl sulfate), it is unlikely that this phenomenon would result in an apparent M_r difference of 12,000 between the liver and hepatoma enzymes. Further, since the migration of the enzymes following electrophoresis under denaturing conditions at pH values ranging from 5.9 to 7.9 are similar,[15] the molecular weight differences do not appear to be due to charge differences between the enzyme. Rather, the loss of a polypeptide as a result of processing in vivo or in the course of enzyme purification might explain the differences in molecular weights reported for poly(A) polymerase from different sources. Further studies are required to resolve this important issue.

III. CHANGES IN POLY(A) POLYMERASE ACTIVITY IN RESPONSE TO VARIOUS STIMULI

In view of the fact that several eukaryotic mRNAs contain poly(A) at their 3′termini and poly(A) appears to stabilize mRNAs,[3] it was logical to determine whether this reaction could be regulated by alterations in poly(A) polymerase activity. Several laboratories have measured the activity of this enzyme from the nuclei and cytosol fractions under a variety of conditions. The most dramatic change in poly(A) polymerase was reported in rabbit heart where the cytoplasmic Mn^{2+}-activated enzyme activity has been shown to increase within 2.5 min after injection of noradrenaline or dibutyryl cAMP to the perfusion buffer; a ninefold stimulation of poly(A) polymerase occurs at 1 min after injection of noradrenaline.[61] If catecholamines are depleted from the heart by treatment with reserpine, as much as a 12-fold increase in the poly(A) polymerase activity has been observed after addition of noradrenaline to the perfusion buffer.[62] The nuclear poly(A) polymerase activity can also respond to steroid hormones. Thus adrenelectomy decreases the poly(A)-synthesizing activity of the nuclear enzyme, which can be restored to control levels by administration of glucocorticoids.[63] Poly(A) polymerase activity in the uterine nuclei increases (sixfold increase in "bound" enzyme and twofold increase in "free" enzyme) following administration of progesterone (1 mg/kg for 5 days) to rabbits.[39] The poly(A)-degrading activity in the uterine nuclei is higher in the progesterone-treated animals, which rules out the possibility that alteration in poly(A) polymerase activity observed after treatment with the hormone is due to an inverse change in poly(A) hydrolytic activity.

Uterine nuclear poly(A) polymerase is altered in a biphasic manner by estradiol following a single i.v. dose of the hormone.[64] A twofold increase observed within 30 min of injection is followed by a decline (40% of control by 1 to 2 hr) and a second marked elevation (two- to threefold) at 12 hr. The latter increase in poly(A) polymerase activity is blocked by prior administration of cycloheximide which seems to suggest that the delayed elevation of poly(A)-synthesizing activity requires *de novo* protein synthesis. In contrast to data obtained with rabbit uterus, no detectable changes in poly(A) polymerase solubilized from quail oviduct were observed after long-term administration of estradiol.[65] These authors have suggested that estradiol decreases the nuclease activity which leads to enhanced polyadenylation in the oviduct by the hormone. Further studies are needed to elucidate the mechanism for the differential responses of the nuclear poly(A) polymerase to the same hormone in different tissues.

The level of poly(A) polymerase activity can also vary according to the age of animals. Thus, enzyme activity in the nuclear lysates from brain and liver of 25-day old rats is significantly greater than in 300-day old animals (26 units of brain enzyme in young rat vs. 1 unit in old rats, 8 units of liver enzyme in young animals vs. 3.3 units in old animals).[66] Since mixing of enzyme preparations from young and old animals produces additive results, the decrease in poly(A) polymerase activity in senescent rats does not appear to be due to a proportional increase in inhibitors of this reaction. Contrary to the data with rat liver and brain, comparison of oviduct nuclear poly(A) polymerase between 70-day old and 950-day old quails does not reveal any significant difference in the enzyme activity between the young and senescent birds.[67] The apparent lack of response of poly(A) polymerase from quail oviduct to either estradiol or aging and the dramatic response of the same enzyme in other tissues to the same hormone[39,64] and to aging[66] suggest that polyadenylation in different tissues might be regulated by different mechanisms, i.e., changes in poly(A)-synthesizing or poly(A)-hydrolytic activity.

Nuclear poly(A) polymerase can also be modified by drugs. Thus, administration of a single dose of phenobarbital[67a] results in approximately twofold increase in the enzyme within 24 hr of drug treatment. This increase in enzyme activity appears to be due to phosphorylation of the enzyme activity (see section on phosphorylation) whereas the rise in the enzyme activity at the later time points (e.g., 48 hr) is due to an increased number of enzyme molecules as determined by radioimmunoassay (see section on immunology of poly(A) polymerase). Although a biphasic stimulation in poly(A) polymerase activity with an intermediate decline is not elicited by phenobarbital, as has been observed with estradiol,[64] the early and late increase in the enzyme activities observed in rabbit uterine nuclei and rat liver nuclei responding to estradiol and phenobarbital, respectively, might be mediated by the same mechanism. The cycloheximide-induced inhibition of the delayed response of the uterine enzyme to the hormone is consistent with the increased number of enzyme molecules observed at later time intervals after phenobarbital treatment. Obviously, labeling of the enzyme followed by immunoprecipitation with specific IgG is required to clearly distinguish *de novo* synthesis of enzyme from the decreased turnover of the enzyme in response to various agents.

Coleman et al.[68] have observed an increase in the activity of cytoplasmic poly(A) polymerase from human lymphocytes in response to phytohemaglutinin. Longacre and Rutter[69] have shown that poly(A) polymerase activity declines during the maturation of erythrocytes which is consistent with the reduction in the general metabolic activity associated with the differentiation process.

There is at least one report which suggests variation in cytoplasmic poly(A) polymerase during fertilization. The level of this enzyme in the four-cell embryo is only about half of that present in the unfertilized egg.[70] This decrease in the enzyme activity is not due to either a postfertilization alteration in intracellular distribution of the enzyme or an increase in poly(A)-hydrolytic activity. Morris and Rutter[71] and Egrie and Wilt[72] did not detect changes in the activity of this enzyme throughout embryogenesis. However the latter investigators[72] observed a rearrangement in the subcellular localization of the enzyme; the enzyme in the unfertilized egg is almost entirely localized in the 100,000 × g supernatant and as embryogenesis proceeds, a progressive increase in the enzyme activity associated with the nuclear fraction occurs with a corresponding decrease in the activity of cytoplasmic enzyme. These studies, however, have not established whether the decrease in cytoplasmic poly(A) polymerase postfertilization is due to activation of cytoplasmic inhibitors of the polyadenylation reaction.

Poly(A) polymerase can also respond to plant hormones. Thus poly(A) polymerase, partially purified from the 20,000 × g fraction of embryo-less half-seeds of wheat, has been shown to respond to the plant hormone gibberellic acid (GA).[73] Clearly, this enzyme is a mixture of microsomal and postmicrosomal poly(A) polymerase. Nevertheless, this appears to be the only report showing the hormonal regulation of poly(A) polymerase in plant cells. The hormonally-induced increase in the enzyme was prevented by cycloheximide or by amino acid analogs which suggests requirement of *de novo* protein synthesis, possibly of the enzyme itself. These data are similar to the second estradiol-induced increase in the nuclear poly(A) polymerase activity observed in uterine nuclei.[64]

Poly(A) polymerase can undergo diurnal variations in activity. The activity of nuclear poly(A) polymerase (the free enzyme) and the hepatic concentration of poly(A)$^+$

RNA show a maximum value in the dark phase.[74] Poly(A) polymerase also shows rapid fluctuations in response to amino acid supply. Overnight starvation has been shown to cause a significant decrease in the activity of nuclear and mitochondrial enzymes.[75] Refeeding the animals with the complete amino acid mixture restores the enzyme activity nearly to the control levels. Since amino acid diet increases the total protein content of the nuclei and mitochondria to a much lesser extent than poly(A) polymerase, this enzyme might be one of the few cellular proteins specifically regulated by amino acid supply. This appears to be the first demonstration of the response of poly(A) polymerase to nutrient supply, and probably the first report of dietary modification of a mitochondrial enzyme from higher organisms. In fact, the most dramatic effect in response to amino acid supply was observed with the mitochondrial enzyme. Subsequently, Matts and Siegel[76] have confirmed the decrease in rat liver nuclear poly(A) polymerase activity following starvation for 48 hr and the subsequent restoration of enzyme activity after refeeding a complete amino acid mixture. However, these investigators observed that the poly(A)-synthesizing activity in the liver nuclei from starved rats was linear for only 10 min whereas the activity from control or refed rats was linear for nearly 30 min and no difference in poly(A) synthetic activity was observed in the linear time range. Apparently, the decrease in poly(A) polymerase activity in the starved group observed after 10 min of incubation is due to elevated poly(A) endonuclease activity. On the contrary, studies in this laboratory[76a] clearly indicate that overnight starvation does not affect the time course of poly(A) synthesis in partially purified preparations. The discrepancy between this data and the above report[76] might have arisen from the use of enzyme concentrations well out of the linear range of the assay in the latter case. Alternatively, the long term starvation (48 vs. 24 hr) used by Matts and Siegel[76] could have activated hydrolytic activity which will be reflected in the poly(A) polymerase assay. Further, liver nuclear poly(A) polymerase can be dephosphorylated following overnight starvation of rats, which is rephosphorylated after administering amino acid diet.[76b] Since the enzyme activity is related to the phosphorylated state of the enzyme (see Section IV) these data could explain, at least in part, the fluctuation of poly(A) polymerase activity in response to amino acid supply. It is not known whether the modification of mitochondrial poly(A) polymerase by dietary manipulation is also caused by dephosphorylation/phosphorylation of the enzyme.

Obviously the relationship of poly(A) polymerase modifications to the in vivo mRNA polyadenylation must be elucidated. It should be recognized that the role and/or extent of polyadenylation of all poly(A)$^+$ mRNAs may not be affected at the same time intervals at which poly(A) polymerase activity is altered in response to physiological or pathological stimuli. Changes in the enzyme might be a very early event which is followed by modification in polyadenylation in vivo at a much later time point. Although these two parameters have not always been determined, there are some indications for a positive correlation between these two events. Thus, the accelerated transfer of mRNA to the cytoplasm after administration of tryptophan[77] or of dietary protein[78] to rats or mice might indeed be related to the increased poly(A) polymerase activity observed after feeding amino acid mixture. A close relationship of poly(A) polymerase alterations to mRNA modification in vivo is observed after ethionine intoxication. Relative to intact polysomes, polysomes disaggregated after ethionine treatment can be seven times more effective as primers for poly(A) polymerase.[79] This could explain augmented incorporation of [^{14}C] adenine into poly(A)$^+$ RNA after ethionine treatment[80] and the extended half-life of mRNA in ethionine-treated animals.[81]

IV. PHOSPHORYLATION OF POLY(A) POLYMERASE AND ITS POTENTIAL REGULATORY ROLE

Investigations in several laboratories have shown that protein phosphorylation is a key event in the regulation of many cellular processes. The fact that phosphorylation of chromatin-associated proteins can lead to increased gene expression has received considerable attention.[82] The close association of nuclear poly(A) polymerase with the chromatin fraction,[35,36] and the rapid response of this enzyme to physiological changes (see previous section) prompted us to investigate whether poly(A) polymerase could be a phosphoprotein and whether this posttranslational modification could alter the enzyme activity. These studies[60, 83] showed that

1. Nuclear poly(A) polymerase is indeed a phosphoprotein.
2. Hepatoma enzyme is much more phosphorylated in vivo than the liver enzyme.
3. Phosphorylation results in enzyme activation; nearly sevenfold increase in enzyme activity is observed at optimal protein kinase concentrations.
4. Phosphorylated enzyme is significantly more stable than poorly phosphorylated enzyme.
5. Phosphorylation does not alter the extent but augments the rate of poly(A) synthesis as a result of increased affinity of enzyme for its polynucleotide primer.
6. Phosphorylation alters the chromatographic characteristics of the enzyme.

Relative to poorly phosphorylated or dephosphorylated enzyme, highly phosphorylated poly(A) polymerase elutes at relatively lower salt concentrations from the phosphocellulose column.[60] Multiple forms of poly(A) polymerase observed in some cases,[3] particularly from the same organelle (e.g., nuclei)[42b,60] could be due to resolution of the highly and poorly phosphorylated forms of enzyme rather than to structurally distinct enzymes.

Phosphorylation of poly(A) polymerase might also regulate mRNA polyadenylation in vivo by controlling the rate of polyadenylation. The close association of a cyclic nucleotide-independent protein kinase (protein kinase NI) with poly(A) polymerase through several stages of purification suggest that the two enzymes might have a concerted action in vivo. Further, phosphorylation of the enzyme in response to physiological stimuli allows the rate of polyadenylation to increase almost immediately without the need for augmented enzyme synthesis. So far, phosphorylation of poly(A) polymerase has been demonstrated only in response to phenobarbital and amino acid supply.[82a] The dramatic responses of poly(A) polymerase at early stages following various stimuli (e.g., cathecholamine-induced increase of poly(A) polymerase activity within 30 sec) might be the result of augmented phosphorylation of the enzyme. It is also likely that the early increase in nuclear poly(A) polymerase activity induced by steroid hormones[63,64] is due to enzyme phosphorylation. This notion is supported by the observed resistance of early increase in poly(A) polymerase activity to cycloheximide.[64] The greater extent of phosphorylation in vivo of hepatoma poly(A) polymerase relative to normal rat liver enzyme[60] as well as the greater affinity of the phosphorylated enzyme for the primer might explain the greater proportion of the chromatin-bound population of the enzyme in hepatoma (40 vs. 1% in liver).[84] It is possible that the rapid rate of polyadenylation by phosphorylated poly(A) polymerase would lead to more efficient mRNA processing and/or a decreased turnover of mRNA which could facilitate gene expression posttranscriptionally. Further studies are needed to substantiate this conclusion.

V. EXONUCLEASE ASSOCIATED WITH PURIFIED NUCLEAR POLY(A) POLYMERASE

Poly(A) polymerase extracted from liver or hepatoma nuclei and purified by chromatographic fractionations on DEAE-Sephadex®, phosphocellulose, hydroxylapatite, QAE-Sephadex®[15] ATP-Sepharose columns,[42] and glycerol density gradient fractionation contains a poly(A) hydrolytic activity. At this stage of purification the enzyme exhibits a single protein band after electrophoresis under denaturing conditions. The poly(A) degrading activity has been shown to be a 3′→5′ exonuclease. It is active in the presence of Mg^{2+} or Mn^{2+} and exhibits a very broad pH optimum (6.4 to 8.0) in the presence of Mn^{2+}. The major product of hydrolysis is AMP. Although it has not been unequivocally established that poly(A)-hydrolytic activity is an intrinsic property of the nuclear poly(A) polymerase molecule, several observations seem to imply that both the synthetic and degradative activities might be catalyzed by a single protein with two distinct catalytic sites. First, the enzyme has gone through six different fractionation steps in addition to initial cell fractionation for isolation of nuclei (which itself removes bulk of the cellular proteins) and the purified enzyme is essentially homogeneous. Second, both activities are retained on ATP-Sepharose columns despite the lack of an ATP requirement in the hydrolytic reaction. Third, unlabeled ATP significantly reduces the poly(A) hydrolytic activity probably due to the shift of the overall reaction to poly(A) synthesis. The "ATP effect" does not appear to be due to its chelation with the divalent metal ions required for the nuclease activity. Fourth, both poly(A)-synthetic and hydrolytic activities are expressed in the presence of Mn^{2+} and at about the same pH. Fifth, if the nuclease were a contaminant, poly(A) synthesis in the presence of ATP should reach a plateau and begin to decline after a short period of incubation; in fact, with tRNA as the primer, the synthetic reaction continues for 24 hr. Finally, and perhaps more important, the extents of ATP incorporation into poly(A) and degradation of poly(A) into AMP by a fixed quantity of the purified enzyme are of the same order. If a minor protein contaminant were responsible for poly(A) degradation, the reaction rates for the synthetic and catabolic reactions would be significantly different. Despite these circumstantial evidences, it is crucial to perform more direct experiments such as the N-terminal analyses. It is plausible that the nuclease and polymerase might be located at different regions of the protein molecule. Although other 3′ exonucleases capable of degrading poly(A) are known to exist,[85,86] these nucleases do not catalyze a synthetic reaction. Polynucleotide phosphorylase can, however, catalyze both synthesis and hydrolysis (3′ to 5′ exonucleolytic cleavage) of poly(A)[87] but this enzyme does not require a primer for the forward reaction and is unlikely to be uninvolved in mRNA polyadenylation in vivo. Moreover, polynucleotide phosphorylase is largely localized in the mitochondria.[87]

What is the role of poly(A)-hydrolytic activity associated with poly(A) polymerase? The obvious function would be in poly(A) turnover which is known to occur in vivo. The shortening or aging of poly(A) associated with mRNA[89,90] and loss or shortening of poly(A) in cells infected with virus[91] might be due to the action of this poly(A) hydrolase. Similarly, the degradation of poly(A) associated with mRNA observed at low concentrations of ATP in cell-free systems used for protein synthesis,[92,93] is most likely catalyzed by the nuclease activity associated with poly(A) polymerase.

Finally, it is not known whether poly(A) polymerases from other cellular compartments contain a nuclease. In one investigation using cytoplasmic poly(A) polymerase from calf thymus no significant nuclease activity was present in purified preparations. Further studies are needed to establish whether the cytoplasmic enzyme from all sources is completely devoid of the poly(A)-hydrolytic activity or whether calf thymus is a special situation. It is worth mentioning that the close association of a nuclease with this polymerizing enzyme is not an isolated case. For example, a 3′-exonuclease activity is known to be associated with highly purified preparations of prokaryotic,[94] eukaryotic,[95] and viral[96] DNA polymerase. Similarly, RNase H activity has been found associated with reverse transcriptase[97,98] and RNA polymerase I from yeast.[99]

VI. IMMUNOLOGY OF POLY(A) POLYMERASE

A. Development of a Radioimmunoassay and Its Applications

Antibodies have been produced in rabbits against poly(A) polymerase purified from nuclei of the rat tumor, Morris hepatoma 3924A. When purified poly(A) polymerase is transferred to diazobenzyloxymethyl (DBM)-filter strips and incubated with the antibodies and then with [^{125}I]-protein A[100] (which complexes with Fc part of IgG), a specific complex is formed between poly(A) polymerase and the antibodies, as detected by autoradiography of the iodinated complex[101] (see Figure 1). The IgG can also inhibit poly(A) polymerase activity. Liver poly(A) polymerase activity is not sensitive to IgG at concentrations which significantly reduce hepatoma enzyme activity (see Figure 2). That the hepatoma enzyme is antigenically distinct from its liver counterpart was further confirmed using a competition radioimmunoassay; purified tumor enzyme is approximately 10 times more efficient than liver enzyme in reducing the amount of iodinated hepatoma poly(A) polymerase recovered in the immune complex. In marked contrast, poly(A) polymerase from fetal liver and from another tumor, mammary carcinoma R3230AC can compete well in the radioimmunoassay (Table 1). These data suggest that the nuclear poly(A) polymerase gene expressed in neoplasia may be another example of the production of fetal antigens reexpressed in the oncogenic state.[102] The embryonic nature of the tumor antigen might explain why the hepatoma enzyme was highly immunogenic; eight out of eight rabbits produced antibodies against the tumor poly(A) polymerase whereas the liver enzyme was a poor immunogen despite numerous attempts.[102a]

Using data from the competition radioimmunoassay for hepatoma poly(A) polymerase, it is estimated that this tumor contains approximately 10^6 molecules per picogram nuclear DNA or 10^7 molecules of nuclear enzyme per cell. In an analogous series of studies, a competition radioimmunoassay for liver enzyme was developed. From these data, it is estimated that rat liver contains 7×10^5 molecules per picogram nuclear DNA or approximately 5×10^6 molecules per liver cell.[102a]

B. Antigenic Relationship of Poly(A) Polymerase to Poly(A) Binding Protein

A major application of the radioimmunoassay developed for poly(A) polymerase is in the identification of proteins that are antigenically related to this enzyme. In one such investigation, the enzyme was shown to be antigenically related to polysomal poly(A)-binding protein from HeLa cells.[103] Increasing levels of purified hepatoma poly(A) polymerase or HeLa cell polysomal poly(A)-binding protein could effectively compete with ^{125}I-labeled enzyme, thereby reducing the amount of radioactivity recovered in the immune complex. No significant difference was observed in the slopes of the competition lines produced by increasing levels of poly(A) polymerase or the binding protein which suggests that these two polypeptides have similar and, perhaps, identical antigenic determinants. This was not a surprising finding in view of the fact that the amino acid composition of the enzyme[15] was similar to that observed for the poly(A)-binding protein.[104] Further, the similarities in the function of the two proteins have raised the possibility that they might even be structurally related.[103] There is, however, an apparent difference in the molecular weight of nuclear poly(A) polymerase (48,000 to 60,000) and polysomal poly(A)-binding protein (75,000 to 78,000). The molecular weight of poly(A) polymerase ranges from 48,000 to 120,000 (depending on

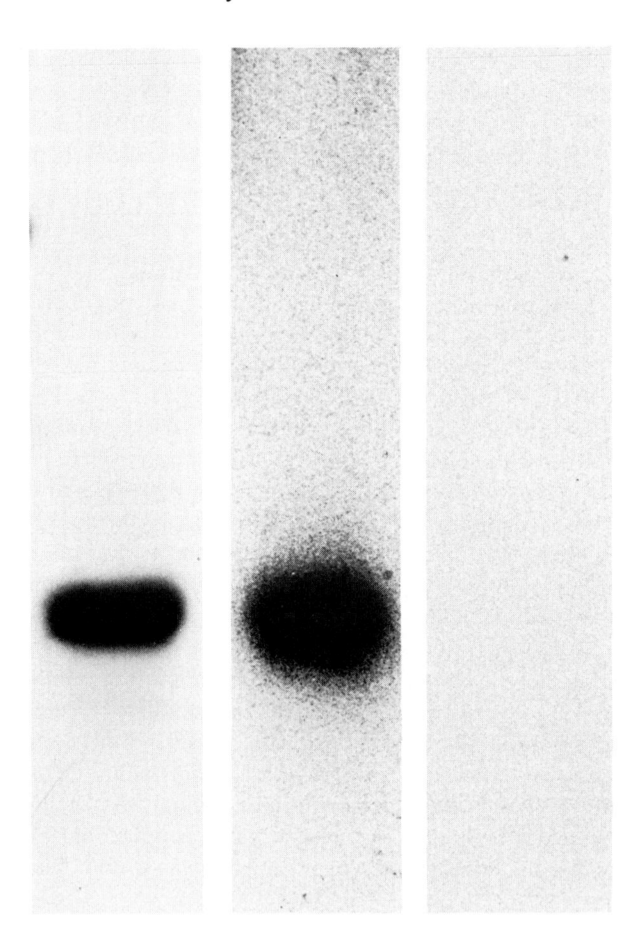

FIGURE 1. Interaction of anti-poly(A) polymerase antibodies with poly(A) polymerase immobilized on DBM-paper. Purified hepatoma poly(A) polymerase was subjected to polyacrylamide gel electrophoresis under denaturing conditions. One gel track was stained for protein with Coomassie® blue (lane A) and two others were used for transfer to DBM-filters. The filters, containing the immobilized enzyme, were then incubated separately with serum from rabbits that had been immunized with either purified hepatoma poly(A) polymerase (lane B) or RNA polymerase I (lane C), followed by incubation with ^{125}I-protein A and autoradiography.[101]

the source of the enzyme) possibly due to the different degree of processing of the enzyme precursor in vivo or in vitro during enzyme purification (see Section II.D). The fairly uniform molecular weight reported for the polysomal poly(A)-binding protein might be due to inhibition of a protease by the detergent used for the isolation of this protein. Obviously structural analysis (e.g., peptide mapping) is required to ascertain whether the two proteins are one and the same entity. It is important to compare the enzyme and the binding protein from the same cellular fraction and perhaps from the same organism.

C. Detection of Antipoly(A) Polymerase Antibodies in the Sera of Tumor-Bearing Animals and of Some Human Cancer Patients

Studies in this laboratory have shown that nuclear poly(A) polymerase from rat liver differs from the corresponding hepatoma enzyme with respect to apparent K_m^{ATP},

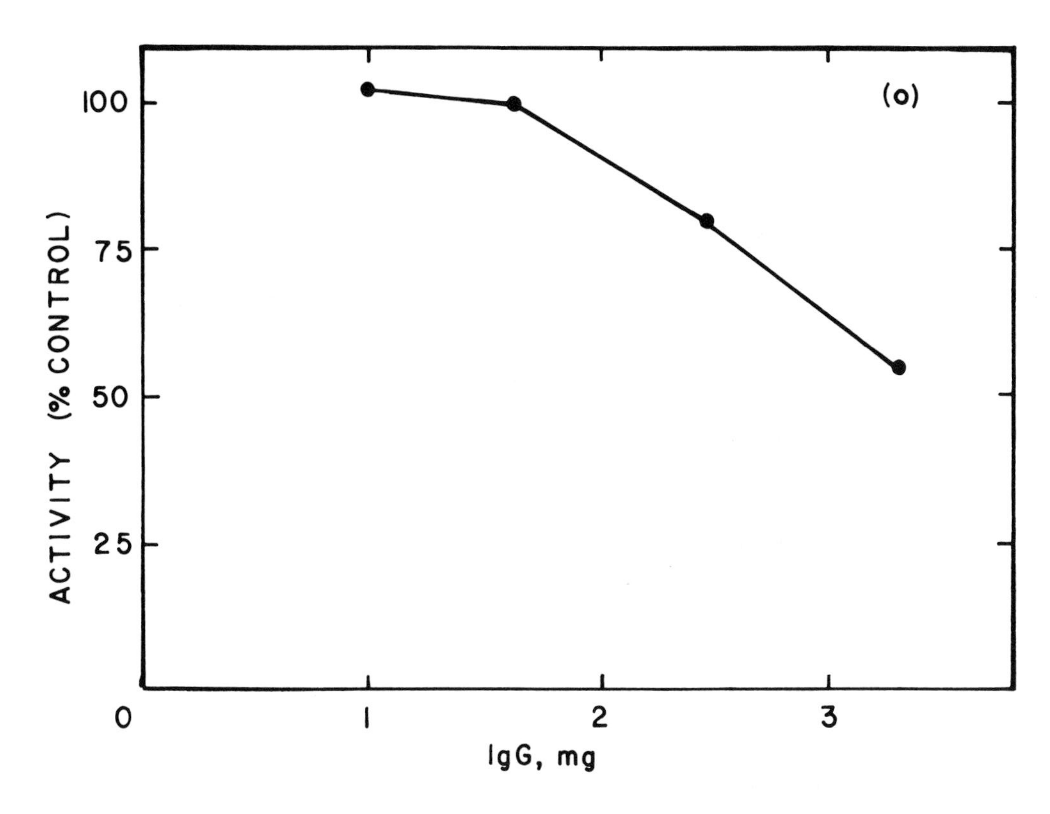

FIGURE 2. Inhibition of poly(A) polymerase activity. A rabbit was injected with purified hepatoma poly(A) polymerase emulsified in Freund's adjuvant or with adjuvant alone (control), sera obtained, and immunoglobulin fractions prepared by $(NH_4)_2SO_4$ precipitation. Varying concentrations of immunoglobulins from a control- or enzyme-injected rabbit were preincubated with 1.4 units (nmol AMP/hr) enzyme activity in the presence of 120 m*M* KCl, 200 μg/mℓ bovine serum albumin, 0.16 m*M* dithiothreitol, 50 m*M* Tris-HCl (pH 8), and 10% glycerol in a volume of 160 μℓ for 30 min at 37°C, followed by 2 hr at 4°C. Cofactors for the poly(A) polymerase assay were then added so that the following final concentrations were obtained in a final volume of 240 μℓ: 50 m*M* Tris-HCl (pH 8); 0.4 m*M* $MnCl_2$; 0.15 m*M* [^{3}H]ATP(4000 cpm/nmol), 0.5 m*M* dithiothreitol, 375 μg/mℓ poly(A), and 100 m*M* KCl. Poly(A) synthesis was carried out for 1 hr at 37°C and reactions processed on DE 81 filters as described earlier.[15] Results obtained with enzyme-injected immunoglobulins are expressed as percent of control reactions containing equivalent quantities of non-immune immunoglobulins. Control values ranged from 1.40 ± 0.08 units (mean ± S.E.M.) at the lowest protein concentration to 0.90 ± 0.10 units at the highest input immunoglobulin. (•) Hepatoma poly(A) polymerase; (o) liver poly(A) polymerase.

primer saturation levels, sensitivity to pH changes, amino acid composition,[15] zinc content,[47] and degree of phosphorylation.[60] Moreover, immunological data (see previous section) have revealed antigenic differences between liver and tumor poly(A) polymerase. Although the peptide maps of the enzyme from these two tissues have not been generated, it is likely that they will also be different. These observations raised the possibility that tumor poly(A) polymerase may be immunogenic in the host and that sera from rats bearing tumor or sera from human cancer patients may contain antipoly(A) polymerase antibodies. Indeed, we could demonstrate the presence of these antibodies in sera from rats which had been inoculated with any one of several slow growing Morris hepatomas (9618A, 5123D, 28A, and 7787) and mammary adenocarcinoma.[101] The production of antibodies was related to the rate and duration of tumor growth. For example, two fast-growing Morris hepatomas 3924A and 7788C did not elicit antibodies against the enzyme in the host. The immense weight of the tumor and indeed, the fatality of the animals bearing these tumors within 6 weeks, are not ideal for the usual immune responses in these animals. In fact, analysis of sera from animals

Table 1
EFFECT OF POLY(A) POLYMERASE FROM VARIOUS SOURCES ON THE COMPETITION RADIOIMMUNOASSAY FOR HEPATOMA 3924A ENZYME[a]

Source of partially purified poly(A) polymerase	Activity	Recovery of [I^{125}] labeled poly(A) polymerase
	(Units × 10^{-3})	(% control)
Hepatoma 3924A	42	60
Fetal liver (day 19)	140	66
Mammary R3230AC	200	66
Adult liver	600	62

[a] All enzymes were extracted from isolated nuclei and subjected to DEAE-Sephadex® chromatography essentially as described previously.[15] Results are expressed as percent control, determined in the absence of competing antigen (600 counts/min above background).

bearing the slow growing Morris hepatoma 9618A or mammary adenocarcinoma R3230AC at different time intervals after inoculation shows that circulating antibodies are detectable only after 6 weeks and the antibody titer increases dramatically (as much as 50-fold) within 16 to 18 weeks (Figure 3).

Interestingly, antibodies to nuclear poly(A) polymerase are also seen in the sera from cancer patients. Initial studies which focused on children and juveniles, showed that sera from patients of this age group consistently produced the antibodies. The specificity of the antibodies was confirmed by the inhibition of poly(A) polymerase activity by these sera as opposed to the control sera. Clearly, extensive studies are required to determine whether the production of anti-poly(A) polymerase antibodies is specific to certain types of cancer. Nevertheless, the presence of antibodies to a nuclear enzyme of known function in the tumor-bearing host offers exciting prospects for future investigations on the role of this protein in neoplasia. For example, it is not known whether a high titer of anti-poly(A) polymerase antibodies can affect tumor growth. Further, the earliest time point of detection of the antibodies during carcinogenesis has not yet been determined. Finally, it is not clear whether poly(A) polymerase itself can be secreted or is present on the cell surface. In all probability, tumor necrosis must occur before the enzyme is exposed to the host immune system.

VII. PROBABLE FUNCTIONS OF POLY(A)

For a detailed discussion on the functional significance of poly(A), the reader is asked to refer to an excellent review article by Brawerman.[2] Only the recent developments on this subject will be discussed in some detail here. Almost all mRNAs, which include both polysomal and mitochondrial species, contain a track of poly(A) at their 3′ termini. A small proportion of the mRNAs appear to lack a detectable terminal oligodenylate sequence under all conditions. However, hybridization[105-108] and translation studies[106,109-111] have not revealed any significant differences in sequence homologies between poly(A)-containing and poly(A)-deficient RNAs. Polyadenylation appears to activate the inactive mRNA population stored in the nonpolysomal cytoplasmic fraction by directing it to the polysomes for translation upon fertilization.[108,112,113] Thus, although poly(A)-containing and -deprived mRNAs do not seem to differ qualitatively, there is a positive correlation between the quantity of mRNA available for translation and the presence of a long terminal poly(A) sequence

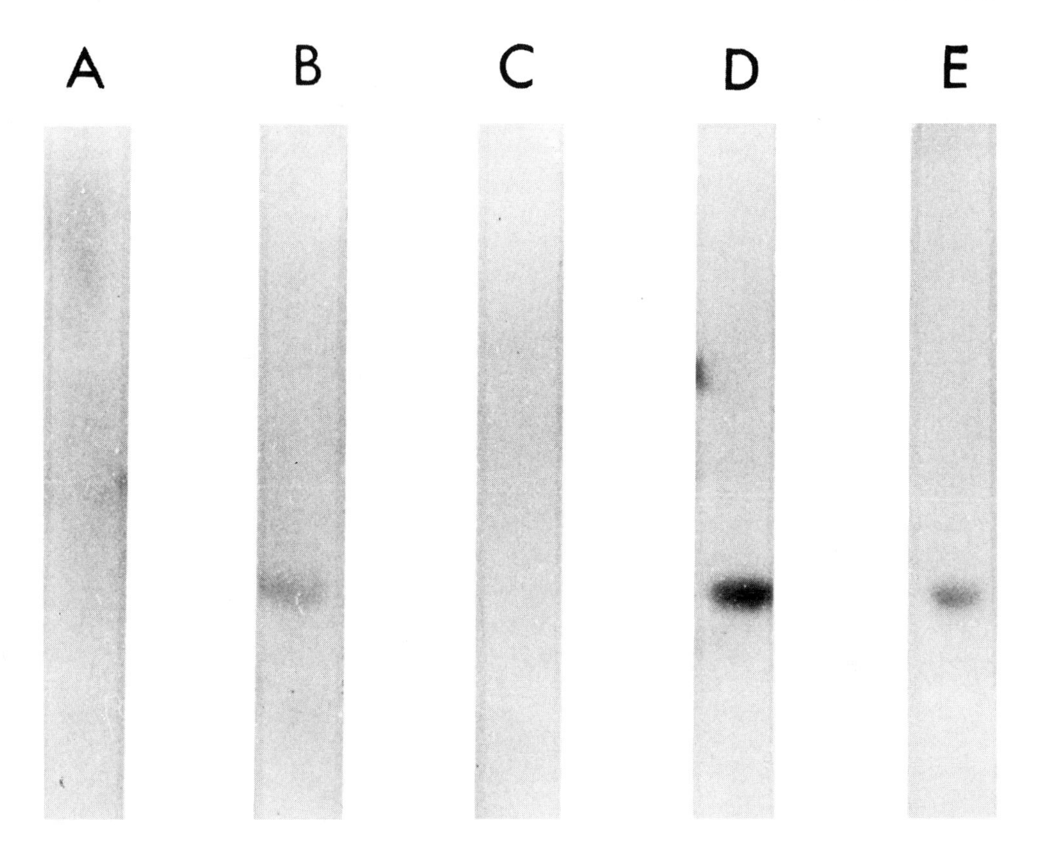

FIGURE 3. Interaction of sera from tumor-bearing rats with immobilized hepatoma poly(A) polymerase. Purified poly(A) polymerase was subjected to polyacrylamide gel electrophoresis under denaturing conditions and transferred to DBM-filters. The filters were then incubated with serum samples from female Fisher strain rats inoculated with mammary adenocarcinoma R3230AC for 7 weeks (lane A, serum diluted 1/5) or 10 weeks (lane B, serum diluted 1/5) prior to sacrifice, or from male Buff/s strain rats inoculated with Morris Hepatoma 9618A for 5 weeks (lane C, serum diluted 1/5) or 16 weeks (lane D, serum diluted 1/20; lane E, serum diluted 1/100) prior to testing. Immune complexes were detected as described in the legend to Figure 1.

in this RNA population. Poly(A) has been postulated to provide a signal for the specific cleavage of mRNA precursors, facilitate transport of some functional mRNA species to the cytoplasm and confer stability to mRNA and consequently enhance its translational capacity. The role of this unique homopolymer in mRNA stabilization has received a good deal of attention. Initially these studies consisted of microinjection of poly(A)$^+$ and deadenylated mRNAs into frog oocytes followed by measurements of the rate of their translation in vivo (for a review, see Reference 2). More recently, this approach has been extended to somatic cells. In one such investigation, HeLa cells were used as the recipient cells for the microinjected poly(A)$^-$ and poly(A)$^+$ mRNA.[114] Following microinjection of either native or deadenylated mixture of rabbit α and β globin mRNA molecules into HeLa cells, the translational capacity of poly(A)$^+$ mRNA, as measured by the synthesis of α and β globin chains, was much higher than that of deadenylated mRNA. Readenylation of the latter RNA in vitro restored its ability to synthesize α and β globin chains after microinjection into the HeLa cells. Although the exact mechanism by which poly(A) stabilizes mRNA has not been elucidated, it seems likely that poly(A) forms a nuclease-resistant hybrid with oligo(U) which is present in poly(A)$^+$ RNA.

Sea urchin embryo mRNAs lacking poly(A)[15] and deadenylated ovalbumin mRNA[116] appear to initiate on the ribosomes at a slower rate than the adenylated mRNAs. Poly(A) has also been shown to be required for viral infectivity.[117] Recently, the requirement of nuclear polyadenylation in mRNA transport has been elegantly demonstrated.[118]

One of the most important functions of poly(A) has been identified in the mitochondria. In human mitochondria, termination codons for some mRNAs are not transcribed from the DNA but are created posttranscriptionally by polyadenylation of the mRNAs.[119] The mitochondrial poly(A) polymerase first identified[23,24] and subsequently characterized[18] in this laboratory might be responsible for this unique reaction.

There has been some suggestion as to the obligatory requirement of polyadenylation in gene splicing[120,121] based largely on the observations that (1) both in vivo[122] and in vitro,[123] polyadenylation precedes splicing, (2) in vivo nonpolyadenylated RNA transcripts are not properly processed, and (3) histone gene is not spliced, and concurrently, histone mRNAs are poorly polyadenylated.[124,126] Bina et al.[127] have proposed a model in which poly(A) facilitates the splicing by promoting the formation of a triple-stranded structure within the mRNA precursor. All these suggestions on the probable role of poly(A) in splicing are circumstantial. Using 3′ deoxyadenosine (Cordycepin) as a specific inhibitor of polyadenylation in vivo, Darnell and colleagues[128] have recently concluded that polyadenylation may not be required at least for splicing of specific adenovirus nuclear RNA molecules from early regions 1b and 2. Although cordycepin is likely to block most of the poly(A) addition to mRNA, it is possible that a short stretch of a few adenylate residues is still synthesized posttranscriptionally.[128] Further, it is not evident whether the lack of requirement of poly(A) in splicing of mRNA precursor is unique to all eukaryotic mRNAs.

VIII. PROBABLE RECOGNITION SITES FOR POLY(A) ADDITION

The probable recognition site for poly(A) polymerase most probably resides in the 3′ region. With the possible exception of chicken ovalbumin gene transcription which terminates near the poly(A) site,[129] most transcription proceeds well beyond the actual termination (or polyadenylation) site (for references, see Fitzgerald and Shenk[130]). It is known that adenovirus late transcripts are cleaved and polyadenylated before the primary transcript is completed.[120] Since an exonucleolytic attack is not possible due to lack of free 3′ OH end, the transcript must be cleaved by an endonuclease. Recently, Fitzgerald and Shenk[130] have identified a major site for poly(A) addition of late SV40 mRNAs. The only unifying feature of more than 30 eukaryotic mRNAs includes the 5-AAUAAA-3′ sequence, which is found 11 to 26 nucleotides from the start of the poly(A) sequence. A 16 pair deletion, which includes this hexanucleotide sequence, prevented poly(A) addition of the SV40 late mRNAs. However, the AAUAAA sequence does not always appear to be the recognition site for polyadenylation. In many instances, this sequence is found at sites that are not amenable to polyadenylation (for references, Fitzgerald and Shenk[130] and Tosi et al.[132]). Thus, other factors are required for providing the signal for endonucleolytic cleavage and polyadenylation of certain mRNAs. Perhaps, an additional or alternate nucleotide sequence is required as the recognition site. For example, chicken lysozyme mRNA[133] and the mouse pancreatic β-amylase mRNA[134] contain an AUUAAA sequence as the probable recognition site rather than the AAUAAA sequence. Alternatively, a combination of primary and secondary structures might be required for recognition by poly(A) polymerase and/or the endonuclease which produces the free 3′OH group. It is also possible that chromatin-associated factors could provide additional specificity, the absence of which might account for the utilization of virtually all classes of RNAs with free 3′OH termini as primers following solubilization of poly(A) polymerase from the chromatin.

IX. CONCLUSIONS AND PERSPECTIVES

Considerable advances have been made in the past decade on several aspects of poly(A) polymerase from a variety of sources. Some of the noteworthy achievements can be summarized as follows: the enzyme has now been well characterized. The nuclear enzyme exists as chromatin-bound and free forms, which appear to catalyze initial poly(A) addition and poly(A) extension reactions, respectively. A mitochondrial poly(A) polymerase was identified and characterized. The probable regulation of enzyme activity has been documented in a number of systems. Clearly, the functional significance of the alterations in enzyme activity should be established. Studies on phosphorylation of poly(A) polymerase and its effect on poly(A) synthesis have provided an important mechanism for controlling the rate of the polyadenylation reaction. It is not known whether poly(A) polymerases in other cellular compartments can also be regulated by phosphorylation. One should, however, bear in mind that any future studies addressing this question must ascertain that poly(A) polymerase used as an acceptor for protein kinase in vitro or in vivo must be in a dephosphorylated state. Further, the extent of addition to and removal of phosphate groups might be related to the metabolic or physiological states of a tissue or cell. The physiological relevance of poly(A) polymerase phosphorylation must await microinjection studies involving phosphorylated and dephosphorylated forms of the enzyme using cells depleted of the enzyme.

The antigenic relationship of poly(A) polymerase to polysomal binding protein must now be extended to answer the obvious question whether they are structurally related. This will provide the enzymatic function of a key cellular protein. The structural relationship of poly(A) polymerase from different cellular compartments also must be established.

Another interesting development has been in regard to a 3'→5' exonuclease activity closely associated with highly purified poly(A) polymerase from rat liver or hepatoma nuclei. Based on several criteria, the two enzyme activities appear to be part of the same protein molecule. However, studies such as N-terminal analysis are needed to unequivocally establish the multifunctional nature of this enzyme molecule. Since the activity of this nuclease seems to depend on the levels of ATP, it is possible that this nucleotide might control the rate of poly(A) synthesis and degradation in vivo. It would be of interest to investigate whether poly(A) polymerases from other cellular fractions also contain this nuclease activity. Highly purified mitochondrial poly(A) enzyme also contains an active 3'→5' exonuclease activity.[136] Although purified cytosolic or microsomal enzymes do not seem to contain a hydrolytic activity, it can be expressed under certain conditions. Further studies are required to determine the nature of this nuclease activity.

The detection of antibodies in the sera of animals bearing tumors and of some cancer patients is a significant observation. Obviously, it is imperative to screen sera from a large number of cancer patients before one can assess the practical application of this observation.

Finally, monoclonal antibodies to poly(A) polymerase must be raised. These would be useful in the rapid purification of poly(A) polymerase, analysis of translation of poly(A) polymerase mRNA in vitro, and in the identification and cloning of the gene for the enzyme. Microinjection of these antibodies into viable cells followed by analysis of mRNA processing might lead to elucidation of the physiological role of poly(A) polymerase as well as the role of mRNA polyadenylation in gene expression.

ACKNOWLEDGMENTS

We are grateful to Ms. Edna J-A. Mayeski for her assistance in the preparation of this chapter. Studies in our laboratory were supported, in part, by the USPHS grant, CA-25078, from the National Cancer Institute.

REFERENCES

1. **Edmonds, M. and Abrams, R.**, Polynucleotide biosynthesis: formation of a sequence of adenylate units from adenosine triphosphate by an enzyme from thymus nuclei, *J. Biol. Chem.*, 235, 1142, 1960.
2. **Brawerman, G.**, Characteristics and significance of the polyadenylate sequence in mammalian messenger RNA, *Prog. Nucl. Acid Res.*, 17, 117, 1976.
3. **Edmonds, M. and Winters, M. A.**, Polyadenylate polymerases, *Prog. Nucl. Acid Res.*, 17, 149, 1976.
4. **Jacob, S. T. and Rose, K. M.**, RNA polymerases and poly(A) polymerase from neoplastic tissues and cells, *Methods Cancer Res.*, 14, 191, 1978.
5. **Moss, B., Rosenblum, E. N., and Paoletti, E.**, Polyadenylate polymerase from vaccinia virions, *Nature (London), New Biol.*, 245, 59, 1973.
6. **Moss, B., Rosenblem, E. N., and Gershowitz, B.**, Characterization of a polyriboadenylate polymerase from vaccinia virions, *J. Biol. Chem.*, 250, 4722, 1975.
7. **Brown, M. K., Dorson, J. W., and Bollum, F. J.**, Terminal riboadenylate transferase: a poly A polymerase in purified vaccinia virus, *J. Virol.*, 12, 203, 1973.
8. **Galet, H. and Prevec, L.**, Polyadenylate synthesis by extracts from L cells infected with vesicular stomatitis virus, *Nature (London), New Biol.*, 243, 200, 1973.

8a. **Villarreal, L. P. and Holland, J. J.**, Synthesis of poly(A) in vitro by purified virious of vesicular stomatitis virus, *Nature (London), New Biol.*, 246, 17, 1973.

8b. **Banerjee, A. K. and Rhodes, D. P.**, In vitro synthesis of RNA that contains polyadenylate by virion-associated RNA polymerase of vesicular stomatitis virus, *Proc. Natl. Acad. Sci. U.S.A.*, 70, 3566, 1973.

9. **Leung, W. C., Leung, M. F. K. L., and Rawls, W. E.**, Distinctive RNA transcriptase, polyadenylic acid polymerase, and polyuridylic acid polymerase activities associated with Pichinde virus, *J. Virol.*, 30, 98, 1979.
10. **Brakel, C. and Kates, J. R.**, Poly(A) polymerase from vaccinia-virus infected cells. Partial purification and characterization, *J. Virol.*, 14, 715, 1974.
11. **Brakel, C. and Kates, J. R.**, Poly(A) polymerase from vaccinia-virus infected cells. II. Product and primer characterization, *J. Virol.*, 12, 724, 1974.
12. **Giron, M., Logeat, F., Forssar, N., and Huppert, J.**, Poly(A) polymerase activity in L cells following encephalomyocarditis virus infection, *J. Gen. Virol.*, 40, 577, 1978.
13. **Winters, M. A. and Edmonds, M.**, A poly(A) polymerase from calf thymus. Purification and properties of the enzyme, *J. Biol. Chem.*, 248, 4756, 1973.
14. **Winters, M. A. and Edmonds, M.**, A poly(A) polymerase from calf thymus. Characterization of the reaction product and the primer requirement, *J. Biol. Chem.*, 248, 4763, 1973.
15. **Rose, K. M. and Jacob, S. T.**, Nuclear poly(A) polymerase from rat liver and a hepatoma, *Eur. J. Biochem.*, 67, 11, 1976.
16. **Grez, M. and Niessing, J.**, Affinity chromatography of poly(A) polymerase on ATP-sepharose, *FEBS Lett.*, 77, 57, 1977.
17. **Ohyama, Y., Fukami, H., and Ohla, T.**, Purification and characterization of a poly(A) polymerase from beef liver nuclei, *J. Biochem. (Tokyo)*, 88, 337, 1980.

17a. **Giron, M. L. and Huppert, G.**, Polyadenylate synthetase des Cellules D'ascite de Souris. Purification et caracterisation de l'enzyme, *Biochim. Biophys. Acta*, 287, 438, 1972.

18. **Rose, K. M., Morris, H. P., and Jacob, S. T.**, Mitochondrial poly(A) polymerase from a poorly differentiated hepatoma: purification and characteristics, *Biochemistry*, 14, 1025, 1975.
19. **Tsiapalis, C. M., Dorson, J. W., and Bollum, F. J.**, Purification of terminal riboadenylate transferase from calf thymus gland, *J. Biol. Chem.*, 250, 4486, 1975.
20. **Sastre, L. and Sebastian, J.**, Purification and properties of a polyadenylate polymerase from *Artemia* dormant embryos, *Biochim. Biophys. Acta*, 661, 54, 1981.
21. **Nevins, J. R. and Joklik, W. K.**, Isolation and partial characterization of the poly(A) polymerases from Hela cells infected with Vaccinia virus, *J. Biol. Chem.*, 252, 6939, 1977.

22. **Niessing, J. and Sekeris, C. E.,** Nucleoside triphosphate polymerizing enzyme associated with ribonucleoprotein particles containing DNA-like RNA, *Nature (London), New Biol.*, 243, 9, 1973.
23. **Jacob, S. T. and Schindler, D. G.,** Polyriboadenylate polymerase solubilized from rat liver mitochondria, *Biochem. Biophys. Res. Commun.*, 48, 126, 1972.
24. **Jacob, S. T. and Schindler, D. G.,** Mitochondrial polyriboadenylate polymerase: relative lack of activity in hepatomas, *Science*, 178, 639, 1972.
25. **Jacob, S. T., Rose, K. M., and Morris, H. P.,** Expression of purified mitochondrial poly(A) polymerase of hepatomas by an endogenous primer from liver, *Biochim. Biophys. Acta*, 361, 312, 1974.
26. **Aujame, L. and Freeman, K. B.** The synthesis of polyadenylic acid-containing ribonucleic acid by isolated mitochondria from ehrlich ascites cells, *Biochem. J.*, 156, 499, 1976.
27. **Cantatore, P., DeGorgi, C., and Saccone, C.,** Synthesis of poly(A) containing RNA in isolated mitochondria from rat liver, *Biochem. Biophys. Res. Commun.*, 70, 431, 1976.
28. **Wilke, N. M. and Smellie, R. M. S.,** Chain extension of ribonucleic acid by enzymes from rat liver cytoplasm, *Biochem. J.*, 109, 485, 1968.
29. **Rose, K. M. and Jacob, S. T.** Poly(A) polymerase activity of rat liver microsomes, *J. Cell Biol.*, 67, 370a, 1975.
30. **Bretthauer, R. K. and Twu, J. S.,** Properties of a polyriboadenylate polymerase isolated from yeast ribosomes, *Biochemistry*, 10, 1576, 1971.
31. **Milchev, G. T. and Hadjilov, A. A.,** Association of poly(A) and poly(U) polymerases with cytoplasmic ribosomes, *Eur. J. Biochem.*, 84, 113, 1978.

31a. **Milchev, G. J., Auramova, Z. U., and Hadjilov, A. A.,** Primer specificity of ribosome-associated poly(A) polymerase from ehrlich ascites cells, *Eur. J. Biochem.*, 103, 109, 1980.

32. **Avramova, Z. V., Milchev, G. I., and Hadjilov, A. A.,** Two distinct poly(A) polymerases isolated from the cytoplasm of ehrlich ascites tumor cells, *Eur. J. Biochem.*, 103, 99, 1980.
33. **Chung, C. W., Mahler, H. R., and Enrione, M.,** Incorporation of adenine nucleotide into ribonucleic acid by cytoplasmic enzyme preparations of chick embryos, *J. Biol. Chem.*, 235, 1448, 1960.
34. **Rose, K. M., Lin, Y-C., and Jacob, S. T.,** Poly(adenylic acid) polymerase: loss of enzyme from rat liver nuclei isolated under isotonic conditions, *FEBS Lett.*, 67, 193, 1976.
35. **Jacob, S. T., Roe, F. J., and Rose, K. M.,** Chromatin-bound and free forms of poly(adenylic acid) polymerase in rat hepatic nuclei, *Biochem. J.*, 153, 733, 1976.
36. **Rose, K. M., Roe, F. J., and Jacob, S. T.,** Two functional states of poly(adenylic acid) polymerase in isolated nuclei, *Biochim. Biophys. Acta*, 478, 180, 1977.
37. **Capone, G. and Drusiam, F.,** Sub-nuclear distribution of poly(A) polymerase activity in rat liver nuclei, *Boll. Soc. Ital. Biol. Sper.*, 57, 246, 1981.
38. **Yu, F. L.,** Rapid inhibition by cycloheximide of rat hepatic nuclear free and engaged poly(A) polymerase activities, *Life Sci.*, 26, 11, 1980.
39. **Orava, M. M., Isomaa, V. V., and Jänne, O.,** Nuclear poly(A) polymerase activities in the rabbit uterus. Regulation by progesterone administration and relation to the activities of RNA polymerases and chromatin template, *Eur. J. Biochem.*, 101, 195, 1979.
40. **Rose, K. M., Bell, L. E., and Jacob, S. T.,** Selective inhibition of initial polyadenylation in isolated nuclei by low levels of cordycepin 5′ triphosphate, *Biochim. Biophys. Acta*, 475, 548, 1977.
41. **Rose, K. M., Bell, L. E., and Jacob, S. T.,** Specific inhibition of chromatin-associated poly(A) synthesis *in vitro* by cordycepin 5′ triphosphate, *Nature (London)*, 267, 178, 1977.
42. **Abraham, A. K. and Jacob, S. T.,** Hydrolysis of poly(A) to adenine nucleotides by purified poly(A) polymerase, *Proc. Natl. Acad. Sci. U.S.A.*, 75, 2085, 1978.

42a. **Sethi, V. S.,** Mechanism of polyadenylate polymerase: formation of enzyme-substrate and enzyme-primer complexes, *FEBS Lett.*, 59, 3, 1975.

42b. **Niessing, J.,** Three distinct forms of nuclear poly(A) polymerase, *Eur. J. Biochem.*, 59, 127, 1975.

43. **Quagliariello, C., Gallerani, R., Gadaleta, G., and Saccone, C.,** Primer directed poly(A) synthesis in rat liver mitochondria, *Biochem. Biophys. Res. Commun.*, 84, 45, 1978.
44. **Jacob, S. T. and Rose, K. M.,** Inhibition of poly(A) polymerase by rifamycin derivatives, *Nuc. Acid Res.*, 1, 1549, 1974.
45. **Rose, K. M., Ruch, P. A., and Jacob, S. T.,** Mechanism of inhibition of RNA polymerase II and poly(adenylic acid) polymerase by the o-n-octyl-oxime of 3-formylrifamycin SV, *Biochemistry*, 14, 3598, 1975.
46. **Hadidi, A. and Sethi, S.,** Polyadenylate polymerase from cytoplasm and nuclei of N.I.H-swiss mouse embryos, *Biochim. Biophys. Acta*, 425, 95, 1976.
47. **Rose, K. M., Allen, M. S., Crawford, I. L., and Jacob, S. T.,** Functional role of zinc in poly(A) synthesis catalyzed by nuclear poly(A) polymerase, *Eur. J. Biochem.*, 88, 29, 1978.
48. **D'Aurora, V., Stern, A. M., and Sigman, D. S.,** 1,10-Phenanthroline-cuprous ion complex, a potent inhibitor of DNA and RNA polymerases, *Biochem. Biophys. Res. Commun.*, 80, 1025, 1978.

48a. **Jacob, S. T. and Rose, K. M.,** unpublished data

48b. **Skaleris, D. A. and Jacob, S. T.,** unpublished data.

49. **Maale, G., Stein, G., and Mans, R.,** Effects of cordycepin and cordycepin triphosphate on polyadenylic and ribonucleic acid-synthesizing enzymes from eukaryotes, *Nature (London)*, 255, 80, 1975.
50. **Horowitz, B., Goldfinger, B., and Marmur, J.,** Effect of cordycepin triphosphate on the nuclear DNA-dependent RNA polymerases and poly(A) polymerase from the yeast, *Saccharomyces cerevisiae, Arch. Biochem. Biophys.*, 172, 143, 1976.
51. **Rose, K. M. and Jacob, S. T.,** Poly(adenylic acid) synthesis in isolated rat liver mitochondria, *Biochemistry*, 15, 5046, 1976.
52. **Hirsch, M. and Penman, S.** The messenger-like properties of the poly(A) plus RNA in mammalian mitochondria, *J. Mol. Biol.*, 83, 131, 1974.
53. **Rose, K. M. and Jacob, S. T.,** Selective inhibition of RNA polyadenylation by Ara-ATP *in vitro:* a possible mechanism for antiviral action of Ara-A, *Biochem. Biophys. Res. Commun.*, 81, 1418, 1978.
54. **Nichols, W. W.,** *In vitro* chromosome breakage induced by arabinosyladenine in human leukocytes, *Cancer Res.*, 24, 1502, 1964.
55. **Brink, J. J. and LePage, G. A.,** 9β-D-arabinofuranosyladenine as an inhibitor of metabolism in normal and neoplastic cells, *Can. J. Biochem.*, 43, 1, 1965.
56. **Rose, K. M. and Jacob, S. T.,** Inhibition of the polyadenylation reaction *in vitro* by polyamines, *Arch. Biochem. Biophys.*, 175, 748, 1976.
57. **Jänne, O., Bardin, C. W., and Jacob, S. T.,** DNA-dependent RNA polymerases I and II from kidney. Effect of polyamines on the *in vitro* transcription of DNA and chromatin, *Biochemistry*, 14, 3589, 1975.
58. **Jacob, S. T. and Rose, K. M.,** Stimulation of RNA polymerases I, II and III from rat liver by spermine, and specific inhibition of RNA polymerase I by higher spermine concentrations, *Biochim. Biophys. Acta*, 425, 125, 1976.
59. **Rose, K. M., Ruch, P. A., Morris, H. P., and Jacob, S. T.,** RNA polymerases from a rat hepatoma. Partial purification and comparison of properties with corresponding liver enzymes, *Biochim. Biophys. Acta*, 432, 60, 1976.
60. **Rose, K. M. and Jacob, S. T.,** Phosphorylation of nuclear poly(A) polymerase. Comparison of liver and hepatoma enzymes, *J. Biol. Chem.*, 254, 10256, 1979.
61. **Casti, A., Corti, A., Reali, N., Mezzetti, G., Orlandini, G., and Caldarera, C. M.,** Modification of major aspects of myocardial ribonucleic acid metabolism as a response to adrenaline. Behaviour of polyadenylate polymerase and ribonucleic acid polymerase, acetylation of histones and rate of synthesis of polyamines, *Biochem. J.*, 168, 333, 1977.
62. **Corti, A., Casti, A., Mezzetti, G., Reali, N., Orlandini, G., and Caldarera, C. M.,** Modifications of major aspects of myocardial ribonucleic acid metabolism as a response of noradrenaline. Action of the hormone on cytoplasmic processing of ribonucleic acid after reserpine treatment, *Biochem. J.*, 168, 341, 1977.
63. **Jacob, S. T., Jänne, O., and Rose, K. M.,** Corticosteroid-induced changes in RNA polymerases, poly(A) polymerase, and chromatin activity in rat liver, in *Regulation of Growth and Differentiated Function in Eukaryote Cells*, Talwar, G. P., Ed., Raven Press, New York, 1975, 369.
64. **Orava, M. M., Isomaa, V. V., and Jänne, O.,** Early changes in nucleoplasmic poly(A) polymerase activity in immature rabbit uterus after estradiol administration, *Steroids*, 36, 689, 1980.
65. **Muller, W. E. G., Totsuka, A., Kroll, M., Nusser, I., and Zahn, R. K.,** Poly(A) polymerase in quail oviduct. Changes during estrogen induction, *Biochim. Biophys. Acta*, 383, 147, 1975.
66. **Richter, R. and Schumm, D. E.,** Age-associated changes in poly(adenylate) polymerase of rat liver and brain, *Mech. Ageing Dev.*, 15, 217, 1981.
67. **Mueller, W. E., Zahn, R. K., Schroder, C. H., and Arendes, J.,** Age-dependent enzymatic poly(A) metabolism in quail oviduct, *Gerontology*, 25, 61, 1979.

67a. **Rose, K. M. and Jacob, S. T.,** unpublished data.

68. **Coleman, M. S., Hutton, J. J., and Bollum, F. J.,** Terminal riboadenylate transferase in human lymphocytes, *Nature (London)*, 248, 407, 1974.
69. **Longacre, S. S. and Rutter, W. J.,** Nucleotide polymerases in the developing avian erytrocyte, *J. Biol. Chem.*, 252, 273, 1977.
70. **Slater, D. W., Slater, I., and Bollum, F. J.,** Cytoplasmic poly(A) polymerase from sea urchin eggs, merogons, and embryos, *Dev. Biol.*, 63, 94, 1978.
71. **Morris, P. W. and Rutter, W. J.,** Nucleic acid polymerizing enzymes in developing *Stronglyocentrotus franciscanus* embryos, *Biochemistry*, 15, 3106, 1976.
72. **Egrie, J. C. and Wilt, F. H.,** Changes in poly(adenylic acid) polymerase activity during sea urchin embryogenesis, *Biochemistry*, 18, 269, 1979.
73. **Berry, M. and Sachar, R. C.,** Hormonal regulation of poly(A) polymerase activity by gibbrellic acid in embryo-less half-seeds of wheat, *FEBS Lett.*, 132, 109, 1981.
74. **Geppinger, G., Broking, P., and Hardeland, R.,** Diurnal rhythmicity of nuclear and cytoplasmic polyadenylated ribonucleic acids and or polyadenylate-dependent polyadenylate polymerase in rat liver, *Int. J. Chronobiol.*, 6, 23, 1979.

75. **Jacob, S. T., Rose, K. M., and Munro, H. N.**, Response of poly(adenylic acid) polymerase in rat liver nuclei and mitochondria to starvation and re-feeding with amino acids, *Biochem. J.*, 158, 161, 1976.
76. **Matts, R. L. and Siegel, F. L.**, Regulation of hepatic poly(A) endonuclease by corticosterone and amino acids, *J. Biol. Chem.*, 254, 11228, 1979.
76a. **Rose, K. M. and Jacob, S. T.**, manuscript in preparation.
76b. **Rose, K. M. and Jacob, S. T.**, unpublished data.
77. **Murty, C. N. and Sidransky, H.** The effect of tryptophan on messenger RNA of the liver of fasted mice, *Biochim. Biophys. Acta*, 262, 328, 1972.
78. **Gaetani, S., Mangheri, E., and Spadoni, M. A.**, Characteristics of newly formed cytoplasmic RNA in liver of protein depleted aminoacid refed rats, *Nutr. Rep. Int.*, 6, 75, 1972.
79. **Dennis, J. and Kisilevsky, R.**, Poly(A) polymerase activity in ethionine-intoxicated rats: the relative effectiveness of disaggregated and intact polyribosomes as primers for poly(A) polymerase, *Can. J. Biochem.*, 58, 236, 1980.
80. **Dennis, J. and Kisilevsky, R.**, Poly(A) containing RNA in the livers of ethionine-treated rats, *Can. J. Biochem.*, 58, 230, 1980.
81. **Endo, Y., Seno, H., Tominaga, H., and Natori, Y.**, Effects of inhibitors of protein or RNA synthesis on turnover rate of messenger RNA in rat liver, *Biochim. Biophys. Acta*, 299, 114, 1973.
82. **Kleinsmith, L. J.**, Do phosphorylated proteins regulate gene activity, in *Chromosomal Proteins and their Role in the Regulation of Gene Expression*, Stein, G. S. and Kleinsmith, L. J., Eds., Academic Press, New York, 1975, 45.
82a. **Rose, K. M. and Jacob, S. T.**, unpublished data.
83. **Rose, K. M. and Jacob, S. T.**, Phosphorylation of nuclear poly(adenylic acid) polymerase by protein kinase: mechanism of enhanced poly(adenylic acid) synthesis, *Biochemistry*, 19, 1472, 1980.
84. **Rose, K. M.**, Nuclear and Mitochondrial Poly(Adenylic Acid) Polymerases from Rat Liver and a Hepatoma: Purification, Properties, Subunit Composition and Levels, Ph.D. thesis, Pennsylvania State University, University Park, 1977.
85. **Razzell, W. E. and Khorana, H. G.**, Studies on polynucleotides. IV. Enzymatic degradation. The stepwise action of venom phosphodiesterase on deoxyribo-oligo-nucleotides, *J. Biol. Chem.*, 234, 2114, 1959.
86. **Lazarus, H. M. and Sporn, M. B.**, Purification and properties of a nuclear exoribonuclease from ehrlich ascites tumor cells, *Proc. Natl. Acad. Sci. U.S.A.*, 57, 1386, 1967.
87. **Grunberg-Manago, M.**, Polynucleotide phosphorylase, *Prog. Nucl. Acid Res. Mol. Biol.*, 1, 93, 1963.
88. **See, Y. P. and Fitt, P. S.**, A study of the localization of polynucleotide phosphorylase within rat liver cells and its distribution among rat tissues and diverse animal species, *Biochem. J.*, 130, 355, 1972.
89. **Mendecki, J., Lee, S., and Brawerman, G.**, Characteristics of the polyadenylic acid segment associated with messenger ribonucleic acid in mouse sarcoma 180 ascites cells, *Biochemistry*, 11, 792, 1972.
90. **Greenberg, J. R. and Perry, R. P.**, The isolation and characterization of steady state labeled messenger RNA from L cells, *Biochim. Biophys. Acta*, 287, 361, 1972.
91. **Galwitz, D., Traub, Y., and Traub, P.**, Fate of histone messenger RNA in mengovirus-infected ehrlich ascites tumor cells, *Eur. J. Biochem.*, 81, 387, 1977.
92. **Abraham, A. K. and Pihl, A.**, Formation of TP from the poly-adenylated region of eukaryotic messenger RNAs, *FEBS Lett.*, 87, 121, 1978.
93. **Abraham, A. K., Pihl, A., and Jacob, S. T.**, Turnover of the poly(A) moiety of mRNA in wheat germ extract, *Eur. J. Biochem.*, 110, 1, 1980.
94. **Gefter, M. L.**, DNA replication, *Annu. Rev. Biochem.*, 44, 45, 1975.
95. **Byrnes, J. J. and Downey, K. M.**, A new mammalian DNA polymerase with 3′ to 5′ exonuclease activity: DNA polymerase delta, *Biochemistry*, 15, 2817, 1976.
96. **Knopf, K. W.**, Properties of herpes simplex virus DNA polymerase and characterization of its associated exonuclease activity, *Eur. J. Biochem.*, 98, 231, 1979.
97. **Molling, K., Bolognesi, D. P., Bauer, W., Busen, W., Plassmann, H. W., and Hausen, P.**, Association of viral reverse transcriptase with an enzyme degrading the RNA moiety of RNA-DNA hybrids, *Nature (London), New Biol.*, 234, 240, 1971.
98. **Verma, I.**, The reverse transcriptase, *Biochim. Biophys. Acta*, 473, 1, 1977.
99. **Huet, J., Wyers, F., Jean-Marie Buhler, Sentenae, A., and Fromageot, P.**, Association of RNase H activity with yeast RNA polymerase A., *Nature (London)*, 261, 431, 1976.
100. **Renart, J., Reiser, J., and Stark, G. R.**, Transfer of proteins from gels to diazobenzyloxymethyl-paper and detection with antisera: a method for studying antibody specificity and antigen structure, *Proc. Natl. Acad. Sci. U.S.A.*, 76, 3116, 1979.
101. **Stetler, D. A., Rose, K. M., and Jacob, S. T.**, Anti-poly(A) polymerase antibodies in sera of tumor-bearing rats and human cancer patients, *Proc. Natl. Acad. Sci. U.S.A.*, 78, 7732, 1981.

102. **Ibsen, K. H. and Fishman, W. H.,** Oncodevelopmental gene expression in cancer, *Biochim. Biophys. Acta,* 560, 243, 1979.
102a. **Rose, K. M. and Jacob, S. T.,** unpublished data.
103. **Rose, K. M., Kumar, A., and Jacob, S. T.,** Poly(A) polymerase and poly(A)-specific mRNA binding protein are antigenically related, *Nature (London),* 279, 260, 1979.
104. **Mazur, G. and Schweiger, A.,** Identical properties of an mRNA-bound protein and a cytosol protein with high affinity for polyadenylate, *FEBS Lett.,* 80, 39, 1978.
105. **Milcarek, C.,** Hela cell cytoplasmic mRNA contains three classes of sequences: predominately poly(A) free, predominately poly(A) containing and bimorphic, *Eur. J. Biochem.,* 102, 467, 1979.
106. **Minty, A. J. and Gross, F.,** Coding potential of non-polyadenylated messenger RNA in mouse Friend cells, *J. Mol. Biol.,* 139, 61, 1980.
107. **Siegal, G. P., Hodgson, C. P., Edler, P. K., Stoddard, L. S., and Getz, M. J.,** Polyadenylate-deficient analogues of poly(A) containing mRNA sequences in cultured AKR mouse embryo cells, *J. Cell Physiol.,* 103, 417, 1980.
108. **Duncan, R. and Humphryes, T.,** Most sea urchin maternal mRNA sequences in every abundance class appear in both polyadenylated and nonpolyadenylated molecules, *Dev. Biol.,* 88, 201, 1981.
109. **Kaufman, Y., Milcarek, C., Berissi, H., and Penman, S.,** Hela cell poly(A) mRNA codes for a subset of poly(A)$^+$ mRNA directed proteins with actin as a major product, *Proc. Natl. Acad. Sci. U.S.A.,* 76, 4801, 1977.
110. **Paterson, B. M. and Bishop, J. O.,** Changes in the mRNA population of chick myoblasts during myogenesis *in vitro, Cell,* 12, 751, 1977.
111 **Brandhorst, B. P., Verma, D. P. S., and Fromson, D.,** Polyadenylated and nonpolyadenylated mRNA fractions from sea urchins code for the same abundant proteins, *Dev. Biol.,* 71, 128, 1979.
112. **Dolecki, G. J., Duncan, R., and Humphreys, T.,** Complete turnover of poly(A) on maternal mRNA of sea urchin embryos, *Cell,* 11, 339, 1977.
113. **Wilt, F. W.,** The dynamics of maternal poly(A)-containing mRNA in fertilized sea urchin eggs, *Cell,* 11, 673, 1977.
114. **Huez, G., Bruck, C., and Cleuter, Y.,** Translational stability of native and deadenylated rabbit globin mRNA injected into Hela cells, *Proc. Natl. Acad. Sci. U.S.A.,* 78, 908, 1981.
115. **Nemer, M., Dubnoff, L. M., and Graham, M.,** Properties of sea urchin embryo messenger RNA containing and lacking poly(A), *Cell,* 6, 171, 1975.
116. **Doel, M. T. and Carey, N. H.** The translational capacity of deadenylated ovalbumin messenger RNA, *Cell,* 8, 51, 1976.
117. **Spector, D. H. and Baltimore, D.,** Requirement of 3′-terminal poly(adenylic acid) for the infectivity of poliovirus RNA, *Proc. Natl. Acad. Sci. U.S.A.,* 71, 2983, 1974.
118. **Egyhazi, E.,** Post-transcriptional polyadenylation is probably an essential step in selection of Balbiani ring transcripts for a cytoplasmic role, *Eur. J. Biochem.,* 107, 315, 1980.
119. **Anderson, S., Bankier, A. T., Barrell, B. G., DeBruijn, M. H. C., Conlson, A. R., Drowin, E. I. D., Nierlich, D. P., Roe, B. A., Sanger, F., Schreier, P. H., Smith, A. J. H., Staden, R., and Young, I. G.,** Sequence and organization of the human mitochondrial genome, *Nature (London),* 290, 457, 1981.
120. **Nevins, J. R. and Darnell, J. E.,** Steps in the processing of Ad2 mRNA: poly(A)$^+$ nuclear sequences are conserved and poly(A) addition preceeds splicing, *Cell,* 15, 1477, 1978.
121. **Lewin, B.,** Alternatives for splicing: recognizing the ends of introns, *Cell,* 22, 324, 1980.
122. **Darnell, J. E.,** Transcription units for mRNA production in eukaryotic cells and their DNA viruses, *Prog. Nucl. Acid Res. Mol. Biol.,* 22, 327, 1979.
123. **Yang, V. W. and Flint, S. J.,** Synthesis and processing of adenoviral RNA in isolated nuclei, *J. Virol.,* 32, 394, 1979.
124. **Kedes, L. H.,** Histone messengers and histone genes, *Cell,* 8, 321, 1976.
125. **Adesnick, M. and Darnell, J. E.,** Biogenesis and characterization of histone messenger RNA in Hela cells, *J. Mol. Biol.,* 397, 406, 1972.
126. **Grunstein, M. and Schedl, P.,** Isolation and sequence analysis of sea urchin (*Lytechinus pictus*) histone H_4 messenger RNA, *J. Mol. Biol.,* 104, 323, 1976.
127. **Bina, M., Feldmann, R. J., and Deeley, R. G.,** Could poly(A) align the splicing sites of messenger RNA precursors, *Proc. Natl. Acad. Sci. U.S.A.,* 77, 1278, 1980.
128. **Zeevi, M., Nevins, J. R., and Darnell, J. E.,** Nuclear RNA is spliced in the absence of poly(A) addition, *Cell,* 26, 39, 1981.
129. **Roop, D. R., Tsai, M. J., and O'Malley, B. W.,** Definition of the 5′ and 3′ ends of transcripts of the ovalbumin gene, *Cell,* 19, 63, 1980.
130. **Fitzgerald, M. and Shenk, T.,** The sequence 5′-AAUAAA-3′ forms part of the recognition site for polyadenylation of late SV 40 mRNA, *Cell,* 24, 251, 1981.
131. **Proudfoot, N. J. and Brownlee, G. G.,** Sequence at the 3′ end of globin mRNA shows homology with immunoglobin light chain mRNA, *Nature (London),* 252, 359, 1974.

132. **Tosi, M., Young, R. A., Hagenbüchle, O., and Schibler, U.,** Multiple polyadenylation sites in a mouse α-amylase gene, *Nucl. Acid Res.,* 9, 2313, 1981.
133. **Jung, A., Sippel, A. E., Grez, M., and Schutz, G.,** Exons encode functional and structural units of chicken lysozyme, *Proc. Natl. Acad. Sci. U.S.A.,* 77, 5759, 1980.
134. **Hagenbüchle, O., Bovey, R., and Young, R. A.,** Tissue-specific expression of mouse α-amylase genes: nucleotide sequence of isozyme mRNAs from pancreatic and salivary gland, *Cell,* 21, 179, 1980.
135. **Benoist, C., O'Hare, K., Breathnack, R., and Chambon, P.,** The ovalbumin gene-sequence of putative control regions, *Nucl. Acid Res.,* 8, 127, 1980.
136. **Jacob, S. T. and Skaleris, D. A.,** unpublished data.

Chapter 6

tRNA NUCLEOTIDYLTRANSFERASE AND THE -C-C-A TERMINUS OF TRANSFER RNA

Murray P. Deutscher

TABLE OF CONTENTS

I. INTRODUCTION

The 3′ terminal trinucleotide sequence, -C-C-A, is present on all tRNA molecules and is required for the acceptor and transfer functions of this nucleic acid.[1] Activities which incorporate nucleotide residues into this terminal sequence have been known since some of the earliest studies on tRNA[2-6] and have been identified in a wide variety of organisms, and also in mitochondria[7,8] and viruses.[9-11] It is now recognized that these activities are due to a single enzyme, termed ATP(CTP)tRNA nucleotidyltransferase, which can complete the -C-C-A sequence when presented with tRNA molecules lacking all, or some, of these terminal residues.[1] In the presence of the appropriate nucleoside triphosphates, tRNA nucleotidyltransferase catalyzes the following reactions:

$$\text{tRNA-N} + 2\text{CTP} + \text{ATP} \rightarrow \text{tRNA-C-C-A} + 3\text{PPi} \quad (1)$$

$$\text{tRNA-N} + \text{CTP} \rightarrow \text{tRNA-C or tRNA-C-C} + \text{PPi} \quad (2)$$

$$\text{tRNA-C} + \text{CTP} \rightarrow \text{tRNA-C-C} + \text{PPi} \quad (3)$$

$$\text{tRNA-C-C} + \text{ATP} \rightarrow \text{tRNA-C-C-A} + \text{PPi} \quad (4)$$

Generally, reaction 3 or 4 is used to assay tRNA nucleotidyltransferase during purification.

tRNA nucleotidyltransferase has found widespread use as a reagent for modifying the 3′ terminus of tRNA molecules.[12] In this regard it has been used for substituting nucleotide analogues within the -C-C-A sequence,[12] for changing the length and composition of the terminal sequence,[13-15] and for synthesizing model tRNA precursors.[16,17] Despite its widespread use as a reagent, the biological role of tRNA nucleotidyltransferase has been less clear.[18] Over the years the putative function of this enzyme has fluctuated from its being required for tRNA biosynthesis and repair, to being required only for tRNA repair, to having different functions in different organisms. However, the continued sequencing of tRNA precursors and tRNA genes, and the isolation of tRNA nucleotidyltransferase mutants, have helped to clarify this situation.

In this article the author will discuss the structural, enzymological, functional, and regulatory aspects of tRNA nucleotidyltransferase. Less emphasis will be placed on the structural and enzymological properties of the enzyme since a detailed review of this area has recently appeared.[18]

II. STRUCTURAL PROPERTIES

In recent years, a number of apparently homogeneous preparations of tRNA nucleotidyltransferase from a variety of sources have become available for study. These include the enzymes from mammals,[20] insects,[21] higher plants,[22] fungi,[23,24] and bacteria.[25-29] tRNA nucleotidyltransferases are usually found in cells in small amounts and purifications of several thousand-fold to as much as 25,000-fold are generally required to obtain pure proteins.[20-29] Evidence for purity of tRNA nucleotidyltransferase preparations has been based mainly on polyacrylamide gel electrophoresis under denaturing conditions. Only in the case of the yeast[23] and liver enzymes[20] has purity been confirmed by sedimentation equilibrium ultracentrifugation.

The size of tRNA nucleotidyltransferases generally lie in the molecular weight range of 40,000 to 50,000 as determined by SDS acrylamide gel electrophoresis and gel filtration. For the rabbit liver enzyme a molecular weight of 48,000 has also been confirmed by sedimentation equilibrium.[20] In two cases, the molecular weight of the purified tRNA nucleotidyltransferase falls outside this range. The enzyme from bakers' yeast has a molecular weight of about 70,000, as determined both by gel electrophoresis and ultracentrifugation,[23] whereas the enzyme purified from houseflies was reported to have a molecular weight of about 30,000.[21] Since all tRNA nucleotidyltransferases are thought to catalyze the identical reaction, it is surprising that the sizes of the yeast and insect enzymes differ by over twofold. This finding is particularly unexpected since the bacterial and mammalian enzymes are apparently quite similar. Further structural analyses of the purified proteins to determine the functional significance, if any, of these size differences should prove worthwhile.

Despite the variation in molecular size, all tRNA nucleotidyltransferases examined are single polypeptide chains.[20,21,23,27-29] In addition, they are devoid of nucleic acid and in the one case studied, the rabbit liver enzyme, display no unusual patterns in amino acid composition.[20] It has been suggested, on the basis of inhibition by chelating agents, that tRNA nucleotidyltransferases may be zinc metalloenzymes.[30,31] However, this point remains to be established conclusively. Thus, although tRNA nucleotidyltransferases must interact with tRNA, two different nucleoside triphosphates, and can accurately synthesize a -C-C-A terminus on all tRNAs lacking this sequence, these enzymes appear relatively uncomplicated structurally. Nevertheless, more detailed examination of tRNA nucleotidyltransferases from a structural viewpoint seems warranted at this time.

III. ENZYMOLOGICAL PROPERTIES

tRNA nucleotidyltransferases have high turnover numbers for enzymes synthesizing a phosphodiester bond, being in the range of 50 to 350 sec^{-1} for various highly-purified enzymes.[20,24,28,32,33] These values are even more impressive when one considers that the enzymes do not catalyze an extended processive reaction, as do nucleic acid polymerases, but most dissociate from tRNA after each AMP addition. The high turnover number for these enzymes may play an important physiological role in preventing the accumulation of defective tRNA in cells.

A. Substrate Specificity

The tRNA nucleotidyltransferases display an interesting range of specificities with respect to their nucleoside triphosphate and nucleic acid substrates. Among the natural nucleoside triphosphates, ATP and CTP are the preferred substrates, being incorporated in such a manner as to ensure synthesis of the -C-C-A sequence.[1] A number of purified tRNA nucleotidyltransferases can also utilize UTP in place of CTP,[21,32,34-37] but only for incorporation into one of the two normal CMP positions, since a U-U sequence is not made.[35,37,38] Even then, the reaction proceeds at less than 10% of the rate of CMP incorporation, and the apparent K_m for UTP is as much as 10- to 40-fold higher than that for CTP.[32,35-37] Inasmuch as CTP is a potent competitive inhibitor of UMP incorporation,[21,35,39] only CMP would be incorporated under normal cellular conditions. In the few cases tested, purified tRNA nucleotidyltransferases are devoid of activity with GTP or ITP.[36,37] In some instances it has been possible to substitute deoxyribonucleoside triphosphates for the natural substrates[40] and the products of these reactions, tRNA-C-C-2′dA and tRNA-C-C-3′dA, have been extremely useful for determining the initial site of tRNA aminoacylation by different aminoacyl-tRNA synthetases.[12]

tRNA nucleotidyltransferases from different sources diverge considerably with respect to the details of their interaction with ATP and CTP. These include substantial differences in apparent K_m values, effects of tRNA on nucleoside triphosphate binding, inhibition or stimulation of incorporation of one nucleotide by the other, and interactions between the nucleoside triphosphate binding sites. A more detailed examination of the nucleoside triphosphate sites of tRNA nucleotidyltransferases has been presented elsewhere.[19]

tRNA nucleotidyltransferases generally display little or no specificity with regard to tRNA substrates, being active with tRNAs from various sources, as well as with tRNAs of different amino acid acceptor activities.[1] Furthermore, both the *Escherichia coli* and rabbit liver enzymes repair individual acceptor species in a population of tRNAs in a completely random fashion,[41] suggesting that there is no discrimination among tRNAs. These results emphasize a particularly intriguing property of tRNA nucleotidyltransferase; namely, that the enzyme must recognize some features of structure which are common to all tRNAs. What these structural features are, and how they mesh with the enzyme, represents an interesting problem in protein-nucleic interaction. One approach to this problem, the use of model acceptor substrates, will be discussed in more detail below.

Despite their ability to incorporate nucleotides on to any tRNA molecules lacking all or part of the -C-C-A sequence, tRNA nucleotidyltransferases show a high degree of discrimination against intact tRNA and other types of nucleic acid. DNA and synthetic polynucleotides are essentially inactive as nucleotide acceptors.[25,36] rRNA, 5S RNA and phage and plant viral RNAs can act as acceptors, with some enzymes, but generally at rates less than 5% of those found with tRNAs.[42-44] In addition to a general recognition of nucleic acid structure, an important determinant of whether a nucleic acid will act as an acceptor with tRNA nucleotidyltransferase is the identity of the 3′ terminal nucleotide. Thus, AMP incorporation into non-tRNA substrates is almost exclusively adjacent to a CMP residue.[43] A dramatic example of this point is the observation that AMP incorporation into *E. coli* 5S RNA, which contains a 3′ terminal uridine residue, is only 6% as rapid as incorporation into wheat germ 5S RNA, which terminates with cytidine.[43] Recognition of the 3′ terminus also plays a role in nucleotide incorporation into tRNA and model acceptors, and will be examined further in Section III.D.

B. Reaction Mechanism

An understanding of the molecular details of nucleotide incorporation by tRNA nucleotidyltransferase also requires an examination of the kinetic mechanism of the enzyme. This type of analysis has been carried out only for the *E. coli*[45] and rabbit liver enzymes.[46] In both bases, reactions were shown to proceed by a sequential mechanism, indicating that both the tRNA and nucleotide substrates are bound to the enzyme prior to the release of any product. These results, taken together with other studies showing no incorporation of [^{14}C]ATP into the protein or release of [^{32}P]PPi from [γ-^{32}P]ATP in the absence of acceptor,[47] strongly suggest that a covalent nucleotide-enzyme is not an intermediate in the reaction. Furthermore, stereochemical analysis of phosphodiester bond formation with ATPαS as substrate are consistent with an S_N2 mechanism in which the 3′ hydroxyl of tRNA-C-C displaces PPi by a nucleophilic attack at the α phosphorus of ATP, obviating the need for an enzyme-AMP intermediate.[48]

The kinetic analyses of these two tRNA nucleotidyltransferases indicated that the rabbit liver enzyme acts by a rapid equilibrium random mechanism,[46] whereas the *E. coli* enzyme is thought to proceed by a random mechanism.[45] At present, it is not clear if these differences are significant or are due to the studies being carried out under different pH conditions. At lower pH values, dissociation of the tRNA product becomes an important determinant of the rate of reaction with the enzyme from either source, implying a similar mechanism under these conditions. However, further studies are required to resolve this issue.

C. Anomalous Incorporation

Under normal conditions, i.e., in the presence of both ATP and CTP, tRNA nucleotidyltransferases can accurately synthesize or complete the -C-C-A terminus of tRNA.[1] The details of how this accurate synthesis is accomplished will be discussed in Section III.E. In contrast, when either ATP or CTP are omitted from reaction mixtures, a series of anomalous reactions take place which lead to the synthesis of sequences other than -C-C-A.[13,34,35,38] Thus, in the absence of ATP, additional CMP residues can be added to tRNA-C-C and even to tRNA-C-C-A.[13,34,35,42] The number of extra CMP residues that can be added varies with the particular tRNA nucleotidyltransferase and with the reaction conditions, but the addition of as many as seven or eight residues has been observed.[38] These anomalous reactions generally proceed at less than 5% the rate of normal CMP incorporation,[13,42] but they can be accelerated by substituting Mn^{2+} for Mg^{2+}.[42] On the other hand, the presence of KCl and, of course, ATP, inhibits the misincorporation of CMP.[42]

In the absence of CTP, misincorporation of AMP also takes place. Under these conditions AMP is incorporated into tRNA-N[13,35,38] and tRNA-C[13,32,34,35,38] in addition to the normal substrate, tRNA-C-C. These reactions generally proceed at rates less than 10% of the normal reaction, although with the rabbit liver enzyme rates as high as 40% are obtained for AMP incorporation into tRNA-C.[36] It is also possible to add several AMP residues to a tRNA acceptor to generate tRNA-C-A-A and tRNA-A-A-A,[13,38] although these reactions proceed at extremely slow rates. Again, these reactions are stimulated by Mn^{2+}, but are inhibited by the presence of low amounts of CTP.[38,39]

Inasmuch as cells contain both ATP and CTP, these misincorporation reactions are probably without physiological significance. However, they have proven extremely useful for synthesizing modified tRNAs which have been employed to study the function of the -C-C-A terminus[13-15] and to isolate tRNA processing enzymes.[49,50]

D. Enzyme - tRNA Recognition

Physical studies and kinetic analyses have shown that tRNA nucleotidyltransferases bind a single molecule of tRNA-N, tRNA-C, or tRNA-C-C[21,51-53] with a dissociation constant in the range of 10^{-6} *M*.[45,46] Although the three types of acceptor tRNAs bind with roughly equal affinities, intact tRNA-C-C-A binds about an order of magnitude more weakly.[24,46,51,54-56] How the addition of a single AMP residue to a tRNA molecule 80 nucleotides long can decrease its affinity for the enzyme is an interesting question which has not yet been resolved. However, it may be related to the finding that with some of these enzymes ATP stimulates CMP incorporation.[39] Perhaps, binding a ligand to the ATP site, including the terminal AMP of tRNA-C-C-A, causes a conformational change in the enzyme which leads to accelerated dissociation of the tRNA molecule.

It has already been noted that since tRNA nucleotidyltransferases act on all tRNA molecules, they must recognize some structural feature common to all tRNAs. Early modification studies of tRNA suggested that many features of tRNA structure were not required for tRNA nucleotidyltransferase action.[1] On the other hand, the fact that the enzyme distinguishes tRNAs from other nucleic acids[25,36] and does not work with denatured tRNA,[57] suggests that the overall structure of tRNA must be important. In addition, studies of AMP incorporation into tRNA-C-A and tRNA-C-U indicated that incorporation into these molecules was less than 5% as rapid as into tRNA-C-C.[43] These results demonstrated that the terminal residue of the tRNA acceptor also plays an important role in enzyme recognition.

Considerable clarification of the recognition problem has come from studies of model acceptor substrates which mimic the 3′ terminus of tRNA. Small compounds such as dinucleoside monophosphates and nucleosides can function as nucleotide acceptors, although with reduced Vmax and increased apparent K_m values.[47,58] These compounds apparently bind to the enzyme in the same position as the 3′ end of tRNA, and display the same acceptor specificities as the macromolecular substrates.[58] Thus, of all the dinucleoside monophosphates and nucleosides tested, only CpC and cytidine were active AMP acceptors.[47,58] Only these two compounds conform exactly to the structure present at the 3′ terminus of the natural acceptor, tRNA-C-C. In contrast, CMP incorporation into the model acceptors is much less specific,[58] as might be expected from the variability found at the 3′ terminus of tRNA-N. Moreover, the best CMP acceptor was ApC,[58] the sequence expected to be found most frequently at the 3′ terminus of tRNA-C. Thus, the use of model acceptors indicates that a complete tRNA molecule is not required for catalysis and that the enzyme specifically recognizes the nucleotide sequence at the 3′ terminus of tRNA for incorporation of the appropriate nucleotide. Presumably, it is this recognition of the 3′ terminus that allows tRNA nucleotidyltransferase to distinguish among tRNA-C-C-A, tRNA-C-C, tRNA-C, and tRNA-N.

It is clear from studies of the model acceptor substrates that the remainder of the tRNA molecule also plays an important role in stabilizing the binding of the acceptor end and also in the efficiency of catalysis. This latter point was demonstrated directly by readdition of the rest of the tRNA molecule to the model system. The presence of the tRNA fragment stimulated nucleotide incorporation as much as 60-fold without changing the apparent K_m values for any of the substrates.[59] Thus, in addition to the reacting end, which is sufficient to trigger catalysis, a second recognition region which is essential for optimum catalytic efficiency, can also be defined. At present, this second recognition region is the rest of the tRNA molecule (tRNA-Np). Whether the whole tRNA molecule is required, or only certain fragments of the overall structure, remains to be elucidated. Nevertheless, this model system permits each of the two recognition regions of the tRNA substrate to be separated and studied independently. Studies such as these should ultimately clarify how the enzyme distinguishes between compounds as similar as CpU and CpC (or between tRNA-C-U and tRNA-C-C).

E. The Active Site and Synthesis of the -C-C-A Sequence

The absence of nucleotides in homogeneous preparations of tRNA nucleotidyltransferase precludes the possibility that accurate synthesis of the -C-C-A sequence is dependent on a nucleic acid template. Rather, synthesis of this sequence must be determined by the specificity and arrangement of donor and acceptor subsites within the active site of the enzyme, and models of this type have been described.[13,36] Experimental support for a subsite model has been provided by the studies with model acceptor substrates described earlier.[39,47,58,59] The high degree of specificity for AMP incorporation into CpC, compared to other dinucleoside monophosphates, defines two subsites which specifically recognize acceptor cytidine residues plus a third subsite recognizing the ATP donor.[58] In addition, the stimulation of nucleotide incorporation into the model acceptors by nonaccepting fragments of tRNA defines another subsite recognizing the rest of the tRNA.[59] However, the structural requirements of the latter subsite have not been well characterized.

Several pieces of evidence serve to distinguish a CMP-donating (or CTP binding) subsite from the one that binds ATP. First of all, in the case of the mammalian enzymes, ATP and ATP analogues stimulate CMP incorporation,[39,42,60] which would not be possible if both ATP and CTP were bound to the same site. Second, under conditions in which the affinity reagent oxidized ATP completely inactivates AMP incorporation, CMP incorporation is much less affected.[39] Third, the binding constants for ATP and CTP differ considerably,[1,19] their ability to stabilize the enzyme against thermal inactivation also differs,[24,55] and their specificities for incorporation into model acceptors differ markedly.[58] Finally, a mutant *E. coli* strain has been isolated[61] which is defective for AMP incorporation, but unaffected for incorporation of CMP.[32]

Other studies,[39] which demonstrate that CTP is a potent inhibitor of AMP incorporation into tRNA-C, and into the model acceptors, CpC and cytidine, indicate that CTP competes with the terminal C residue on the various acceptors. These data strongly suggest that the CMP-donating subsite overlaps, or is identical, with the subsite which recognizes the terminal C acceptor moiety. This, of course, would be expected since once a CMP residue, derived from CTP, has been added to tRNA-C to generate tRNA-C-C, this residue would become the acceptor for incorporation of the incoming AMP.

All these results are most easily accommodated into a model of the active site which contains four donor and acceptor subsites arranged in tandem that recognize tRNA (subsite 1), the terminal residues of tRNA (subsites 2 and 3), and the donors, ATP and CTP (subsites 2, 3 and 4) (Figure 1). Accurate synthesis of the -C-C-A sequence comes about because tRNA acceptors are positioned in the active site adjacent to the appropriate donor subsites in a manner that is determined by the terminal structure of the acceptor and the availability of nucleoside triphosphates. Thus, tRNA-N and tRNA-C would be positioned adjacent to a CMP-donating subsite because the high affinity of CTP for its binding site(s) would prevent these tRNAs from binding adjacent to the AMP-donating site (Figure 1A and 1B). However, when CTP is absent, these tRNAs become AMP acceptors because they can bind adjacent to the ATP site with some frequency (Figure 1D). The higher level of AMP incorporation into tRNA-C, compared to tRNA-N, is presumably due to the terminal C residue increasing the frequeny of tRNA-C binding in the AMP acceptor position. When tRNA-C-C is the substrate, the presence of the two terminal CMP residues directs binding of this acceptor to the position adjacent to the AMP-incorporating site for synthesis of tRNA-C-C-A (Figure 1C). In the absence of ATP, tRNA-C-C-C is made because of a low level of CMP incorporation from the AMP-donating site (subsite 4). However, under normal cellular conditions, i.e., high ATP, this reaction would be insignificant.

This model satisfactorily accounts for various features of -C-C-A synthesis catalyzed by tRNA nucleotidyltransferase and for the anomolous incorporations observed in the absence of one of the nucleoside triphosphates. One feature of the model which remains to be clarified is whether both of the sites which recognize the terminal CMP residues of the acceptor (subsites 2 and 3) are also CMP-donating sites. If only one of these sites (subsite 3) is a donor site, it would necessitate a translocation of the tRNA-C product back one position to allow binding and incorporation of the second CMP. If both sites are also donor sites, no movement of the tRNA chain would be required to synthesize tRNA-C-C.

IV. FUNCTIONAL ROLE

The existence of an enzyme that could accurately synthesize the -C-C-A sequence of tRNA in vitro naturally led to the conclusion that this enzyme carries out the same function in vivo. Thus, it has been generally assumed that the -C-C-A terminus of tRNA is synthesized in a simple posttranscriptional process involving tRNA nucleotidyltransferase, ATP and CTP.[1] Likewise, the enzyme has also been thought to play a role in the repair of tRNA molecules which have lost part of their -C-C-A sequence by the end-turnover process known to occur in vivo in many systems.[1]

These generally accepted functions for tRNA nucleotidyltransferase were initially thrown into doubt with the isolation and sequencing[62] of a precursor to *E. coli* su_{III} $tRNA^{Tyr}$. In this molecule the -C-C-A sequence was already present and was followed by additional residues at the 3′ end, leading to the conclusion that the -C-C-A sequence

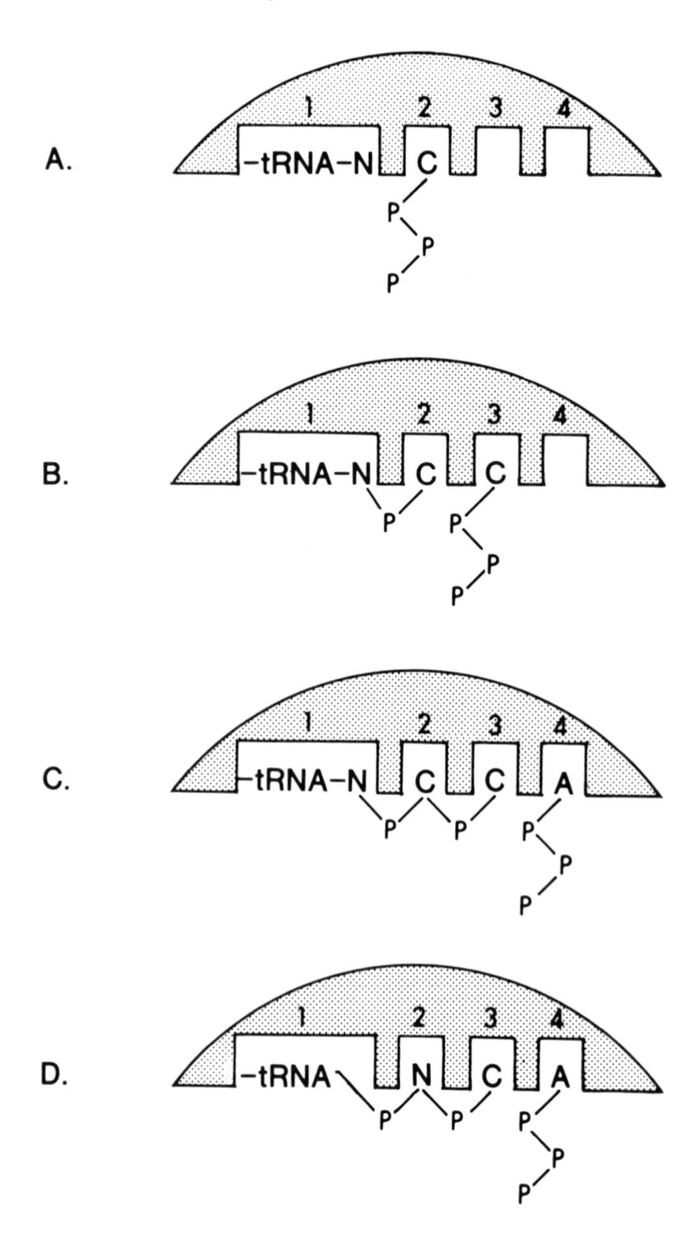

FIGURE 1. Model for tRNA nucleotidyltransferase catalyses. The active site contains four subsites (numbered 1 to 4) which recognize tRNA-N (subsite 1), each of the terminal CMP residues and/or CTP (subsites 2 and 3), and ATP (subsite 4). Depending on the structure of the tRNA and the availability of nucleoside triphosphates, the tRNA acceptor will positioned adjacent to the appropriate donor subsite to ensure accurate synthesis of the -C-C-A sequence. A. CMP incorporation into tRNA-N. Since it is not proven that subsite 2 also binds CTP, CMP incorporation into tRNA-N may come from subsite 3 followed by translocation of the tRNA-C product back one position to reopen subsite 3. B. CMP incorporation into tRNA-C. C. AMP incorporation into tRNA-C-C. D. Anomalous AMP incorporation into tRNA-C in the absence of CTP.

Table 1
-C-C-A SEQUENCES IN tRNA PRECURSORS

-C-C-A Present[a]	Ref.	-C-C-A Absent[b]	Ref.
Escherichia coli $tRNA^{Tyr}_{SuIII}$	62	T4 $tRNA^{Pro}$-$tRNA^{ser}$	68
E. coli $tRNA^{Leu}_{1}$	63	T4 $tRNA^{Gln}$	69
E. coli $tRNA^{Gly}_{2}$-$tRNA^{Thr}$	64	T4 $tRNA^{Ile}$	67
E. coli $tRNA^{Asp}_{1}$	65	*Saccharomyces cerevisiae* $tRNA^{Tyr}$	70
E. coli $tRNA^{Ser}_{3}$-$tRNA^{Arg}_{2}$	66	*S. cerevisiae* $tRNA^{Arg}_{3}$	71
T4 $tRNA^{Thr}$	67	*S. cerevisiae* $tRNA^{Trp}$	72
		Schizosaccharomyces pombe $tRNA^{Ser}$-$tRNA^{Met}$	73
		Drosophila melanogaster $tRNA^{Arg}_{2}$	74
		Bombyx mori $tRNA^{Ala}_{2}$	75, 76

[a] Listed are only those precursors in which the -C-C-A sequence is followed by additional residues.

[b] All the eukaryotic precursors are transcripts synthesized in *Xenopus* nuclear extracts in vitro or synthesized from cloned genes injected into *Xenopus* oocytes.

is encoded within the transcriptional unit of the tRNA. On the other hand, since the end-turnover of tRNA in vivo is largely limited to the terminal AMP residue,[1] it seemed unlikely that the sole function of tRNA nucleotidyltransferase in vivo would be to add a single AMP residue to defective tRNAs, especially since the enzyme has such a high turnover number, and end-turnover is a relatively slow process. Although complete resolution of this paradox for *E. coli* has not been forthcoming, work in the last several years has, nevertheless, greatly clarified our understanding of the role of this enzyme.

A. tRNA Biosynthesis

Clarification of the role of tRNA nucleotidyltransferase in tRNA biosynthesis in a variety of systems has come about from sequencing of tRNA precursors and tRNA genes, from the isolation of mutants deficient in the enzyme, and from development of systems for studying tRNA biosynthesis and processing. The combination of information from each of these areas has led to our current picture of how the -C-C-A sequence finds its way into tRNA molecules.

1. Sequences of tRNA Precursors and tRNA Genes

Isolation and sequencing of tRNA precursors has been accomplished for several different systems, including bacteria, phage-infected bacteria, yeast, and higher eukaryotes (Table 1). In some cases it has been possible to use the sequence information to determine whether the -C-C-A sequence is part of the initial transcript.[62-76] Thus, in those instances in which a monomeric or dimeric precursor contains the -C-C-A sequence followed by additional residues, or those in which the 5′ tRNA of a dimeric precursor contains -C-C-A, it is clear that these residues are not a product of tRNA nucleotidyltransferase action, but of the transcriptional process itself. This is true of several tRNA precursors from *E. coli*[62-66] and bacteriophage-infected *E. coli*[67-69] (see Table 1). In other instances of prokaryotic and eukaryotic tRNA precursors, although the -C-C-A sequence is present at the 3′ terminus of the precursor, it is not followed by additional residues. In this situation it is not possible to use the sequence information unambiguously to determine whether the -C-C-A residues are a product of transcription, or are added later, since 3′ processing may have taken place. In fact, it is clear, from sequences of tRNA genes, that in some cases such a -C-C-A sequence at the 3′ terminus of a tRNA precursor arises by transcription, whereas in others it is due to posttranscriptional addition. Therefore, reliance on precursor sequence information alone could lead to erroneous conclusions.

In contrast to the possible ambiguity resulting from finding a 3′ terminal -C-C-A in a tRNA precursor, the absence of all or part of this sequence is a good indication that posttranscriptional addition of these residues is required to generate a functional tRNA. No instance of a bacterial tRNA precursor lacking the -C-C-A sequence has yet been found. However, precursors to four of the eight new tRNAs found after infection of *E. coli* with bacteriophage T4 are lacking a complete -C-C-A sequence[67-69] (Table 1). Likewise, a number of tRNA precursors from eukaryotic cells have been shown to lack the -C-C-A sequence[70-76] (Table 1); however, in all cases except one, the eukaryotic precursors are transcripts synthesized in vitro in nuclear extracts of *Xenopus* oocytes.[71-76] In one instance, yeast $tRNA^{Tyr}$, the precursor was synthesized in *Xenopus* oocytes after microinjection of the cloned gene.[70]

Obviously, the most direct means of determining whether the -C-C-A sequence is encoded in the tRNA gene, or must be added posttranscriptionally, is to sequence the gene. In recent years, the ease of cloning tRNA genes has led to the sequencing of a large number of these genes from a variety of systems (Table 2). These include: tRNA genes from *E. coli* present as tRNA operons,[77] tRNA genes cotranscribed with structural genes,[78,79] and tRNA genes cotranscribed with rRNA, both as spacers[80,81] or distal tRNAs;[82,83] the bacteriophage T4 tRNA gene cluster;[84,85] eukaryotic tRNA genes from organisms ranging from yeast to human;[86-95] and tRNA genes from the eukaryotic organelles, mitochondria, and chloroplasts.[96-101]

Sequencing of this variety of tRNA genes has led to the conclusion that prokaryotic tRNA genes, at least as exemplified by *E. coli*, contain the information for the complete -C-C-A sequence encoded in the sequence of the gene.[77-83] Whether this conclusion is valid for all prokaryotes, or even for nonlaboratory strains of *E. coli*, remains to be determined. In contrast, all eukaryotic tRNA genes examined lack information for the -C-C-A sequence[73-75,86-95] and it must be added posttranscriptionally, presumably by tRNA nucleotidyltransferase. Even tRNA genes in mitochondria and in chloroplasts, which are thought to have some homology to *E. coli* tRNA genes, are missing the -C-C-A sequence.[96-101] The one exception to this simple division between prokaryotes and eukaryotes has been found for the tRNA genes of bacteriophage T4.[84,85] In this instance, four of the eight tRNA genes encode the -C-C-A sequence and four lack it. Even more surprisingly, of the four genes that do not encode the -C-C-A sequence, two lack the whole sequence and two lack only the last two residues.

The physiological significance of the differences in 3′ terminal structure between tRNA genes of prokaryotes and eukaryotes, and among T4 tRNA genes is not known. Possibly, it may represent an example of an evolving nucleotide sequence. Random mutations in the extraneous 3′ residues that would normally be removed during processing may have over time converted these nucleotides to -C-C-A in some systems, such as *E. coli.* If encoding of the -C-C-A sequence in the gene conferred some advantage to these cells, such as more rapid tRNA processing or lower energy requirements, these genetic changes would be fixed. In fact, just such an advantage has been observed when comparing the rates of synthesis of bacteriophage T2 and T4 $tRNA^{Ser}$. T2 $tRNA^{Ser}$, which has a transcriptionally derived -C-C-A, is synthesized more rapidly than its T4 counterpart even though both mature tRNAs have the same nucleotide sequence.[102] Why laboratory strains of *E. coli* would have attained such a dramatic change to complete encoding of -C-C-A, whereas this has not occurred in any eukaryotic system, is not clear. Perhaps, some facet of the many years of growth under laboratory conditions has contributed to this change.

2. *tRNA Nucleotidyltransferase Mutants*

The absence of a -C-C-A sequence in certain tRNA genes and tRNA precursors and the requirement for posttranscriptional addition of these residues, of course, does not prove that tRNA nucleotidyltransferase is the enzyme involved in this process. Some information bearing on this problem, at least in prokaryotes, has been obtained from study of *E. coli* mutants deficient in tRNA nucleotidyltransferase.[61]

Table 2
-C-C-A SEQUENCES IN tRNA GENES

-C-C-A Present	Ref.	-C-C-A Absent	Ref.
Escherichia coli-tRNA operons	77	Bacteriophage T4	84, 85
-spacer tRNA in rRNA transcript	80, 81	*Saccharomyces cerevisiae*	73, 86-89
-distal tRNA in rRNA transcript	82, 83	*Drosophila melanogaster*	74, 90-92
-cotranscribed with mRNA	78, 79	*Bombyx mori*	75
Bacteriophage T4	84, 85	*Xenopus laevis*	93
		Human	94
		Neurospora crassa	95
		Mitochondria-*S. cerevisiae*	96
		-*N. crassa*	97
		-*Aspergillus nidulans*	98
		-Human	99
		Chloroplasts-maize	100
		-*Euglena gracilis*	101

tRNA nucleotidyltransferase mutants were isolated after *N*-methyl-*N'*-nitro-*N*-nitrosoguanidine mutagenesis of *E. coli*.[61] Since no selection procedure for these mutants existed, colonies containing lowered levels of the enzyme were identified by a rapid screening procedure which involved assaying tRNA nucleotidyltransferase in cells made permeable to tRNA and nucleoside triphosphates by treatment with toluene.[103] Using this procedure, six clones containing less than 20% of normal levels of enzyme activity were isolated.[61] The mutation in each of the strains was located at the same position on the *E. coli* chromosome, defined as the *cca* locus, at 66 min on the genetic map.[104] Purification of tRNA nucleotidyltransferase from one of the mutant strains demonstrated that the *cca* locus is probably the structural gene for the enzyme.[32] Construction of partial diploid strains with F′ factors covering the *cca* locus indicated that the mutations in these various strains were recessive.[104] Studies with the partial diploids also demonstrated that the level of tRNA nucleotidyltransferase activity was affected by the gene dosage and that the level of the enzyme in *E. coli* was probably not regulated.[104] The *cca* locus cotransduced with a nearby marker, *tolC*, at about a 50% frequency,[104] which allowed transfer of the *cca* mutation to a variety of other strains.

tRNA nucleotidyltransferase activity in the various mutant strains varied from about 2% to about 20% of normal levels based on AMP incorporation.[61,104] In most of the mutant strains AMP and CMP-incorporating activities were decreased to a similar extent,[61] supporting the previous conclusion based on enzyme purification that a single protein was responsible for synthesis of the complete -C-C-A sequence.[106] In one case, however, AMP-incorporating activity was depressed to about 20% of normal, whereas CMP-incorporating activity was unaffected.[61] This mutant enzyme was extensively purified and compared to the wild type enzyme.[32] The mutant retained its unusual nucleotide specificity throughout purification. Studies of this mutant enzyme have provided supporting evidence for the conclusion that tRNA nucleotidyltransferase contains distinct AMP- and CMP-incorporating subsites.

Studies with the various tRNA nucleotidyltransferase mutants demonstrated that the enzyme is essential for the normal growth of *E. coli* and that the growth rate depends on the level of enzyme present.[105] Since a growth defect was associated only with those strains containing less than 15% of normal activity, these data suggested that tRNA nucleotidyltransferase is normally present in large excess over its requirements.[105] Most importantly, the *cca* mutant strains also were affected in their tRNAs.[61,107] Again, depending on the level of tRNA nucleotidyltransferase activity remaining, as many as 10 to 15% of the tRNA molecules had defective 3′ termini.[61] Since the defect was localized mainly to the 3′ terminal AMP residue (<2% of the terminal CMP was missing), it was concluded that the mutant strains were affected in end-turnover of tRNA, but not in tRNA biosynthesis. This conclusion was reenforced by studies with the antibiotics, chloramphenicol and rifampicin,[107] which inhibit end-turnover and tRNA biosynthesis, respectively.[108] The results supported the idea that all the defective tRNAs, except possibly that for $tRNA^{Cys}$, arose by end-turnover which was not repaired in the -C-C-A mutants.[107] Another report has suggested that the -C-C-A terminus of $tRNA^{Cys}$ has unusual metabolic properties and that its defective termini may also arise by end-turnover.[109] The end-turnover process of tRNA will be discussed in more detail in Section IV.B.

Additional studies with the *cca* mutants showed that the *E. coli* tRNA suppressors one to six were made in normal amounts in the mutant strains,[110] so that tRNA nucleotidyltransferase is presumably not required for the biosynthesis of these tRNAs. Thus, no evidence for an involvement of the enzyme in *E. coli* tRNA biosynthesis was obtained from any studies with the *cca* mutant strains. Obviously, these results are entirely consistent with what is now known about the presence of the -C-C-A sequence in the genes for *E. coli* tRNAs.

3. Bacteriophage tRNAs

In contrast to the situation in *E. coli*, considerable information has accumulated which indicates that tRNA nucleotidyltransferase is required for the biosynthesis of certain tRNAs specified by the T-even bacteriophages. Evidence has already been presented that four of the eight T4 tRNA genes do not encode the -C-C-A sequence.[84,85] Since a phage-specific tRNA nucleotidyltransferase is not induced upon infection of *E. coli* with T-even phages,[111] synthesis of the -C-C-A sequence on those tRNA precursors lacking it must involve the host enzyme. The first indication that this idea was correct was the finding that expression of the serine-inserting psu_1^+-amber suppressor tRNA and the glutamine-inserting psu_2^+-*ochre* suppressor tRNA was greatly depressed in *cca* mutant strains.[105] In addition, the level of suppressor activity remaining was related to the level of tRNA nucleotidyltransferase present in the various mutants.[105]

Detailed examination of tRNA synthesis after infection of a *cca* mutant has shown that two of the eight T2 tRNAs, four of the eight T4 tRNAs and one of the six T6 tRNAs lack an intact -C-C-A sequence (Table 3).[112,113] In addition, the biosynthetic origin of the -C-C-A sequence in tRNAs from different phages can differ. For example, the -C-C-A sequence of T4 $tRNA^{ser}$ and $tRNA^{Ile}$ are synthesized by tRNA nucleotidyltransferase, whereas these residues for T2 $tRNA^{Ser}$ and T6 $tRNA^{Ile}$ are added during transcription.[112,113] Why these tRNA genes have evolved differently in the different phages is not clear.

Analysis of the pathways of T4 tRNA precursor processing have shown that in some instances addition of the -C-C-A residues can be important for sequential maturation of the precursor. Thus, in the case of pre-$tRNA^{Pro+Ser}$, a dimeric precursor lacking -C-C-A at its 3′ terminus accumulates in *cca* strains, and addition of the -C-C-A sequence to the precursor is required for cleavage by ribonuclease P in vivo.[114,115] In addition, purified tRNA nucleotidyltransferase can add -C-C-A in vitro to the dimeric precursor that accumulates in a *cca* mutant.[116] On the other hand, removal of the 5′ tail from immature $tRNA^{Pro}$ by RNase P does not require -C-C-A addition.[102] Likewise, RNase P cleavage of the pre-$tRNA^{Thr+Ile}$ does not require prior addition of -C-C-A to the 3′ terminus of the precursor.[67] Thus, although ordered processing of phage precursors may occur, addition of -C-C-A does not appear to be obligatory for RNase P action on all precursors.

One puzzling feature of phage tRNA synthesis in *cca* strains is the finding that of the four tRNAs that require tRNA nucleotidyltransferase for maturation, three can be found in a partially processed form in the mutant.[102,112] Immature T4 $tRNA^{Gln}$, T2 $tRNA^{Pro}$, and T4 $tRNA^{Ile}$ are all found to contain two terminal CMP residues despite the fact that for the former two tRNAs, only one CMP is encoded, and for the latter tRNA, neither CMP is encoded in the gene (in one preparation immature $tRNA^{Ile}$ was found to lack terminal CMP residues).[84,85] In contrast, the dimeric precursor to $tRNA^{Ser}$ or the small amount of immature $tRNA^{Ser}$ (RNase P cleaved) that is formed in vivo have not been found with any 3′ terminal CMP residues.[102,112] Most likely, the CMP residues are added to the various precursors by some residual tRNA nucleotidyltransferase left in the best mutant strain, despite the fact that in vitro assays indicated less CMP-incorporating (0.6%) than AMP-incorporating (1.6%) activity remaining.[107] Undoubtedly, the in vitro assays do not accurately reflect the situation in vivo in this regard. Nevertheless, it is curious that no CMP residues are found on the pre-$tRNA^{Pro+Ser}$ or immature $tRNA^{Ser}$.[102,112] Perhaps, addition of CMP residues, by the small amount of activity remaining in vivo, is much more efficient with the immature $tRNA^{Gln}$, $tRNA^{Pro}$, and $tRNA^{Ile}$, which exist in monomeric form even in *cca* strains, than with $tRNA^{Ser}$ which is predominantly dimeric.[102,112]

Since *E. coli* tRNAs are synthesized with the -C-C-A sequence already present and followed by additional residues, whereas some phage tRNAs are lacking all or part of

Table 3
3′ TERMINI OF BACTERIOPHAGE tRNAS PRESENT IN A *cca* CELL[a]

	Bacteriophage		
RNA Species[b]	T2	T4	T6
tRNA^{Gln}	G-C-C	G-C-$\underline{C}$[c]	G-C-C
tRNA^{Leu}	A-C-C-A	A-C-C-A	A-C-C-A
tRNA^{Gly}	U-C-C-A	U-C-C-A	U-C-C-A
tRNA^{Pro}	A-C-$\underline{C}$	A-C-$\underline{C}$	—[d]
tRNA^{Ser}	G-C-C-A	G	—
tRNA^{Thr}	—	A-C-C-A	—
tRNA^{Ile}	—	A-[$\underline{C}$-C]	n.d.
tRNA^{Arg}	—	A-C-C-A	A-C-C-A
tRNA α	A-C-C-A	—	—
tRNA δ-2	A-C-C-A	—	—

[a] Taken from References 112 and 113.
[b] One of the specific T2 tRNAs and one of the specific T6 tRNAs were not analyzed and are not shown.
[c] Underlined residues are absent from the gene and are presumably added by residual tRNA nucleotidyltransferase. For tRNA^{Ile} the extra CMP residues are not found in all preparations.
[d] A dash indicates a particular tRNA species is not made.

the -C-C-A sequence, it raises the interesting question of how these different tRNAs are processed nucleolytically at their 3′ termini. The extra residues on *E. coli* tRNAs are thought to be removed exonucleolytically, probably by RNase D.[117] This enzyme does not remove residues from the -C-C-A terminus during processing, as would be expected from the fact that tRNA nucleotidyltransferase is not required for the biosynthesis of *E. coli* tRNAs. On the other hand, the enzyme responsible for removing the extraneous residues from the T4 tRNAs missing an intact -C-C-A sequence (tRNA^{Pro}, tRNA^{Ser}, and tRNA^{Ile}; tRNA^{Gln} does not require nucleolytic trimming prior to -C-A addition) is not known. This enzyme would have to remove the extraneous residues, but not go into the tRNA sequence itself, or tRNA nucleotidyltransferase would not be able to repair the 3′ terminus. An *E. coli* mutant is known, strain BN,[118] which is unaffected for the synthesis of *E. coli* tRNAs and for the five of eight T4 tRNAs which do not require nucleolytic trimming pior to -C-C-A synthesis, but which cannot process tRNA^{Pro}, tRNA^{Ser}, and tRNA^{Ile}.[119] It had been suggested that strain BN is deficient in a 3′ ribonuclease,[119,120] but attempts to isolate this enzyme have so far been unsuccessful.[117] It is curious that *E. coli* would contain a nuclease that is apparently not required for the host, but is needed for processing some of the phage-specific tRNAs. Other work has suggested the possibility that a host enzyme, RNase D, is modified after bacteriophage infection to allow it to process these phage tRNAs.[121] However, further studies are required to test this possibility.

4. *Eukaryotic tRNAs*

Since eukaryotic tRNA genes apparently do not encode the -C-C-A sequence,[86-101] and initial transcripts of the genes also lack these residues,[70-76] it has been assumed that the 3′ terminal sequence is added posttranscriptionally by tRNA nucleotidyltransferase. Inasmuch as eukaryotic mutants deficient in tRNA nucleotidyltransferase have not been isolated, evidence for the involvement of this enzyme in tRNA biosynthesis has been indirect; namely that systems which synthesize tRNA contain tRNA nucleotidyltransferase. However, since tRNA nucleotidyltransferase is ubiquitous, and all cells synthesize tRNA, this reasoning is obviously circular. Perhaps, somewhat more convincing is the finding that mitochondria also contain a tRNA nucleotidyltransferase, apparently distinct from the cellular enzyme.[7] Since mitochondrial tRNA genes also do not contain information for the -C-C-A sequence,[96-99] an enzyme of this type would be required to process tRNA precursors. Clearly, isolation of a eukaryotic tRNA nucleotidyltransferase mutant would be extremely useful for proving that this enzyme is required for addition of -C-C-A during tRNA biosynthesis.

Another important question related to the role of tRNA nucleotidyltransferase in eukaryotic tRNA biosynthesis has to do with the subcellular site of -C-C-A addition. Some evidence has been obtained which suggests that the -C-C-A sequence is added in the nucleus. The tRNA precursors which accumulate in the yeast mutant, ts136, already contain the mature 3′ terminal sequence.[122,123] Since this mutant is thought to be affected in transport of RNA from the nucleus to the cytoplasm,[124] it has been assumed that the -C-C-A addition is a nuclear event. However, it has not been shown that the tRNA precursors which accumulate in this strain are, in fact, located in the nucleus. Additional evidence that 3′ processing occurs in the nucleus has come from studies with nuclear extracts of *Xenopus laevis* oocytes.[71-76] Germinal vesicle extracts from oocytes have been the main system used to study tRNA transcription and processing in vitro. This system is capable of synthesizing tRNA precursors, and processing them to mature tRNAs. Since -C-C-A is added in this nuclear system, it again suggests that this process normally occurs in the nucleus. Unfortunately, germinal vesicles are generally contaminated with cytoplasmic material making a clear interpretation difficult.

The best evidence that the -C-C-A sequence is added in the nucleus has come from experiments in which a cloned yeast $tRNA^{Tyr}$ gene was injected into intact oocytes.[125] A precursor tRNA made from this gene and containing the 3′ terminal residues, was shown to be present in the nucleus (and the cytoplasm). Also, direct microinjection of tRNA-C-C into oocyte nuclei and cytoplasm showed that both compartments could repair the tRNA, suggesting that tRNA nucleotidyltransferase is present in both locations.[125] Direct assay of tRNA nucleotidyltransferase, by both AMP and CMP incorporation, in cytoplasm and in mechanically isolated nuclei from *Xenopus* oocytes, has shown that about one quarter of tRNA nucleotidyltransferase activity is present in the nuclear compartment and at a higher specific activity than the cytoplasm. Thus, at least in the oocyte system, tRNA nucleotidyltransferase activity is present in the nuclei. These results contrast with earlier studies of rat liver which indicated that nuclei from this source are devoid of tRNA nucleotidyltransferase.[7] A continuation of this work has shown that rat liver nuclei and *Xenopus* liver nuclei, isolated under a variety of conditions, contain essentially no tRNA nucleotidyltransferase activity.[126] The explanation for the difference between oocytes and liver is not known. It may represent a difference in the two types of cells, or it may be due to the use of mechanical isolation for the oocyte nuclei, since the usual aqueous isolation procedures often lead to leakage of enzymes from the organelle. It would be worthwhile to explore other nuclear isolation procedures and to develop other tRNA-synthesizing systems to determine whether nuclear processing of tRNA precursors is a general phenomenon.

B. End-Turnover of tRNA

It has been known for some time that the 3′ terminal nucleotides of tRNA undergo a turnover process in vivo independent of the rest of the tRNA molecule.[1] Although the physio logical significance of this process is not completely clear, the phenomenon is widespread, having been demonstrated in *E. coli*,[108,127] yeast,[128] liver,[129,130] reticulocytes,[131] and possibly even mitochondria.[132] Generally, end-turnover of the terminal AMP residue is much more rapid than turnover of the terminal CMPs,[127-131] but even for AMP the turnover amounts to only about 20% per generation.[108,128] End-turnover can be suppressed by the presence of the protein synthesis inhibitor, chloramphenicol.[108]

Some clarification of our understanding of the turnover process has come from studies of *E. coli* tRNA nucleotidyltransferase mutants.[61] As noted before, *cca* mutants contain tRNA molecules with defective 3′ termini.[61,107] In the mutant with the lowest level of the enzyme, about 10 to 15% of the tRNA molecules were lacking a terminal AMP residue and about 2% were missing CMP.[61,107] The wide disparity between the amount of AMP and CMP missing from the 3′ terminus of tRNA initially suggested that the defective tRNA molecules arose by end-turnover and this was confirmed by studies with chloramphenicol.[107] In the presence of the antibiotic, continued turnover was inhibited and defective molecules were slowly repaired by the residual tRNA nucleotidyltransferase present in the mutant cells. The level of defective tRNA in a *cca* strain increased in stationary phase cells, but otherwise was unaffected by changes in growth conditions or transfer of the mutation to other genetic backgrounds.[107]

In stringent *E. coli* strains the presence of the *cca* mutation led to greatly decreased growth rates[105] and elevated levels of ppGpp.[133] In *relA* (relaxed) mutants the same *cca* allele had relatively little effect on growth of the strain[105] and of course, ppGpp levels were unaffected.[133] Thus, the response of stringent and relaxed strains of *E. coli* to the same level of defective tRNA was markedly different. At the present time it is not clear whether the slow growth of stringent cells was a consequence of elevated ppGpp, or vice versa; i.e., the elevated ppGpp was due to the slow growth. Most likely, the second possibility is correct since the elevation of ppGpp was not of the magnitude of a true stringent response.[133] However, why defective tRNA leads to slow growth only in stringent strains remains to be determined. Nevertheless, these studies do explain why tRNA nucleotidyltransferase is required for the normal growth of *E. coli* even though the enzyme does not appear to be needed for tRNA biosynthesis.

Of considerable interest to our understanding of the end-turnover process was the finding that defective tRNA molecules were not envely distributed over the whole tRNA population.[107] Rather, tRNAs specific for certain amino acids (aspartic acid, cysteine, glycine, leucine, and methionine) were 20 to 60% defective, and other tRNAs were unaffected.[107] The defect in tRNAs for a particular amino acid was not restricted to specific isoacceptors, but affected all the tRNAs for a given amino acid equally.[107] Why certain tRNAs are affected by end-turnover more than others is not understood. However, it is apparently not due to differential repair of tRNA molecules by the limiting amount of tRNA nucleotidyltransferase present in mutant cells since in vitro studies, at least, indicate that all tRNA molecules are repaired randomly.[41] Whether a similar random repair occurs in vivo is, of course, an open question.

On the other hand, it is clear that the level of aminoacylation of a particular tRNA plays an important role in determining whether a tRNA will participate in end-turnover (Table 4).[107] Mutant cells harboring both the *cca* mutation and a temperature-sensitive valyl-tRNA synthetase mutation (valS^{ts}) contain extremely high levels of defective tRNA^{Val} (as much as 75%), whereas cells containing either mutation alone have normal tRNA^{Val}.[107] This experiment shows that when a tRNA is not charged with an amino acid, its 3′ terminus becomes susceptible to nucleolytic degradation. If tRNA nucleotidyltransferase is also not present to repair the defective tRNA molecules, then defective tRNAs will accumulate. In fact, in this experiment a significant fraction of the defective tRNA molecules were also lacking a CMP residue. Although these results do not explain why certain tRNAs are more defective than others, they do suggest that those tRNAs which are defective in *cca* strains spend a higher fraction of the time in

Table 4
EFFECT OF AMINOACYLATION ON THE STATE OF THE -C-C-A TERMINUS OF tRNA^{Val}

Strain	Temperature (°C)	Defective termini (%)
cca$^+$, *valS*^{ts}	30	<7
	42	7
cca, valS	30	<7
	42	<7
cca, valS^{ts}	30	39 (50)[a]
	42	61 (75)

[a] Values in parentheses represent tRNA repair carried out with CTP, in addition to ATP.

Data from Deutscher, M. P., Lin, J. J-C., and Evans, J. A., *J. Mol. Biol.*, 117, 1081, 1977.

the uncharged state rendering them sensitive to the turnover machinery. These data also offer an explanation for the inhibition of end-turnover by chloramphenicol.[108] In the presence of this antibiotic, protein synthesis is inhibited, and tRNAs accumulate in the aminoacylated form, rendering them resistant to turnover.

These studies also raise the interesting question of the identity of the nuclease responsible for the cleavage of the 3′ terminal AMP residue. It has been suggested that the exonuclease, RNase II, is involved in this process.[108] However, studies of terminal AMP exchange with purified RNase II and tRNA nucleotidyltransferase indicate that the involvement of this nuclease in the turnover process is highly unlikely since no exchange occurred.[134] RNase II is known to hydrolyze RNA chains processively,[135] thus preventing tRNA nucleotidyltransferase from binding and repairing the defective RNA chain. In contrast, active AMP exchange takes place in the presence of RNase D,[134] an exonuclease known to hydrolyze by a random mechanism.[117] These results do not prove that RNase D is involved in end-turnover, only that the process is most likely due to an exonuclease with a random mode of hydrolysis. In fact, faster-growing revertants of *cca* strains have been isolated which contain about one half as much defective tRNA as the original *cca* parent.[133] Since these revertants still have the same low level of tRNA nucleotidyltransferase as the *cca* parent, it has been suggested that the cells have been affected in the nuclease removing AMP.[133] However, these strains have normal levels of RNase D,[136] suggesting that either another unknown nuclease is involved in turnover, or that other factors besides just the nuclease and tRNA nucleotidyltransferase influence the levels of defective tRNA.

V. REGULATORY ASPECTS

A. Possible Control Functions of tRNA Nucleotidyltransferase and the -C-C-A Terminus of tRNA

The fact that the 3′ terminus turns over in vivo raises the important question of whether this process serves any regulatory function. Obviously, a simple and rapid means of turning off protein synthesis would be to remove the terminal adenylate moiety from tRNA, but not repair it. The studies described earlier demonstrated that exposure of the 3′ terminal nucleotides of tRNA, when the molecule is not aminoacylated, is an important determinant of the turnover process,[107] but how these uncharged tRNA molecules arise is not clear. The turnover rate is too slow to be coupled one-to-one with peptide bond formation or peptide chain release.[1] It could represent the occasional damage of tRNA molecules that happen to be uncharged at any given instant, such as might occur on the ribosome after the peptide chain has been transferred or that might occur prior to recharging by an aminoacyl-tRNA synthetase after release from the ribosome. Under these circumstances the turnover process may have no metabolic significance, but is simply a manifestation of an incorrect nucleolytic cleavage by a ribonuclease which normally carries out another function. As long as sufficient tRNA nucleotidyltransferase were present to rapidly repair the damaged molecules, the cells would suffer no ill effects. In this model, tRNA nucleotidyltransferase would be a general scavenger for defective tRNA molecules, a role which fits well with its high turnover number and its presence in large excess, at least in prokaryotic cells.

On the other hand, several observations are consistent with a possible control function for end-turnover of tRNA. A number of systems have been found which normally contain significant levels of defective tRNA. These include stationary phase yeast[137] and *E. coli*,[107] spores from *Bacillus*,[138,139] and *Clostridium*[140] species, unfertilized sea urchin eggs,[141] nonlactating bovine mammary gland,[142] lupin seeds,[143] and full-term human placenta.[144] All of these systems are either dormant or in a phase of greatly decreased proliferation. One additional system, skeletal muscle from polymyopathic hamsters, also contains defective tRNA.[145] In several of these systems the level of tRNA nucleotidyltransferase was examined and found to be normal.[138,143,144] Thus, the increase in defective tRNA could be due to increased removal of terminal nucleotides by elevated nuclease activity or higher levels of uncharged tRNA. Alternatively, some other component of the synthesizing system could be limiting.

Examples of each of these mechanisms has been observed. In stationary phase *E. coli* increased levels of RNase are known to be present.[107] In dormant spores and seeds the primary deficit appears to be decreased levels of ATP which affect aminoacylation levels and the ability to repair the defective tRNA.[138,140,143] Whether these phenomena represent a regulatory process to shut off protein synthesis or simply a secondary effect of another metabolic process is not clear. However, if regulation is involved, the control is more likely on the removal of nucleotides from the 3′ terminus than on the tRNA nucleotidyltransferase since no change in enzyme level has been observed in these systems.

Thus far, only one example is known of variations in tRNA nucleotidyltransferase activity in response to a developmental change. tRNA nucleotidyltransferase activity increases about twofold after fertilization of *X. laevis* eggs and then decreases about threefold after gastrulation.[146] Unfortunately, no correlation between these changes in enzyme activity and tRNA synthesis was observed in this system and its physiological significance remains a mystery.

One additional example of a role for tRNA nucleotidyltransferase which, perhaps, is related to a regulatory process, is the observation that *E. coli cca* mutants are greatly affected in their ability to support the normal growth of T-even phages.[61] In *cca* mutants, in a variety of genetic backgrounds, the burst size of T-even phages is decreased about 90%.[147] No effect on T-odd or RNA phages was found. The presence of the *cca* mutation did not effect the total yield of phage particles, but only their viability, such that 90% of the phage that were released during infection of a *cca* strain were not able to carry out a successful second cycle of infection, although they could adsorb and inject DNA normally.[147] Since T4 tRNAs are not required for successful infection[148] and the amount of defective host tRNA does not change after infection,[18] these results raise the possibility that tRNA nucleotidyltransferase may have another function during infection.

B. Why -C-C-A?

The -C-C-A sequence is known to be required for the aminoacylation of tRNA and for peptide bond formation by peptidyltransferase.[1] However, tRNAs are not the only RNA molecules to contain this 3′ terminal sequence. Included among this group are bacteriophage RNAs,[149] plant viral RNAs,[150] two stable, low-molecular weight RNAs specified by bacteriophage T4 (termed band C or 1 and band D or 2),[84,85] and the small nuclear RNA, U2.[151] In addition to containing a -C-C-A sequence, two of these RNAs, band C RNA[85] and turnip yellow mosaic virus (TYMV) RNA,[152] are also known to utilize tRNA nucleotidyltransferase in vivo for completion of the 3′ terminus. One wonders how these diverse RNAs have evolved to all contain the identical 3′ terminal sequence, and what physiological function of the non-tRNA molecules is being carried out which requires them to maintain this sequence. It is known that plant viral RNAs can accept amino acids in vitro,[150] and that TYMV RNA can be aminoacylated in *X. laevis* oocytes in vivo after endonucleolytic cleavege which generates a tRNA-like structure.[152] However, aminoacylation of other RNAs in this group has not been demonstrated.

Alternatively, the -C-C-A sequence may serve a protective function for the non-tRNAs containing these residues. tRNA molecules are relatively stable in vivo, and if any 3′ exonucleolytic cleavage does occur it is immediately repaired by tRNA nucleotidyltransferase. Perhaps, these RNAs evolved a 3′ -C-C-A sequence as protection against cleavage by cellular exonucleases. Since in most cells exonucleases act from the 3′ terminus, the presence of a -C-C-A sequence and the existence of an active enzyme which can repair this sequence, could act as a barrier against extensive exonucleolytic degradation. Such a mechanism might provide a simple means for an infectious RNA to be stabilized in a foreign cytoplasm.

REFERENCES

1. **Deutscher, M. P.,** Synthesis and functions of the -C-C-A terminus of transfer RNA, *Prog. Nucl. Acid Res. Mol. Biol.,* 13, 51, 1973.
2. **Heidelberger, C., Haibers, E., Leibman, K. C., Yakagi, Y., and Potter, V. R.,** Specific incorporation of adenosine-5′-phosphate-^{32}P into ribonucleic acid in rat liver homogenates, *Biochim. Biophys. Acta,* 20, 445, 1956.
3. **Canellakis, E. S.,** On the mechanism of incorporation of adenylic acid from ATP into ribonucleic acid by soluble mammalian enzyme systems, *Biochim. Biophys. Acta,* 25, 217, 1957.
4. **Edmonds, M. and Abrams, R.,** The incorporation of 8-^{14}C adenine into calf thymus nuclei *in vitro, Biochim. Biophys. Acta,* 26, 226, 1957.
5. **Herbert, E.,** The incorporation of adenine nucleotides into RNA of cell-free systems from liver, *J. Biol. Chem.,* 231, 975, 1958.
6. **Hecht, L. I., Zamecnik, P. C., Stephenson, M. L., and Scott, J. F.,** Nucleoside triphosphates as precursors of ribonucleic acid end groups in a mammalian system, *J. Biol. Chem.,* 233, 954, 1958.

7. **Mukerji, S. K. and Deutscher, M. P.**, Subcellular localization and evidence for a mitochondrial tRNA nucleotidyltransferase, *J. Biol. Chem.*, 247, 481, 1972.
8. **Bouhnik, J., Michel, O., and Michel, R.**, Caractérization de la tRNA nucleotidyltransferase dans les mitochondries musculaires de rat normal ou thyroidectomisé, *Biochimie*, 55, 1179, 1973.
9. **Kolakafsky, D.**, tRNA nucleotidyltransferase and tRNA in Sendai virus, *J. Virol.*, 10, 555, 1972.
10. **Thomassen, M. J., Kingsley-Lechner, E., Ohe, K., and Wu, A. M.**, tRNA nucleotidyltransferase activity in virions of type-C viruses, *Biochem. Biophys. Res. Commun.*, 72, 258, 1976.
11. **Faras, A. J., Levinson, W. E., Bishop, J. M., and Goodman, H. M.**, Identification of a tRNA nucleotidyltransferase and its substrates in virions of avian RNA tumor viruses, *Virology*, 58, 126, 1974.
12. **Sprinzl, M. and Cramer, F.**, The -C-C-A end of tRNA and its role in protein biosynthesis, *Prog. Nucl. Acid Res. Mol. Biol.*, 22, 1, 1979.
13. **Rether, B., Gangloff, J., and Ebel, J. P.**, Studies of tRNA nucleotidyltransferase from baker's yeast. 2. Replacement of the terminal CCA sequence in yeast $tRNA^{Phe}$ by several unusual sequences, *Eur. J. Biochem.*, 50, 289, 1974.
14. **Tal, J., Deutscher, M. P., and Littauer, U. Z.**, Biological activity of *Escherichia coli* $tRNA^{Phe}$ modified in its -C-C-A terminus, *Eur. J. Biochem.*, 28, 478, 1972.
15. **Kirschenbaum, A. H. and Deutscher, M. P.**, Amino acid acceptor activity of tRNA-C-C-C-A, *Biochem. Biophys. Res. Commun.*, 70, 258, 1976.
16. **Deutscher, M. P. and Ghosh, R. K.**, Preparation of synthetic tRNA precursors with tRNA nucleotidyltransferase, *Nucl. Acids Res.*, 5, 3821, 1978.
17. **Schmidt, F. J. and McClain, W. H.**, An *Escherichia coli* ribonuclease which removes an extra nucleotide from a biosynthetic intermediate of bacteriophage T4 proline transfer RNA, *Nucl. Acids Res.*, 5, 4129, 1978.
18. **Deutscher, M. P., Foulds, J., Morse, J. W., and Hilderman, R. H.**, Synthesis of the CCA terminus of transfer RNA, *Brookhaven Symp. Biol.*, 26, 124, 1974.
19. **Deutscher, M. P.**, tRNA nucleotidyltransferase, in *The Enzymes*, Vol. 15, 3rd ed., Boyer, H., Ed., Academic Press, New York, in press.
20. **Deutscher, M. P.**, Purification and physical and chemical properties of rabbit liver tRNA nucleotidyltransferase, *J. Biol. Chem.*, 247, 450, 1972.
21. **Poblete, P., Jedlicky, E., and Litvak, S.**, Purification and properties of tRNA nucleotidyltransferase from *Musca domestica*, *Biochim. Biophys. Acta*, 476, 333, 1977.
22. **Cudny, H., Pietrzak, M., and Kaczkowski, J.**, Isolation and purification of tRNA nucleotidyltransferase from *Lupinus luteus* seeds, *Planta*, 142, 23, 1978.
23. **Sternbach, H., von der Haar, F., Schlimme, E., Gaertner, E., and Cramer, F.**, Isolation and properties of tRNA nucleotidyltransferase from yeast, *Eur. J. Biochem.*, 22, 166, 1971.
24. **Rether, B., Bonnet, J., and Ebel, J. P.**, Studies on tRNA nucleotidyltransferase from baker's yeast 1. Purification of the enzyme, *Eur. J. Biochem.*, 50, 281, 1974.
25. **Carre, D. S., Litvak, S., and Chapeville, F.**, Purification and properties of *Escherichia coli* CTP (ATP) tRNA nucleotidyltransferase, *Biochim. Biophys. Acta*, 224, 371, 1970.
26. **Best, A. N. and Novelli, G. D.**, Studies with tRNA adenylyl (cytidylyl) transferase from *Escherichia coli* B I. purification and kinetic properties, *Arch. Biochem. Biophys.*, 142, 527, 1971.
27. **Miller, J. P. and Philipps, G. R.**, Transfer RNA nucleotidyltransferase from *Escherichia coli* II. purification, physical properties and substrate specificity, *J. Biol. Chem.*, 246, 1274, 1971.
28. **Schofield, P. and Williams, K. R.**, Purification and some properties of *Escherichia coli* tRNA nucleotidyltransferase, *J. Biol. Chem.*, 252, 5584, 1977.
29. **Leineweber, M. and Philipps, G. R.**, Comparison of tRNA nucleotidyltransferase from *Escherichia coli* and *Lactobacillus acidophilus*, *Hoppe-Seyler's Z. Physiol. Chem.*, 359, 473, 1978.
30. **Valenzuela, P., Morris, R. W., Faras, A., Levinson, W., and Rutter, W. J.**, Are all nucleotidyltransferases metalloenzymes? *Biochem. Biophys. Res. Commun.*, 53, 1036, 1973.
31. **Williams, K. R. and Schofield, P.**, Evidence for metalloenzyme character of tRNA nucleotidyltransferase, *Biochem. Biophys. Res. Commun.*, 64, 262, 1975.
32. **McGann, R. G. and Deutscher, M. P.**, Purification and characterization of a mutant tRNA nucleotidyltransferase, *Eur. J. Biochem.*, 106, 321, 1980.
33. **Sternbach, H., Sprinzl, M., Hobbs, J. B., and Cramer, F.**, Affinity labeling of tRNA nucleotidyltransferase from baker's yeast with $tRNA^{Phe}$ modified on the 3′ terminus, *Eur. J. Biochem.*, 67, 215, 1976.
34. **Best, A. N. and Novelli, G. D.**, Studies with adenylyl (cytidylyl) transferase from *Escherichia coli* B II. regulation of AMP and CMP incorporation into tRNApCpC and tRNApC, *Arch. Biochem. Biophys.*, 142, 539, 1971.
35. **Carre, D. and Chapeville, F.**, Study of the *Escherichia coli* tRNA nucleotidyltransferase. Specificity of the enzyme for nucleoside triphosphates, *Biochimie*, 56, 1451, 1974.

36. **Deutscher, M. P.,** Catalytic properties of two purified rabbit liver tRNA nucleotidyltransferases, *J. Biol. Chem.,* 247, 459, 1972.
37. **Sprinzl, M., Sternbach, H., von der Haar, F., and Cramer, F.,** Enzymatic incorporation of ATP and CTP analogues into the 3′ end of tRNA, *Eur. J. Biochem.,* 81, 579, 1977.
38. **Deutscher, M. P.,** Extents of normal and anomalous nucleotide incorporation catalyzed by tRNA nucleotidyltransferase, *J. Biol. Chem.,* 247, 469, 1972.
39. **Masiakowski, P. and Deutscher, M. P.,** Dissection of the active site of rabbit liver tRNA nucleotidyltransferase, specificity and properties of subsites for donor nucleoside triphosphates, *J. Biol. Chem.,* 255, 11240, 1980.
40. **Sprinzl, M., Scheit, K. H., Sternbach, H., von der Haar, F., and Cramer, F.,** In vitro incorporation of 2′-deoxyadenosine and 3′-deoxyadenosine into yeast $tRNA^{Phe}$ using tRNA nucleotidyltransferase and properties of $tRNA^{Phe}$-C-C-2′dA and $tRNA^{Phe}$-C-C-3′dA, *Biochem. Biophys. Res. Commun.,* 51, 881, 1973.
41. **Deutscher, M. P. and Evans, J. A.,** Transfer RNA nucleotidyltransferase repairs all transfer RNAs randomly, *J. Mol. Biol.,* 109, 593, 1977.
42. **Deutscher, M. P.,** Properties of the poly(c) polymerase activity associated with rabbit liver tRNA nucleotidyltransferase, *J. Biol. Chem.,* 248, 3108, 1973.
43. **Deutscher, M. P.,** Anomalous adenosine monophosphate incorporation catalyzed by rabbit liver tRNA nucleotidyltransferase, *J. Biol. Chem.,* 248, 3116, 1973.
44. **Prochiantz, A., Bénicourt, C., Carre, D., and Haenni, A-L.,** tRNA nucleotidyltransferase-catalyzed incorporation of CMP and AMP into RNA-bacteriophage genome fragments, *Eur. J. Biochem.,* 52, 17, 1975.
45. **Williams, K. R. and Schofield, P.,** Kinetic mechanism of tRNA nucleotidyltransferase from *Escherichia coli, J. Biol. Chem.,* 252, 5589, 1977.
46. **Evans, J. A. and Deutscher, M. P.,** Kinetic analysis of rabbit liver tRNA nucleotidyltransferase, *J. Biol. Chem.,* 253, 7276, 1978.
47. **Masiakowski, P. and Deutscher, M. P.,** The dinucleoside monophosphate, CpC, is a model acceptor substrate for rabbit liver tRNA nucleotidyltransferase, *FEBS Lett.,* 77, 261, 1977.
48. **Eckstein, F., Sternbach, H., and von der Haar, F.,** Stereochemistry of internucleotide bond formation by tRNA nucleotidyltransferase from baker's yeast, *Biochemistry,* 16, 3429, 1977.
49. **Ghosh, R. K. and Deutscher, M. P.,** Purification of potential 3′ processing nucleases using synthetic tRNA precursors, *Nucl. Acids Res.,* 5, 3831, 1978.
50. **Ghosh, R. K. and Deutscher, M. P.,** The purification of 3′ processing nucleases using synthetic tRNA precursors, in *tRNA: Biological Aspects,* Söll, D., Abelson, J. N., and Schimmel, P. R., Eds., Cold Spring Harbor Laboratory, Cold Spring Harbor, New York, 1980, 59.
51. **Morris, R. W. and Herbert, E.,**Purification and characterization of yeast nucleotidyltransferase and investigation of enzyme-tRNA complex formation, *Biochemistry,* 9, 4819, 1970.
52. **Hondo, H.,** Formation of the complex between CCA-enzyme and transfer RNA, *Biochim. Biophys. Acta,* 195, 587, 1969.
53. **Carre, D. S., Litvak, S., and Chapeville, F.,** Studies of *Escherichia coli* tRNA nucleotidyltransferase, interactions of the enzyme with tRNA, *Biochim. Biophys. Acta,* 361, 185, 1974.
54. **Miller, J. P. and Philipps, G. R.,** Transfer RNA nucleotidyltransferase from *Escherichia coli* III Kinetic analysis, *J. Biol. Chem.,* 246, 1280, 1971.
55. **Miller, J. P. and Philipps, G. R.,** Studies on thermal inactivation of tRNA nucleotidyltransferase from *Escherichia coli, Biochemistry,* 10, 1001.
56. **Deutscher, M. P. and Masiakowski, P.,** Binding of tRNA nucleotidyltransferase to Affi-Gel Blue: rapid purification of the enzyme and binding studies, *Nucl. Acids Res.,* 5, 1947, 1978.
57. **Lindahl, T., Adams, A., Geroch, M., and Fresco, J. R.,** Selective recognition of the native conformation of tRNAs by enzymes, *Proc. Natl. Acad. Sci. U.S.A.,* 57, 178, 1967.
58. **Masiakowski, P., and Deutscher, M. P.,** Dissection of the active site of rabbit liver tRNA nucleotidyltransferase, specificity and properties of the tRNA and acceptor subsites determined with model acceptor substrates, *J. Biol. Chem.,* 255, 11233, 1980.
59. **Masiakowski, P., and Deutscher, M. P.,** Separation of functionally distinct regions of a macromolecular substrate, *J. Biol. Chem.,* 254, 2585, 1979.
60. **Anthony, D. D., Starr, J. L., Kerr, D. S., and Goldthwait, D. A.,** Evidence for separate enzymatic sites for incorporation of AMP and CMP, *J. Biol. Chem.,* 238, 690, 1963.
61. **Deutscher, M. P. and Hilderman, R. H.,** Isolation and partial characterization of *Escherichia coli* mutants with low levels of tRNA nucleotidyltransferase, *J. Bacteriol.,* 118, 621, 1974.
62. **Altman, S. and Smith, J. D.,** Tyrosine tRNA precursor molecule polynucleotide sequence, *Nature (London), New Biol.,* 233, 35, 1971.
63. **Schedl, P., Roberts, J., and Primakoff, P.,** In vitro processing of *E. coli* tRNA precursors, *Cell,* 8, 581, 1976.

64. **Chang, S. and Carbon, J.**, The nucleotide sequence of a precursor to the glycine- and threonine-specific Transfer RNAs of *Escherichia coli, J. Biol. Chem.*, 250, 5542, 1975.
65. **Shimura, Y., Sakano, H., and Nagawa, F.**, Specific ribonucleases involved in processing of tRNA precursors of *Escherichia coli, Eur. J. Biochem.*, 86, 267, 1978.
66. **Sakano, H. and Shimura, Y.**, Characterization and *in vitro* processing of transfer RNA precursors accumulated in a temperature-sensitive mutant of *Escherichia coli, J. Mol. Biol.*, 123, 287, 1978.
67. **Guthrie, C. and Scholla, C. A.**, Asymmetric maturation of a dimeric transfer RNA precursor, *J. Mol. Biol.*, 139, 349, 1980.
68. **Barrell, B. G., Seidman, J. G., Guthrie, C., and McClain, W. H.**, Transfer RNA biosynthesis: the nucleotide sequence of a precursor to serine and proline transfer RNAs, *Proc. Natl. Acad. Sci. U.S.A.*, 71, 413, 1974.
69. **Guthrie, C.**, The nucleotide sequence of the dimeric precursor to glutamine and leucine transfer RNAs coded by bacteriophage T4, *J. Mol. Biol.*, 95, 529, 1975.
70. **Melton, D. A., DeRobertis, E. M., and Cortese, R.**, Order and intracellular location of the events involved in the maturation of a spliced tRNA, *Nature (London)*, 284, 143, 1980.
71. **Schmidt, O., Mao, J., Ogden, R., Beckmann, J., Sakano, H., Abelson, J., and Söll, D.**, Dimeric tRNA precursors in yeast, *Nature (London)*, 287, 750, 1980.
72. **Ogden, R. C., Beckman, J. S., Abelson, J., Kang, H. S., Söll, D., and Schmidt, O.**, In vitro transcription and processing of a yeast tRNA gene containing an intervening sequence, *Cell*, 17, 399, 1979.
73. **Mao, J., Schmidt, O., and Söll, D.**, Dimeric tRNA precursors in *S. pombe, Cell*, 21, 509, 1980.
74. **Silverman, S., Schmidt, O., Söll, D., and Hovemann, B.**, The nucleotide sequence of a cloned *Drosophila* arginine tRNA gene and its *in vitro* transcription in *Xenopus* germinal vesicle extracts, *J. Biol. Chem.*, 254, 10290, 1979.
75. **Garber, R. L. and Gage, L. P.**, Transcription of a cloned *Bombyx mori* $tRNA^{Ala}_{2}$ gene: nucleotide sequence of the tRNA precursor and its processing *in vitro, Cell*, 18, 817, 1979.
76. **Hagenbüchle, O., Larson, D., Hall, G. I., and Sprague, K. U.**, The primary transcription product of a silkworm alanine tRNA gene: identification of in vitro sites of initiation, termination and processing, *Cell*, 18, 1217, 1979.
77. **Nakajima, N., Ozeki, H., and Shimura, Y.**, Organization and structure of an *E. coli* tRNA operon containing seven tRNA genes, *Cell*, 23, 239, 1981.
78. **Hudson, L., Rossi, J., and Landy, A.**, Dual function transcripts specifying tRNA and mRNA, *Nature (London)*, 294, 422, 1981.
79. **Altman, S., Modell, P., Dixon, G. H., and Wosnick, M. A.**, An *E. coli* gene coding for a protamine-like protein, *Cell*, 26, 299, 1981.
80. **Sekiya, T. and Nishimura, S.**, Sequence of the gene for isoleucine tRNA, and the surrounding region in a ribosomal RNA operon of *Escherichia coli, Nucl. Acids Res.*, 6, 575, 1979.
81. **Young, R. A., Macklis, R., and Steitz, J. A.**, Sequence of the 16S-23S spacer region in two ribosomal RNA operons of *Escherichia coli, J. Biol. Chem.*, 254, 3264, 1979.
82. **Duester, G. L. and Holmes, W. M.**, The distal end of the ribosomal RNA operon rrnD of *Escherichia coli* contains a $tRNA^{Thr}_{1}$ gene, two 5S rRNA genes and a transcription terminator, *Nucl. Acids Res.*, 8, 3793, 1980.
83. **Sekiya, T., Mori, M., Takahashi, N., and Nishimura, S.**, Sequence of the distal $tRNA^{Asp}_{1}$ gene and the transcription termination signal in the *Escherichia coli* ribosomal RNA operon rrnF (or G), *Nucl. Acids Res.*, 8, 3809, 1980.
84. **Fukada, K. and Abelson, J.**, DNA sequence of a T4 transfer RNA gene cluster, *J. Mol. Biol.* 139, 377, 1980.
85. **Mazzara, G. P., Plunkett, G., III, and McClain, W. H.**, DNA sequence of the transfer RNA region of bacteriophage T4: implications for transfer RNA synthesis, *Proc. Natl. Acad. Sci. U.S.A.*, 78, 889, 1981.
86. **Goodman, H. M., Olson, M. V., and Hall, B. D.**, Nucleotide sequence of a mutant eukaryotic gene: the yeast tyrosine-inserting ochre suppressor SUP4-0, *Proc. Natl. Acad. Sci. U.S.A.*, 74, 5453, 1977.
87. **Valenzuela, P., Venegas, A., Weinberg, F., Bishop, R., and Rutter, W. J.**, Structure of yeast phenylalanine-tRNA genes: an intervening DNA segment within the region coding for the tRNA, *Proc. Natl. Acad. Sci. U.S.A.*, 75, 190, 1978.
88. **Page, G. S. and Hall, B. D.**, Characterization of the yeast $tRNA^{Ser}_{2}$ gene family: genomic organization and DNA sequence, *Nucl. Acids Res.*, 9, 921, 1981.
89. **Venegas, A., Quiroga, M., Zaldivar, J., Rutter, W. J., and Valenzuela, P.**, Isolation of yeast $tRNA^{Leu}$ genes, *J. Biol. Chem.*, 254, 12306, 1979.
90. **DeFranco, D., Schmidt, D., and Söll, D.**, Two control regions for eukaryotic tRNA gene transcription, *Proc. Natl. Acad. Sci. U.S.A.*, 77, 3365, 1980.
91. **Hershey, N. D. and Davidson, N.**, Two *Drosophila melanogaster* $tRNA^{Gly}$ genes are contained in a direct duplication at chromosomal locus 56F, *Nucl. Acids Res.*, 8, 4899, 1980.

92. **Robinson, R. R. and Davidson, N.**, Analysis of a *Drosophila* gene cluster: two tRNA^{Leu} genes containing intervening sequences, *Cell*, 23, 251, 1981.
93. **Müller, F. and Clarkson, S. G.**, Nucleotide sequence of genes coding for tRNA^{Phe} and tRNA^{Tyr} from a repeating unit of *X. laevis* DNA, *Cell*, 19, 345, 1980.
94. **Santos, T. and Zasloff, M.**, Comparative analysis of human chromosomal segments bearing monallelic dispersed tRNA$^{met}_{i}$ genes, *Cell*, 23, 699, 1981.
95. **Selker, E. and Yanofsky, C.**, A phenylalainine tRNA gene from *Neurospora crassa:* conservation of secondary structure involving an intervening sequence, *Nucl. Acids Res.*, 8, 1033, 1980.
96. **Berlani, R. E., Pentella, C., Macino, G., and Tzagaloff, A.**, Assembly of the mitochondrial membrane system: isolation of mitochondrial tRNA mutants and characterization of tRNA genes of *Saccharomyces cerevesiae, J. Bacteriol.*, 141, 1086, 1980.
97. **Heckman, J. E. and RajBhandary, U. L.**, Organization of tRNA and rRNA genes in *N. crassa* mitochondria: intervening sequence in the large rRNA gene and strand distribution of the RNA genes, *Cell*, 17, 583, 1979.
98. **Köchel, H. G., Lazarus, C. M., Basak, N., and Küntzel, H.**, Mitochondrial tRNA gene clusters in *Aspergillus nidulans:* organization and nucleotide sequence, *Cell*, 23, 625, 1981.
99. **Anderson, S., Bankier, A. T., Barrell, B. G., deBruijn, M. H. L., Coulson, A. R., Drovin, J., Eperon, I. C., Nierlich, D. P., Roe, B. A., Sanger, F., Schreier, P. H., Smith, A. J. H., Staden, R., and Young, I. G.**, Sequence and organization of the human mitochondrial genome, *Nature (London)*, 290, 457, 1981.
100. **Schwarz, S., Jolly, S. O., Steinmetz, A. A., and Bogorad, L.**, Overlapping divergent genes in the maize chloroplast chromosome and in vitro transcription of the gene for tRNA^{His}, *Proc. Natl. Acad. Sci. U.S.A.*, 78, 3423, 1981.
101. **Graf, L., Kossel, H., and Stutz, E.**, Sequencing of the 16S-23S spacer in a ribosomal RNA operon of *Euglena gracilus* chloroplast DNA reveals two tRNA genes, *Nature (London)*, 286, 908, 1980.
102. **Seidman, J. G., Barrell, B. G., and McClain, W. H.**, Five steps in the conversion of a large precursor RNA into bacteriophage proline and serine transfer RNAs, *J. Mol. Biol.*, 99, 733, 1975.
103. **Deutscher, M. P.**, Preparation of cells permeable to macromolecules by treatment with toluene: studies of tRNA nucleotidyltransferase, *J. Bacteriol.*, 118, 633, 1974.
104. **Foulds, J., Hilderman, R. H., and Deutscher, M. P.**, Mapping of the locus for *Escherichia coli* tRNA nucleotidyltransferase, *J. Bacteriol.*, 118, 628, 1974.
105. **Deutscher, M. P., Foulds, J., and McClain, W. H.**, tRNA nucleotidyltransferase plays an essential role in the normal growth of *Escherichia coli* and in the biosynthesis of some bacteriophage T4 tRNAs, *J. Biol. Chem.*, 249, 6696, 1974.
106. **Deutscher, M. P.**, A single enzyme catalyzes the incorporation of AMP and CMP into transfer RNA, *J. Biol. Chem.*, 245, 4225, 1970.
107. **Deutscher, M. P., Lin, J. C., and Evans, J. A.**, Transfer RNA metabolism in *Escherichia coli* cells deficient in tRNA nucleotidyltransferase, *J. Mol. Biol.*, 117, 1081, 1977.
108. **Cannon, M.**, Further studies concerning end-group instability of soluble RNA of *Escherichia coli in vivo, Biochim. Biophys. Acta*, 129, 221, 1966.
109. **Mazzara, G. P. and McClain, W. H.**, Cysteine transfer RNA of *Escherichia coli:* nucleotide sequence and unusual metbolic properties of the 3′ C-C-A terminus, *J. Mol. Biol.* 117, 1061, 1977.
110. **Morse, J. W. and Deutscher, M. P.**, Apparent non-involvement of tRNA nucleotidyltransferase in the biosynthesis of *Escherichia coli* suppressor tRNAs, *J. Mol. Biol.*, 95, 141, 1975.
111. **Hilderman, R. H. and Deutscher, M. P.**, tRNA nucleotidyltransferase activity in bacteriophage T4-infected *Escherichia coli, Biochem. Biophys. Res. Commun.*, 54, 205, 1973.
112. **McClain, W. H., Seidman, J. G., and Schmidt, F. J.**, Evolution of the biosynthesis of the 3′ terminal C-C-A residues in T-even bacteriophage transfer RNAs, *J. Mol. Biol.*, 119, 519, 1978.
113. **Moen, T. L., Seidman, J. G., and McClain, W. H.**, A catalogue of transfer RNA-like molecules synthesized following infection of *Escherichia coli* by T-even bacteriophages, *J. Biol. Chem.*, 253, 7910, 1978.
114. **Seidman, J. G. and McClain, W. H.**, Three steps in the conversion of large precursor RNA into serine and proline transfer RNAs, *Proc. Natl. Acad. Sci. U.S.A.*, 72, 1491, 1975.
115. **Schmidt, F. J., Seidman, J. G., and Bock, R. M.**, tRNA biosynthesis, substrate specificity of ribonuclease P, *J. Biol. Chem.*, 251, 2440, 1976.
116. **Schmidt, F. J.**, A novel function of *Escherichia coli* tRNA nucleotidyltransferase, *J. Biol. Chem.*, 250, 8399, 1975.
117. **Cudny, H. and Deutscher, M. P.**, Apparent involvement of ribonuclease D in the 3′ processing of tRNA precursors, *Proc. Natl. Acad. Sci. U.S.A.*, 77, 837, 1980.
118. **Maisurian, A. N. and Buyanovskaya, E. A.**, Isolation of an *Escherichia coli* strain restricting bacteriophage suppressor, *Mol. Gen. Genet.*, 120, 227, 1973.
119. **Seidman, J. G., Schmidt, F. J., Foss, K., and McClain, W. H.**, A mutant of *Escherichia coli* defective in removing 3′ terminal nucleotides from some transfer RNA precursor molecules, *Cell*, 5, 389, 1975.

120. **Schmidt, F. J. and McClain, W. H.**, An *Escherichia coli* ribonuclease which removes an extra nucleotide from a biosynthetic intermediate of bacteriophage T4 proline transfer RNA, *Nucl. Acids Res.*, 5, 4129, 1978.
121. **Cudny, H., Roy, P., and Deutscher, M. P.**, Alteration of *Escherichia coli* RNase D by infection with bacteriophage T4, *Biochem. Biophys. Res. Commun.*, 98, 337, 1981.
122. **Hopper, A. K., Banks, F., and Evangelides, V.**, A yeast mutant which accumulates precursor tRNAs, *Cell*, 14, 211, 1978.
123. **Knapp, G., Beckmann, J. S., Johnson, P. F., Fuhrmann, S. A., and Abelson, J.**, Transcription and processing of intervening sequences in yeast tRNA genes, *Cell*, 14, 221, 1978.
124. **Hutchinson, H. T., Hartwell, L. H., and McLaughlin, C. S.**, Temperature sensitive yeast mutant defective in RNA production, *J. Bacteriol.*, 99, 807, 1969.
125. **Melton, D. A., DeRobertis, E. M., and Cortese, R.**, Order and intracellular location of the events involved in the maturation of a spliced tRNA, *Nature (London)*, 284, 143, 1980.
126. **Solari, A., Allende, J. E., and Deutscher, M. P.**, unpublished results, 1981.
127. **Rosset, R. and Monier, R.**, Étude du renouvellement de l'AMP terminal chez *Escherichia coli* en fonction du taux de croissance, *Biochim. Biophys. Acta*, 108, 385, 1965.
128. **Rosset, R. and Monier, R.**, Renouvellement de l'AMP terminal chez *Saccharomyces cerevisiae*, *Biochim. Biophys. Acta*, 108, 376, 1965.
129. **Scholtissek, C.**, End-turnover of rat liver soluble RNA *in vivo*, *Biochim. Biophys. Acta*, 61, 499, 1962.
130. **Landin, R. M. and Moulé, Y.**, Particularités de renouvellement du RNA soluble *in vivo* dans le foie de rat, *Biochim. Biophys. Acta*, 129, 249, 1966.
131. **Holt, C. E., Joel, P. B., and Herbert, E.**, Turnover of terminal nucleotides of soluble RNA in intact reticuloyctes, *J. Biol. Chem.*, 241, 1819, 1966.
132. **Fukamachi, S. Bartoov, B., and Freeman, K. B.**, Synthesis of RNA by isolated rat liver mitochondria, *Biochem. J.*, 128, 299, 1972.
133. **Deutscher, M. P., Foulds, J., and Setlow, P.**, *relA* overcomes the slow growth of *cca* mutants, *J. Mol. Biol.*, 117, 1095, 1977.
134. **Cudny, H., Zaniewski, R., and Deutscher, M. P.**, *Escherichia coli* RNase D, catalytic properties and substrate specificity, *J. Biol. Chem.*, 256, 5633, 1981.
135. **Nossal, N. G. and Singer, M. F.** The processive degradation of individual polyribonucleotide chains 1. *E. coli* ribonuclease II, *J. Biol. Chem.*, 243, 913, 1968.
136. **Zaniewski, R. and Deutscher, M. P.**, unpublished results, 1980.
137. **Geige, R. and Ebel, J. P.**, Intégrité de la séquence pCpCpA terminale des tRNA extraits de levures a differents stades de laur croissance, *Biochim. Biophys. Acta*, 161, 125, 1968.
138. **Setlow, P., Primus, G., and Deutscher, M. P.**, Absence of 3′ terminal residues from tRNA of dormant spores of *Bacillus megaterium*, *J. Bacteriol.*, 117, 126, 1974.
139. **Vold, B.**, Degree of completion of 3′-terminus of tRNA of *Bacillus subtilis* 168 at various developmental stages and asporogenous mutants, *J. Bacteriol.*, 117, 1361, 1974.
140. **Hausenbauer, J. M., Waites, W. M., and Setlow, P.**, Biochemical properties of *Clostridium bifermentans* spores, *J. Bacteriol.*, 129, 1148, 1977.
141. **Timourian, H.**, Protein synthesis in sea urchin eggs I. Fertilization-induced changes in subcellular fractions, *Dev. Biol.*, 16, 594, 1967.
142. **Herrington, M. D. and Hawtrey, A. O.**, Evidence for the absence of the terminal adenine nucleotide at the amino acid acceptor end of tRNA in non-lactating mammary gland and its inhibitory effect on the aminoacylation of rat liver tRNA, *Biochem. J.*, 116, 405, 1970.
143. **Dziegielewski, T., Kedzierski, W., and Pawelkiewicz, J.**, Levels of aminoacyl-tRNA synthetases, tRNA nucleotidyltransferase and ATP in germinating lupin seeds, *Biochim. Biophys. Acta*, 64, 37, 1979.
144. **Baliga, S., Hubert, C., Murphy, A., Meadow, F., Dourmashkin, P., and Munro, H.**, Status of tRNA charging, trinucleotide acceptor sequence and tRNA nucleotidyltransferase activity in the human placenta, *Can. J. Biochem.*, 54, 609, 1976.
145. **Bester, A. J. and Gervers, W.**, Evidence for defective tRNA in polymyopathic hamsters and its inhibitory effect on protein synthesis, *Biochem. J.*, 132, 203, 1973.
146. **Paradiso, P. and Schofield, P.**, Changes in tRNA nucleotidyltransferase activity during embryonic development of *Xenopus laevis*, *Exp. Cell Res.*, 100, 9, 1976.
147. **Morse, J. W. and Deutscher, M. P.**, A physiological role for tRNA nucleotidyltransferase during bacteriophage infection, *Biochim. Biophys. Res. Commun.*, 73, 953, 1976.
148. **Wilson, J. H.**, Function of bacteriophage T4 transfer RNAs, *J. Mol. Biol.*, 74, 753, 1973.
149. **Weissmann, C., Billeter, M. A., Goodman, H. M., Hindley, J., and Weber, H.**, Structure and function of phage RNA, *Annu. Rev. Biochem.*, 42, 303, 1973.
150. **Genevaux, M., Pinck, M., and Duraton, H. M.**, Amino acid accepting structures in tymovirus RNAs, *Ann. Microbiol.*, 127A, 47, 1976.

151. **Shibata, H., Ro-Choi, T. S., Reddy, R., Choi, Y. C., Henning, D., and Busch, H.,** The primary nucleotide sequence of nuclear U2 RNA, *J. Biol. Chem.,* 250, 3909, 1975.
152. **Toshi, S., Haenni, A. L., Hubert, E., Huez, G., and Marbaix, G.,** *In vivo* aminoacylation and processing of turnip yellow mosaic virus RNA in *Xenopus laevis* oocytes, *Nature (London),* 275, 339, 1978.

INDEX

A

B

C

D

E

F

G

H

I

L

M

N

O

P

S

V

W

X

Y

Z